The Role of Antioxidants in Longevity and Age-Related Diseases

Bee Ling Tan • Mohd Esa Norhaizan

The Role of Antioxidants in Longevity and Age-Related Diseases

Springer

Bee Ling Tan
Department of Nutrition
Faculty of Medicine and Health Sciences
Universiti Putra Malaysia
Serdang, Selangor, Malaysia

Mohd Esa Norhaizan
Department of Nutrition
Faculty of Medicine and Health Sciences
Universiti Putra Malaysia
Serdang, Selangor, Malaysia

ISBN 978-3-030-82858-5 ISBN 978-3-030-82859-2 (eBook)
https://doi.org/10.1007/978-3-030-82859-2

This Springer imprint is published by the registered company Springer Nature Switzerland AG
The registered company address is: Gewerbestrasse 11, 6330 Cham, Switzerland

Preface

The average life expectancy has increased worldwide over the last decades. Despite the increasing life expectancy reflecting positive human development, new challenges are arising. Aging can be viewed as the progressive loss of organ and tissue function over time. It is characterized as the onset of major underlying causes for age-related diseases such as cancer, neurodegenerative disorders, cardiovascular disease (CVD), type 2 diabetes, arthritis, osteoporosis, stroke, and Alzheimer's disease. It has been recognized that chronic inflammation is a prominent underlying factor for age-related diseases. Oxidative stress is linked to the development of age-related diseases. The vast majority of published work has revealed the potential preventive and/or therapeutic roles of antioxidant in aging and age-related diseases. Ample evidence has suggested that antioxidant can modulate autoxidation by inhibiting the formation of free radicals or disrupting the propagation of free radicals and thus increase health longevity, enhance immune function, and decrease oxidative stress. In fact, oxidative damage is highly dependent on the acquired or inherited defects in enzymes involved in the redox-modulated signaling pathways. In view of the dramatic and recent increase of age-related diseases among the elderly in developed countries, the discovery of interventions that target the aging aspects of these diseases would have a significant public health impact. Based on our knowledge, the literature reported on antioxidants with their biological mechanisms that mediate age-related diseases has not well been compiled in the form of a book/brief. In fact, this disintegrated knowledge needs to be compiled together to deliver information at one point. Hence, this brief/book attempts to clarify several issues linked to antioxidant, aging, and age-related diseases, namely changes in organ systems over the lifespan, age-related oxidative stress-induced redox imbalance, inflammaging, implications of inflammation in aging and age-related diseases, antioxidant and age-related diseases, and future prospects. By summarizing all the literature in a cohesive and lucid manner in one book/brief, it would provide a comprehensive representation of the information on the mode of actions of antioxidants involved in the prevention of age-related diseases for the readers and allied stakeholders. A better understanding of how chronic inflammation contributes to aging and age-related diseases would provide a new intervention targeting inflammation.

Serdang, Selangor, Malaysia
Bee Ling Tan

Contents

Chapter 1
Introduction and Background

Aging is a biological, universal, and unavoidable process that affecting almost all multicellular organisms and probably common in unicellular organisms such as bacteria, yeast, and protozoa (Książek 2010; Lin and Austriaco 2014; Teulière et al. 2020). It is characterized as degenerative in nature, concomitantly with progressive loss of function, and lastly increased mortality rate (Thoma et al. 2020). Among the mammalian species, humans have the longest lifespan and consume more energy throughout their lifespan per weight basis (Ben-Haim et al. 2018). Despite numerous hypotheses that have been proposed to explain the molecular mechanisms of aging, previous studies suggest that aging is caused by an accumulation of molecular damage, giving rise to a unified theory of aging (Niedernhofer et al. 2018; Ogrodnik et al. 2019). The free radical theory of aging has long been established to explain the aging process (Harman 1956). This theory proposed that aging is a consequence of the failure of some defensive mode of action to respond to the reactive oxygen species (ROS)-induced damage, especially at the mitochondria (Theurey and Pizzo 2018).

In the context of thermodynamic, all aerobic organisms are susceptible to the action of a common oxidant, namely oxygen (Matschke et al. 2019). The redox potential of the $O_2/2H_2O$ redox system is positive compared to other biologically relevant redox systems (Gutteridge and Mitchell 1999). Hence, the oxidation by oxygen in organic compounds showed a negative free enthalpy and response spontaneously (Yang et al. 2018). In this regard, structures or organic compounds consisted of them are thermodynamically unstable in an oxygen-containing atmosphere (Granzow 1978). Molecular oxygen in its triplet basal state is unreactive due to the spin restriction (Sadowska-Bartosz and Bartosz 2014). Subsequently, the formation of free radicals or ROS demonstrates potentially deleterious oxidative reactions of oxygen (Sies and Jones 2020). The previous study has revealed that utilization of oxygen by mitochondria of aerobic organisms can lead to the production of reactive radicals, for example, hydrogen peroxide (H_2O_2), superoxide ($O_2^{\bullet-}$), as well as hydroxyl radical ($HO^{\bullet}$) (Shim and Kim 2013; Xie et al. 2019; Li Puma et al. 2020).

B. L. Tan, M. E. Norhaizan, *The Role of Antioxidants in Longevity and Age-Related Diseases*, https://doi.org/10.1007/978-3-030-82859-2_1

Furthermore, nitric oxide ($NO^•$) produced by mitochondria may further implicate the aging process and age-related diseases (Opatrilova et al. 2018; Somasundaram et al. 2019; Lieu et al. 2020). Another potent source of oxidant production is via phagocytes (Stawski et al. 2019), in which they produce hypochlorous acid (HOCl), $NO^•$, $HO^•$, H_2O_2, and $O_2^{•-}$ (Misztal et al. 2019). Indeed, HOCl is an inflammatory mediator as well as a strong chlorinating and oxidizing compound that can produce other reactive metabolites including nitrogen dioxide ($NO_2^•$) and nitryl chloride (NO_2Cl), in the presence of nitrite (Rayner et al. 2018). A study found that activated human polymorphonuclear neutrophils can convert nitrite into NO_2Cl and $NO_2^•$ metabolites, and thereby leading to the formation of harmful compounds (Whiteman et al. 2003). The deleterious effects of chlorinating compounds, ROS, and reactive nitrogen species (RNS) may subsequently contribute to the aging process (Saxena and Batra 2020; Brunetta et al. 2020; Semerád et al. 2020).

Aerobic organisms are protected against oxidative challenges by sophisticated antioxidant defense systems (Rezayian et al. 2019). Oxidative stress triggered by oxidant species occurs when the rate constant of free radicals is higher than the antioxidant defense mechanisms or when the antioxidant defenses are depleted (Simioni et al. 2018). The defense mechanisms decline with age and accelerate oxidative damage and thereby leading to deterioration in physiological function (Ridhima et al. 2018). Indeed, the oxidative damage of biomolecules is increased with age (Kuzmic et al. 2019). This damage has been hypothesized to play a critical role in cellular biochemical senescence (Moldogazieva et al. 2019). A study found that the antioxidant enzymes such as catalase and superoxide dismutase (SOD) are overexpressed in transgenic *Drosophila*, implied that increased antioxidant enzymes could increase maximum and average lifespan and decrease oxidative damage (Milisav et al. 2019). Furthermore, the study also found that aged species have a greater likelihood to generate a significant amount of ROS compared to the younger counterparts (Hekimi et al. 2016). Data from animal studies have shown a strong inverse correlation between the maximum lifespan and the rate of mitochondrial oxidant production in different species (Munro and Pamenter 2019). A study further revealed that animals with high mitochondrial metabolism have a short lifespan due to the high oxygen consumption and thereby leading to higher oxidant production (Venditti and Di Meo 2020). In support of this, Lin et al. (2019) found a strong correlation between SOD activity and maximum lifespan in *Caenorhabditis elegans*. This finding implies that the increase of SOD activity may protect against superoxide radicals (Case 2017), suggesting that lifespan may partly depend on the activity of this enzyme (Leite et al. 2020).

Free radical theory of aging is commonly explaining the oxidative damage theory of aging, which is a primary facet of intrinsic biological instability of living systems (Yegorov 2020). This theory shows that free radicals or other ROS are produced in the course of metabolism and increased concomitantly due to the action of different exogenous factors and the damage of biomolecules, and accumulation of this damage may result in aging and age-related diseases (Luo et al. 2020).

The average life expectancy of the world population has increased rapidly in the recent decades, with an average of 72.0 years in 2016 (World Health Organization

2020a). The world population aged 60 years is projected to live another 20.5 years on average in 2016 (World Health Organization 2020b). Based on the demographics of the global population between 2000 and 2050, the population aged more than 60 years is expected to increase from 605 million to 2 billion people (World Health Organization 2014). The predicted increase in life expectancy would potentially lead to biological and cognitive degeneration, for instance, psychological impairment, physical frailty, and cognitive decline (Garcia et al. 2018; Marzetti et al. 2019).

In the twenty-first century, age-related diseases have become the greatest health threat (Tan and Norhaizan 2019). The interplay among co-factors, antioxidants, and free radicals is crucial in the modulation of age-related diseases, aging, and health (Tan et al. 2018). Age-related diseases are linked to the structural changes in mitochondria, concomitantly with the modifications of biophysical properties of the membranes, for instance, reduced fluidity, modifications in the electron transport chain (ETC) complexes activities, and thereby leading to energy imbalance and mitochondrial failure (Liu et al. 2020). This perturbation destroys the mitochondrial function and cellular homeostasis and increases susceptibility to oxidative stress (Madreiter-Sokolowski et al. 2018). Older people are vulnerable to oxidative stress due to a decrease in the efficiency of endogenous antioxidant systems (Zhang et al. 2015). In particular, organs including hearts and brains, with limited respiration levels and high amounts of oxygen consumption, are susceptible to these phenomena, suggesting that it may contribute to the high prevalence of neurodegenerative disease and vascular disease in the elderly (Takahashi and Takahashi 2013; Cervellati et al. 2020).

Oxidative stress is regarded as a result of an imbalance between anti- and prooxidant species (Dogru et al. 2018). It plays a prominent role in the development of age-related diseases such as osteoporosis, vascular disease, cancer, osteoporosis, dementia, diabetes, and arthritis (Tan et al. 2018). Increased ROS has been linked to the progression and onset of aging (Moldogazieva et al. 2019). Despite ROS may not be an essential factor for aging, they are more likely to aggravate the progression of age-related diseases through interaction with mitochondria and oxidative damage (Kowalska et al. 2020; Luo et al. 2020). The balance between detrimental and beneficial effects of ROS is preserved in the cells by mediating the complex array of enzymatic and non-enzymatic detoxification mechanisms (Bast and Haenen 2015). Antioxidants can counteract the damage induced by ROS in cells, and thus protecting the physiological targets, for instance, proteins, DNA, and lipids (He et al. 2017). Emerging evidence has revealed that natural products possess a unique complex of bioactive compounds. These compounds can improve the immune system and decrease oxidative stress (Alvarez-Arellano et al. 2020; Romiti et al. 2020). Indeed, the interplay between the overall redox system and antioxidants in humans is very complex and is highly dependent on the acquired or inherited defects in enzymes involved in the redox signaling pathways (Łuczaj et al. 2017). Therefore, the role of diet/food high in antioxidant molecules that combat oxidative stress and promote healthy aging is worth discussing further.

According to the best of the authors' knowledge, the literature on antioxidants and their mode of action that modulate age-related diseases has not well been compiled in the form of a brief/book. Indeed, this disintegrated information needs to be compiled together to deliver knowledge at one point. Therefore, this brief/book explores several issues related to antioxidant, aging, and age-related diseases, including changes in organ systems over the lifespan, age-related oxidative stress-induced redox imbalance, inflammaging, implications of inflammation in aging and age-related diseases, the role of antioxidants on health and age-related diseases, and future prospects.

References

Alvarez-Arellano L, González-Garcia N, Salazar-García M et al (2020) Antioxidants as a potential target against inflammation and oxidative stress in attention-deficit/hyperactivity disorder. Antioxidants 9:176

Bast A, Haenen GRMM (2015) Chapter 2 Nutritional antioxidants: it is time to categorise. In: Lamprecht M (ed) Antioxidants in sport nutrition. CRC Press, Boca Raton

Ben-Haim MS, Kanfi Y, Mitchell SJ et al (2018) Breaking the ceiling of human maximal life span. J Gerontol 73:1465–1471

Brunetta HS, Holwerda AM, van Loon LJC et al (2020) Mitochondrial ROS and aging: understanding exercise as a preventive tool. J Sci Sport Exerc 2:15–24

Case AJ (2017) On the origin of superoxide dismutase: an evolutionary perspective of superoxide-mediated redox signaling. Antioxidants 6:82

Cervellati C, Trentini A, Pecorelli A et al (2020) Inflammation in neurological disorders: the thin boundary between brain and periphery. Antioxid Redox Signal 33:191–210

Dogru M, Kojima T, Simsek C et al (2018) Potential role of oxidative stress in ocular surface inflammation and dry eye disease. Investig Ophthalmol Vis Sci 59:DES163–DES168

Garcia MA, Saenz JL, Downer B et al (2018) Age of migration differentials in life expectancy with cognitive impairment: 20-year findings from the Hispanic-EPESE. Gerontologist 58:894–903

Granzow A (1978) Flame retardation by phosphorus compounds. Acc Chem Res 11:177–183

Gutteridge JMC, Mitchell J (1999) Redox imbalance in the critically ill. Br Med Bull 55:49–75

Harman D (1956) Aging: a theory based on free radical and radiation chemistry. J Gerontol 11:298–300

He L, He T, Farrar S et al (2017) Antioxidants maintain cellular redox homeostasis by elimination of reactive oxygen species. Cell Physiol Biochem 44:532–553

Hekimi S, Wang Y, Noë A (2016) Mitochondrial ROS and the effectors of the intrinsic apoptotic pathway in aging cells: the discerning killers! Front Genet 7:161

Kowalska M, Peikut T, Prendecki M et al (2020) Mitochondrial and nuclear DNA oxidative damage in physiological and pathological aging. DNA Cell Biol 39:1410–1420

Książek K (2010) Let's stop overlooking bacterial aging. Biogerontology 11:717–723

Kuzmic M, Galas S, Lecomte-Pradines C et al (2019) Interplay between ionizing radiation effects and aging in C. *elegans*. Free Radic Biol Med 134:657–665

Leite NR, de Araújo LCA, dos Santos da Rocha P (2020) Baru pulp (*Dipteryx alata* Vogel): fruit from the Brazilian Savanna protects against oxidative stress and increases the life expectancy of *Caenorhabditis elegans* via SOD-3 and DAF-16. Biomol Ther 10:1106

Li Puma LC, Hedges M, Heckman JM et al (2020) Experimental oxygen concentration influences rates of mitochondrial hydrogen peroxide release from cardiac and skeletal muscle preparations. Am J Phys 318:R972–R980

Lieu M, Traynham CJ, de Lucia C et al (2020) Loss of dynamic regulation of G protein-coupled receptor kinase 2 by nitric oxide leads to cardiovascular dysfunction with aging. Am J Phys 318:H1162–H1175

Lin SJ, Austriaco N (2014) Aging and cell death in the other yeasts, *Schizosaccharomyces pombe* and *Candida albicans*. FEMS Yeast Res 14:119–135

Lin C, Zhang X, Su Z et al (2019) Carnosol improved lifespan and healthspan by promoting antioxidant capacity in *Caenorhabditis elegans*. Oxidative Med Cell Longev 2019, Article ID 5958043, 13 pages

Liu YJ, McIntyre RL, Janssens GE et al (2020) Mitochondrial fission and fusion: a dynamic role in aging and potential target for age-related disease. Mech Ageing Dev 186:111212

Łuczaj W, Gęgotek A, Skrzydlewska E (2017) Antioxidants and HNE in redox homeostasis. Free Radic Biol Med 111:87–101

Luo J, Mills K, le Cessie S et al (2020) Ageing, age-related diseases and oxidative stress: what to do next? Ageing Res Rev 57:100982

Madreiter-Sokolowski CT, Sokolowski AA, Waldeck-Weiermair M et al (2018) Targeting mitochondria to counteract age-related cellular dysfunction. Genes 9:165

Marzetti E, Picca A, Marini F et al (2019) Inflammatory signatures in older persons with physical frailty and sarcopenia: the frailty "cytokinome" at its core. Exp Gerontol 122:129–138

Matschke V, Theiss C, Matschke J (2019) Oxidative stress: the lowest common denominator of multiple diseases. Neural Regen Res 14:238–241

Milisav I, Ribarič S, Poljsak B (2019) Antioxidant vitamins and ageing. In: Harris J, Korolchuk V (eds) Biochemistry and cell biology of ageing: part I Biomedical science, Subcellular biochemistry, vol 90. Springer, Singapore, pp 1–23

Misztal T, Golaszewska A, Tomasiak-Lozowska MM et al (2019) The myeloperoxidase product, hypochlorous acid, reduces thrombus formation under flow and attenuates clot retraction and fibrinolysis in human blood. Free Radic Biol Med 141:426–437

Moldogazieva NT, Mokhosoev IM, Mel'nikova TI et al (2019) Oxidative stress and advanced lipoxidation and glycation end products (ALEs and AGEs) in aging and age-related diseases. Oxidative Med Cell Longev 2019, Article ID 3085756, 14 pages

Munro D, Pamenter ME (2019) Comparative studies of mitochondrial reactive oxygen species in animal longevity: technical pitfalls and possibilities. Aging Cell 18:e13009

Niedernhofer LJ, Gurkar AU, Wang Y et al (2018) Nuclear genomic instability and aging. Annu Rev Biochem 87:295–322

Ogrodnik M, Salmonowicz H, Gladyshev VN (2019) Integrating cellular senescence with the concept of damage accumulation in aging: relevance for clearance of senescent cells. Aging Cell 18:e12841

Opatrilova R, Kubatka P, Caprnda M et al (2018) Nitric oxide in the pathophysiology of retinopathy: evidences from preclinical and clinical researches. Acta Ophthalmol 96:222–231

Rayner BS, Zhang Y, Brown BE et al (2018) Role of hypochlorous acid (HOCl) and other inflammatory mediators in the induction of macrophage extracellular trap formation. Free Radic Biol Med 129:25–34

Rezayian M, Niknam V, Ebrahimzadeh H (2019) Oxidative damage and antioxidative system in algae. Toxicol Rep 6:1309–1313

Ridhima W, Riya G, Pawan MK (2018) Oxidative stress and accelerated aging in neurodegenerative and neuropsychiatric disorder. Curr Pharm Des 24:4711–4725

Romiti GF, Corica B, Raparelli V et al (2020) The interplay between antioxidants and the immune system: a promising field, still looking for answers. Nutrients 12:1550

Sadowska-Bartosz I, Bartosz G (2014) Effect of antioxidants supplementation on aging and longevity. Biomed Res Int 2014, Article ID 404680, 17 pages

Saxena R, Batra J (2020) Chapter 8 Arthritis as a disease of aging and changes in antioxidant status. In: Preedy VR, Patel VB (eds) Aging: oxidative stress and dietary antioxidants, 2nd edn. Academic, London, pp 83–94

Semerád J, Filip J, Ševců A et al (2020) Environmental fate of sulfidated nZVI particles: the interplay of nanoparticle corrosion and toxicity during aging. Environ Sci Nano 7:1794–1806

Shim S-Y, Kim H-S (2013) Oxidative stress and the antioxidant enzyme system in the developing brain. Korean J Pediatr 56:107–111

Sies H, Jones DP (2020) Reactive oxygen species (ROS) as pleiotropic physiological signalling agents. Nat Rev Mol Cell Biol 21:363–383

Simioni C, Zauli G, Martelli AM et al (2018) Oxidative stress: role of physical exercise and antioxidant nutraceuticals in adulthood and aging. Oncotarget 9:17181–17198

Somasundaram V, Basudhar D, Bharadwaj G et al (2019) Molecular mechanisms of nitric oxide in cancer progression, signal transduction, and metabolism. Antioxid Redox Signal 30:1124–1143

Stawski R, Walczak K, Perdas E et al (2019) Decreased integrity of exercise-induced plasma cell free nuclear DNA–negative association with the increased oxidants production by circulating phagocytes. Sci Rep 9, Article number: 15970

Takahashi K, Takahashi M (2013) Exogenous administration of coenzyme Q10 restores mitochondrial oxygen consumption in the aged mouse brain. Mech Ageing Dev 134:580–586

Tan BL, Norhaizan ME (2019) Carotenoids: how effective are they to prevent age-related diseases? Molecules 24:1801

Tan BL, Norhaizan ME, Liew W-P-P et al (2018) Antioxidant and oxidative stress: a mutual interplay in age-related diseases. Front Pharmacol 9:1162

Teulière J, Bhattacharya D, Bapteste E (2020) Ancestral germen/soma distinction in microbes: expanding the disposable soma theory of aging to all unicellular lineages. Ageing Res Rev 60:101064

Theurey P, Pizzo P (2018) The aging mitochondria. Genes 9:22

Thoma A, Akter-Miah T, Reade RL et al (2020) Targeting reactive oxygen species (ROS) to combat the age-related loss of muscle mass and function. Biogerontology 21:475–484

Venditti P, Di Meo S (2020) The role of reactive oxygen species in the life cycle of the mitochondrion. Int J Mol Sci 21:2173

Whiteman M, Rose P, Halliwell B (2003) Inhibition of hypochlorous acid-induced oxidative reactions by nitrite: is nitrite an antioxidant? Biochem Biophys Res Commun 303:1217–1224

World Health Organization (2014) World Health Statistics – large gains in life expectancy. http://www.who.int/mediacentre/news/releases/2014/world-health-statistics-2014/en/. Accessed on Dec 2014

World Health Organization (2020a) Life expectancy. https://www.who.int/gho/mortality_burden_disease/life_tables/situation_trends/en/. Accessed 20 Apr 2020

World Health Organization (2020b) Life tables. https://www.who.int/gho/mortality_burden_disease/life_tables/life_tables/en/. Accessed 20 Apr 2020

Xie X, He Z, Chen N et al (2019) The roles of environmental factors in regulation of oxidative stress in plant. Biomed Res Int 2019, Article ID 9732325, 11 pages

Yang Y, Banerjee G, Brudvig GW et al (2018) Oxidation of organic compounds in water by unactivated peroxymonosulfate. Environ Sci Technol 52:5911–5919

Yegorov YE (2020) Healthy aging: antioxidants, uncouplers and/or telomerase? Mol Biol 54:311–316

Zhang H, Davies KJA, Forman HJ (2015) Oxidative stress response and Nrf2 signaling in aging. Free Radic Biol Med 88(Part B):314–336

Chapter 2
Changes in Organ Systems over the Lifespan

Aging is characterized by the impairment of the regulatory process that provides functional integration between organs and cells (Srivastava 2017). Increasing age caused an interindividual variability of the physiological changes (Kedlian et al. 2019). Aging is not only a single entity but a collective term as the sum of cumulative local effects at the tissue, cellular, and molecular levels (Ashapkin et al. 2019). Indeed, aging is the effect of these underlying changes but not the cause. Despite an all-encompassing definition of aging is not possible; some characteristics have been identified (Mangoni and Jackson 2004). The most consistent is the functional and time-associated loss units (Chen 2019). This unit is the smallest structure that capable to perform certain physiological activities of the organs including neurons, alveoli, and nephrons (Sgarbieri and Pacheco 2017). Another characteristic is a disruption of the regulatory process that provides functional integration between organs and cells (Jeffries et al. 2019). As a result, the ability to counteract external stresses and maintain homeostasis is reduced (Tan et al. 2020). Loss of the functional reserve is linked to the increase in vulnerability and decrease in viability (Vila 2019). Aging is not solely a progressive decline in function but results in physiological and anatomical changes, and thus may lead to the decompensation of the system when they progress beyond a threshold (Larsson et al. 2019).

2.1 Changes in Nervous System

The brain has a limited capacity for repair. During aging, neurons are progressively reduced and cannot be replaced (Li et al. 2019). Aging is linked to numerous neurological diseases, in which the capacity to communicate and transmit signals in the brain decreased (Hertelendy et al. 2019). Reduced brain function is the biggest challenge facing by the elderly such as dementia and Alzheimer's disease (Nelson et al.

B. L. Tan, M. E. Norhaizan, *The Role of Antioxidants in Longevity and Age-Related Diseases*, https://doi.org/10.1007/978-3-030-82859-2_2

2016; Acosta et al. 2017). Several neurodegenerative diseases, for instance, Parkinson's disease or sudden devastation of a stroke are also increasing with age (Prange et al. 2019; Yousufuddin and Young 2019). Parkinson's and Alzheimer's diseases are progressive neurodegenerative diseases related to aging (Harrison et al. 2019; Trist et al. 2019). Alzheimer's disease is characterized by progressive deterioration of cognition along with the decreased in daily living activities and changed in behaviors (Takeda 2019). Among the neurodegenerative diseases, Alzheimer's disease is the most common type of senile and pre-senile dementia (Armstrong 2016). It causes tissue loss and nerve cell death throughout the brain, in which the cortex in the brain shrivels up and destroys the area involved in remembering, planning, and thinking (Jonathan et al. 2015). Indeed, the shrinkage of the nerve cell is severe in the hippocampus, the area of the cortex that plays an important role in the formation of new memories (Theofilas et al. 2017). Alzheimer's disease leads to an overall misbalance in the elderly by changes in behavior-like depression and personality, aggressiveness, irritability, mood swings, social withdrawal, apathy, and memory loss (Scott et al. 2020; Wu et al. 2020). Defects in protein processing also may cause the deposition of amyloid plaques (Wang et al. 2020). Notably, there is a steady increase in the symptoms of Alzheimer's disease with age, with nearly doubling the incidence for every 5 years of life after 65 years old (Qiu et al. 2009). Such finding implies that there is not a defined syndrome, in which the late-onset could be labeled as a natural aging of the brain (Murman 2015).

2.2 Changes in Cognitive Function

The hippocampus is a brain area deals with memory as well as the initial target for aging-associated physiological and structural impairments (Bettio et al. 2017) that mediate spatial memory decline (Subramaniapillai et al. 2019). Neurotransmission by glutamate, acetylcholine, dopamine, and γ-aminobutyric acid (GABA) interacts to modulate learning and memory abilities (Cai and Ford 2018; Ravichandran et al. 2018; Mohebi et al. 2019). Glutamate is converted to GABA facilitated by glutamate decarboxylase (GAD) (Zhao and Gammie 2014). GAD is widely used for monitoring neuronal function during aging (Jacobson et al. 2008), while choline acetyltransferase (ChAT) is extensively utilized as a marker for cholinergic function in learning and memory (Hawley et al. 2015). The dopamine output of the striatum also plays an important role in age-related cognitive decline (Darvas and Palmiter 2011). Treatment with L-Dopa has been demonstrated to benefit age-related memory loss through enhanced dopamine output (Zhao et al. 2017). Indeed, enhancement of neuronal plasticity during the pre-symptomatic period could be a trophic response to the initial neuronal loss (Mazarakis et al. 2005). Likewise, a study in middle-aged gerbils showed a higher level of brain-derived neurotrophic factor (BDNF) in the hippocampus compared to aged and young groups (Hwang et al. 2006). Compared to young and aged rats, the striatum of middle-aged rats has a higher extracellular dopamine concentration (Segovia et al. 1999). Kim et al. (2005)

have also been reported to enhance GluR-1 positive neurons in the frontal cortex of middle-aged rats.

In the context of behavioral response to plasticity in striatal dopamine and GABA, the middle-aged rats showed less anxiety compared to aged groups (Jacobson et al. 2008). A study by Bessa et al. (2005) further supported that middle-aged rats were shown to have less anxious behaviors compared to the aged rats by illustrating the avoidance of open arm entries. Despite limited evidence reported on the role of dopamine in anxiety during brain aging, the previous study has demonstrated that anxiety observed in Parkinson's patients could be attributed to the diminish of striatal dopamine output (Richard 2005).

The most common cognitive decline with aging is memory impairment, especially in short-term, episodic, and spatial memory (Poddar et al. 2019). In a study by Jacobson et al. (2008) focusing on age-associated neuronal degeneration and plasticity in rat models using behavioral tests, the data showed that aged rats performed more working memory errors in the radial arm maze (RAM), made more entries into incorrect arms in the T-maze test, and were more slowly to find the platform in the Morris water maze (MWM), implied that poor performance of aged rats in hippocampus-dependent tasks (Arias-Cavieres et al. 2017).

Mild cognitive impairment (MCI) is a clinical prodrome, which may evolve to overt dementia in advanced age (Mielke et al. 2017). Neurodegeneration may occur before a clear decrease in function, a model that has been depicted in other forms of neurodegenerative situation, for instance, Parkinson's disease (Zeng et al. 2018). Indeed, this condition may also happen in pathological brain aging. Older people showed a greater difficulty with memory tasks, particularly involving strategic, effortful, and complex memory tasks (Lee et al. 2018). Such finding indicates that aging reduces the memory functions resulted from impairments of basic cognitive functions (Harman and Martín 2020). The mechanisms underlying the age-related changes are not fully understood, but it is hypothesized that lack of cognitive control, loss of inhibitory functions, decreased processing resources, and general slowing contributes to these phenomena (Luo and Craik 2008).

Based on the general slowing, aging is accompanied by a reduction in processing speed and thereby leading to a decline in cognitive functions such as memory performance (Shim et al. 2020). Processing speed plays a crucial role in several cognitive functions. In this regard, complex tasks involving multiple types of processing may gradually slow down, indicating that greater age-related declines (Monteiro et al. 2020). Nonetheless, substantial evidence highlights that this view is primarily derived from path analysis data. It has remained obscure how the hypothesis accounts for some experimental analyses. For instance, (1) age-related declines have been demonstrated in several tasks including free recall that does not need a speed component (Dunlosky and Salthouse 1996); (2) unlimited processing time does not decrease memory problems in the elderly but improves performance in young people (Luo and Craik 2008); and (3) aging is linked to differential effects on tasks which do not involve different levels of processing (Luo and Craik 2008).

During aging, older people experience a reduction in cognitive function such as memory loss (Appel et al. 2020). According to Flexner et al. (1962, 1963, 1967), changes in the amount of protein synthesis in the brain may affect some of the cognitive effects of aging. When more and more brain-specific proteins and the putative roles were discovered, it became evident that the individual protein expression could decrease or increase with age, and the rates of change across the lifespan may differ between proteins (Tong et al. 2020). In view of the discovery that *de novo* protein synthesis is required for the establishment of long-term information storage, substantial studies were performed to evaluate the translation of individual proteins in restricted areas of the brain, and in response to artificial or cognitive stimulation (Wegmann et al. 2019). Indeed, the regulation of transcription undergoes age-related changes, and that decreases or increases in translation, transcription, or both determine the expression of certain proteins at any stage during the lifespan (Baehr et al. 2017).

Several studies found that the amounts of brain protein synthesis increased during the development of rodents, in which it reaches maximum amounts during the first 6 months of age and gradually decline (Dwyer et al. 1980; Ekstrom et al. 1980; Fando et al. 1980; Ingvar et al. 1985; Smith et al. 1995; Filion and Laughrea 1985; Richardson 1981; Richardson and Birchenall-Sparks 1983; Richardson et al. 1983). Likewise, the same effect was also observed in white leghorn chicken (Yang et al. 1977). Nevertheless, not all studies demonstrated such a link. The previous study evaluated the exact time courses of the decline, in which some studies reported a greater rate of decrease in translation during adulthood (Ekstrom et al. 1980; Fando et al. 1980; Ingvar et al. 1985; Smith et al. 1995), and other study showed the greatest rate of decline during senescence (Dwyer et al. 1980). Despite the protein synthesis reduced was observed in several cells and organ tissues including muscle, kidney, and liver, the rates and time courses of the decline are varied among each other (Webster 1985). Although there remain several disagreements on the timing of the acceleration decrease in brain tissues, the data showed that translation is declined with age (Hipp et al. 2019). It is worth highlighting that most of the data evaluated rates of translation in the brains of animals are not manipulated or cognitively stimulated before measurement, or in non-perturbed *in vitro* preparations. Therefore, the findings found from the foregoing studies demonstrate resting rates of synthesis that may or may not reflect rates occurring when the system is stimulated. In another study, Ingvar et al. (1985) evaluated the amount of protein synthesis in a broad spectrum of brain structures during aging, and after a decade followed this study to calculate this level accurately (Smith et al. 1995). The data from the animal study revealed that age-related declines were observed in brain structures, cerebellar white matter, internal capsule, and olfactory cortex that having a role in the extrapyramidal motor, visual, and auditory systems (Muñoz-Moreno et al. 2018). In particular, the protein synthesis is affected in extrapyramidal motor and sensory systems during aging; however, an alternate interpretation of this finding is that activity in sensorimotor circuitry might be decreased when the reduced locomotion in aged animals. Despite Ingvar et al. (1985) suggested that areas required for higher cognitive function are relatively spared, the age-related protein synthesis is reduced in the locus coeruleus, nucleus accumbens, and hippocampal dentate gyrus region, the regions that modulate memory and learning processes.

Many studies reported that protein expression in specific brain structures can be changed during the aging process (Gelfo et al. 2018). In general, the transcriptional activities involved in regulating synaptic function, neural plasticity, and mitochondrial function are downregulated during aging; whereas the genes that mediate the inflammatory/immune responses and stress are upregulated (Schimanski and Barnes 2010). The translation of certain proteins is preceded by transcription of the mRNAs that encode several proteins, in which it can undergo age-related modifications (Browning and Bailey-Serres 2015).

Normal aging is often accompanied by memory impairment (Rosenzweig and Barnes 2003; Burke and Barnes 2006). Despite numerous forms of memory are left intact by the aging process, age-related deficits have been identified in several types of hippocampus-dependent memory such as trace eyeblink conditioning (Weiss and Disterhoft 2015), fear memory (Zhan et al. 2018), and spatial memory (Yamamoto et al. 2019). Notably, this hippocampal function-dependent memory deficit was found across phylogeny, for instance, human, monkey, dog, rabbit, rat, and mouse (Murray et al. 2018). The previous study showed the increased decline rate of spatial memory in aged rats concomitantly with the rapid decay of long-term potentiation (LTP) at the perforant path-granule cell synapse (Barnes and McNaughton 1985). A study by Winocur (1988a) further supported that aged rodents have a greater likelihood of some types of long-term memory deficits compared to short-term memory. The data showed that aged rats demonstrate selective deficits 21 days after training when evaluated for recall of a passive avoidance response (Winocur 1988b). A study by Foster (1999) further demonstrated that a "saw-toothed" manner of performance in aged rats when several swim trials are performed each day for several concurrent days in the Morris water maze.

Several studies reported by Frankland et al. (2004) and Brightwell et al. (2005) evaluated the age-associated changes in cAMP-responsive element-binding protein (CREB) activity in relation to memory impairment in aging. CREB is a transcriptional activity that is required for hippocampal memory consolidation. CREB is active when phosphorylated (Gonzalez and Montminy 1989), and thus can be upregulated through the induction of LTP (Impey et al. 1996; Schulz et al. 1999). The previous study showed that pCREB upregulates the protein expression that promotes the long-lasting forms of memory and synaptic plasticity (Bailey et al. 1996; Yin and Tully 1996). Some studies have revealed that pCREB activity is decreased in aged rodents (Morris and Gold 2012; An et al. 2018).

Epigenetic mechanisms may also affect the age-related changes in synaptic plasticity and memory (Castellano et al. 2012; Kosik et al. 2012; Kwapis et al. 2018). Alteration methylation of the Arc gene has also been observed in aged rats (Penner et al. 2011). It was evident that changes in histone acetyltransferases (HATs) can improve synaptic plasticity and memory in animal models of neurological disorders and normal groups (Karisetty et al. 2020). HATs are co-activator that regulated the CRE-dependent transcription via CREB (Piccirillo et al. 2019). Taken together, these findings indicate that the expression of protein synthesis-dependent synaptic strengthening is reduced in aged rodents. An alteration of signaling pathways and calcium homeostasis could influence the recruitment of the protein synthesis process and induction of synaptic potentiation (Schimanski and Barnes 2010).

2.3 Changes in Skin

Human skin is constantly exposed to environmental pollutants (Poljšak and Fink 2014), solar radiation (Ali et al. 2020), air (Poljšak and Dahmane 2012), as well as chemical and mechanical agents (Rudolf and Cervinka 2011; Wong et al. 2011), and thus inducing the production of ROS (Hepel and Andreescu 2015; Panich et al. 2016). Indeed, ROS are often of little harm if intracellular modes of action that decrease the damaging effects are worked effectively (Trouba et al. 2002). The most important mechanisms involved in this process including nonenzymatic and antioxidative enzymatic defenses as well as repair processes (Herrling et al. 2006; Prasad and Pospišil 2011; Rastogi and Pospísil 2011). Furthermore, the repair processes and endogenous antioxidative mechanisms are reduced with advancing age (Tan et al. 2018). Thus, the skin tends to become finely wrinkled, drier, less elastic, and thinner (Langton et al. 2019). Changes of skin are partly due to elastin and collagen. Less elastin and collagen are produced when people are getting older. Subsequently, the skin tears more easily (Vanzi and Toma 2018).

Several factors developed extrinsic skin damage, for instance, environmental pollution, overeating, poor nutrition, alcohol intake, several psychological and physical stress, ionizing radiation, and exposure to UV radiation (Farage et al. 2008; Naidoo and Birch-Machin 2017). UV triggered the production of ROS in the skin to develop oxidative stress when the formation exceeds the antioxidant defense ability of the target cells (Lin et al. 2019). Acute exposure to UV radiation promotes protein oxidation and decreases the catalase activity in the skin (Svobodová et al. 2011). Among all the environmental factors, UV radiation is accounting for more than 80% in skin aging and the development of skin cancer (Amaro-Ortiz et al. 2014). The predominant mode of action by which UV radiation initiates molecular responses in the human skin is through the mediation of photochemical generation of ROS primarily via the formation of singlet oxygen ($^1O_2$), hydroxyl radical ($OH^•$), H_2O_2, and superoxide anion ($O_2^{-•}$) (Pospíšil et al. 2019). UV radiation penetrates the skin and subsequently reaches the cells, which is absorbed by DNA and thus resulted in the formation of photo-products that inactivate the functions of DNA (Ibarz et al. 2015). UV radiation can modulate the skin damage via two different mechanisms including (1) photosensitization mechanisms, in which the light is absorbed by exogenous (endogenous) sensitizers that are excited to the triplet states; and (2) direct absorption of the incident light by cellular components and leading to the excited state formation for a subsequent chemical reaction (Zaheer et al. 2016; Brem et al. 2017; de Jager et al. 2017). In general, the excited photosensitizers can trigger cellular damage via two mechanisms including (1) energy transfer from oxygen to produce reactive excited state, singlet oxygen (Type II) (Bacellar et al. 2015); and (2) hydrogen abstraction and electron transfer processes to produce free radicals (Type I) (Fukuzumi et al. 2003; Amić et al. 2017). The previous study showed that eukaryotic and prokaryotic cells exposed to UV can cause morphological transformation, mutation, chromosome changes, and cell death (Nawkar et al. 2013). In addition to nuclear DNA, UV radiation can trigger oxidative damage to mitochondrial DNA (mtDNA) (Van Houten et al. 2016). Indeed, aging is a multifactorial phenomenon characterized by increased susceptibility to functional decline and cellular loss, in which mtDNA damage response and mutations play a pivotal role (Höhn et al. 2017). The previous study has revealed that sunlight passing through the skin is not only can lead to DNA damage in white cells circulating through the skin capillaries (Brenner and Hearing 2008), but it can cause the damage of dermal mtDNA (Stout and Birch-Machin 2019). Singlet oxygen produced by UVA light has been reported to cause strand breaks in the mtDNA and leading to the deletions of mtDNA (Schuch et al. 2017). A study reported by Hahn and Zuryn (2019) has revealed that mtDNA serves as the most crucial target of endogenous ROS production because it lies in the inner mitochondrial membrane, which is close to the ETC that produced the freest radicals. Emerging evidence has suggested that mitochondria exert effective DNA repair mechanisms in the last few decades (Kowluru and Mishra 2018). The predominant DNA repair pathway has been reported to actively take place in mammalian mitochondria including the base excision repair pathway (Jang et al. 2019). The data showed that mtDNA is higher in sun-exposed skin compared to protected skin (Berneburg et al. 1999; Birch-Machin et al. 1998). In this regard, avoidance of sporadic and excessive cumulative sun exposure is vital in decreasing skin aging and the risk of skin cancer (Watson et al. 2016). The most frequent mutation is the 4977-base pair deletion or known as common deletion, which is increased in photoaged skin (Berneburg et al. 2004). Collectively, the DNA damage caused by ROS is not a rare phenomenon because the human cell is estimated to sustain about 10^5 oxidative hits per day (Poljšak and Dahmane 2012). Since the DNA is functionally stable, thus the incidence of cancer is expected to be low, by taking into account the high frequency of oxidative hits.

2.4 Changes in Cardiovascular System

Atherosclerosis is regarded as a scourge of increased age, albeit its onset can be traced from a young age (Sarbacher and Halper 2019). In particular, young individuals have a healthy intima with a thin layer of endothelial cells attached with glycocalyx on its luminal surface (Halper 2018). The endothelial cells are tightly linked to the internal elastic lamina and basement membrane (Halper 2018). Nonetheless, increasing age enhanced a repeating cycle of endothelial repair and injury leading to the thickening of tunica intima that may subsequently cause plaque formation and atherosclerosis (Sarbacher and Halper 2019). A significant change has been found in the function and structure of the cardiovascular system in the elderly, for instance, decreased elasticity (Fhayli et al. 2019). Increasing age is related to an elevation of the collagen cross-linking and collagen levels, activation of the fragmentation of the internal elastic membrane, stimulation of vascular smooth muscle hypertrophy, and increased intimal thickness (Lacolley et al. 2017). These microscopic changes are accompanied by the increased arterial stiffness and thickness and progressive elongation and dilatation of arteries. Arterial stiffening is linked to aging in Western societies even in the absence of cardiovascular disease (CVD) (Kitzman et al. 1988; Vaitevicius et al. 1993). It has demonstrated an increase in pulse wave velocity, widening of blood pressure, and elevation of systolic blood pressure (Kim and Kim 2019; Papaioannou et al. 2019). These changes create early reflected pressure waves that alter the pressure waveform (London et al. 2019). Such effects lead to an elevation of late systolic pressure peak and thus promote central vascular systolic blood pressure in the elderly (Malm et al. 2020). Further, the elevation of arterial stiffness promotes the afterload and end-systolic wall stress and subsequently may contribute to the progression of left ventricular hypertrophy (Table 2.1).

Increased impedance to left ventricular ejection and high systolic arterial pressure results in interstitial fibrosis and left ventricular hypertrophy (Mancusi et al. 2017). Aging reduced the myocardial relaxation rate. The left ventricle becomes stiffer and takes longer to relax and fill in diastole, and thereby increasing time of atrial contraction that contribute to normal left ventricular end-diastolic volume (Nichols 2005). When the pulse wave reaches the iliac bifurcation, it is reflected and transmitted back to the aorta. Likewise, the backward transmission wave is also accelerated (Weisfeldt 1998).

Aging is linked to an increased in sinoatrial node conduction time and a decrease in the intrinsic heart rate (Moghtadaei et al. 2016). There is a 40–50% decreased in the total number of nuclei and cells in the myocardium in individuals between 20 and 90 years old (Olivetti et al. 1991), suggesting that an age-related drop out of cells with compensatory hypertrophy of the remaining cells. Further, left ventricular

Table 2.1 Cardiovascular changes related to the aging process and their clinical responses

Cardiovascular structure and function	Outcomes	Clinical responses
Diastolic function	↓ Early diastolic filling	↓ E velocity
	↑ Late diastolic filling	Prolong E deceleration time
	↓ Myocardial diastolic velocities	↑ A velocity
Systolic function	Preserved	–
Cellular changes	↓ Sarcoplasmic reticulum Ca^{2+} ATPase protein level	Prolong of myocardial relaxation
	↓ Reuptake of Ca^{2+} into the sarcoplasmic reticulum	↑ Isovolumetric relaxation time
Myocardium	Loss of myocytes	–
	↑ Thickness left ventricular wall ratio to chamber size	
Arterial wall	↑ Thickness	↑ Systolic blood pressure
	↓ Compliance	↑ Pulse wave velocity
		Widening of pulse pressure

Source: Oxenham and Sharpe (2003)

cell hypertrophy also impairs left ventricular ejection (Milani et al. 2011). Strikingly, the central arterial stiffness is increased with age (Hughes et al. 2014). In the context of left ventricular function load, increased stiffness produces greater impedance to the left ventricular ejection (Weisfeldt 1998; Bell et al. 2017). Aging may stiffen the central aorta, and thereby result in the left ventricle ejecting the same amount of blood into the aorta (Weisfeldt 1998). This phenomenon may accelerate the velocity of blood movement to the arterial system. Acceleration in pulse wave velocity is then extended to the whole arterial tree (Weisfeldt 1998; Avramovski et al. 2018). The rate of transmission backward and forward is steady in individuals aged 20 years old and the systole is completed when the reflected wave returns to the heart (Weisfeldt 1998). The previous study has shown an effect in promoting the aortic diastolic pressure when the aortic valve is shut off (Weisfeldt 1998; Pagoulatou and Stergiopulos 2017). Indeed, it does not elevate the cardiac work and therefore regulates the aortic blood pressure during diastole. In individuals aged 40 years old, there is a small effect on blood pressure during systole, despite the reflected wave is returned well before the aortic valve is closed (Weisfeldt 1998). Likewise, the reflected wave in elderly people (80 years old) is also returned well before the aortic valve is shut off. This increases the arterial blood pressure and systolic left ventricular and thereby elevates the burden of the left ventricle in ejecting the blood (Weisfeldt 1998). Indeed, the reflected wave exerts an additional effect by reducing the diastolic aortic pressure to support the coronary flow. Reduced entire circulatory system response to β-adrenergic activation is one of the predominant changes in the cardiovascular system with age (Strait and Lakatta 2012). However, no alteration was found in response to α-stimulation. In particular, there is only a selective reduction of β-sympathetic in response with age (Weisfeldt 1998).

Compared to the young adults, elderly had a relatively constant or even higher circulating epinephrine and norepinephrine level (Fleg et al. 1985). Reduced cardiac response in the elderly is more likely due to the increased of activation, and thereby leading to desensitization as observed in heart failure (Fleg and Strait 2012). The age-related changes are mediated via a few mechanisms such as reduced intracellular calcium mobilization from the intracellular calcium stores of the cardiac muscle cells and decreased sympathetic response at the receptor levels (Weisfeldt 1998).

2.5 Changes in Renal System

The renal mass is reduced with age (Özcan et al. 2018), implied the decrease of nephrons (Denic et al. 2017). Increased age can also alter the intra-renal vascular activity, comprising of hyalinization of the vascular tuft, and thereby leading to the decreased blood flow in the afferent arterioles in the cortex (Weinstein and Anderson 2010). However, increased age did not show any changes in the medullary vasculature (Ljungqvist and Laggergren 1962). Indeed, the glomerular filtration rate and plasma flow were decreased with age (Abdulkader et al. 2017). Intriguingly, reduced glomerular filtration rate did not concomitant promote the creatinine plasma levels, indicating that creatinine is not a reliable indicator of glomerular filtration rate in elderly individuals (Mangoni and Jackson 2004).

Under the physiological state, the acid-base balance is maintained. However, a decreased response to stress could be due to an inability to cope with the acid loads, and thereby leading to the defective renal tubular in secretion of ammonium ions (Batlle et al. 2001). In addition, the ability to concentrate the urine is decreased during water deprivation (Sands 2012). Such finding is more likely due to a reduced number of nephrons to counteract with elevation perfusion of juxtamedullary glomeruli or increased solute load to produce medullary washout (Denic et al. 2016). The previous study has stated that decreased thirst was observed in elderly individuals during water deprivation (Phillips et al. 1984), although the considerable increases in plasma osmolality. This finding could be due to the increased activity of the renin-angiotensin-aldosterone axis and decreased sensation of mouth dryness (Phillips et al. 1984; O'Neill and McLean 1992).

2.6 Changes in Gastrointestinal System

The predominant changes occurred during aging are the secretion of pepsin and hydrochloric acid, which are reduced under the basal state. This alteration could be due to the changes in the neural and hormonal regulatory system and the enzymes secreting cells (Blechman and Gelb 1999). Indeed, the gastric emptying in young individuals is similar to that of the elderly (Gainsborough et al. 1993). Webster and Leeming (1975) and Husebye and Engedal (1992) did not find any changes in motility and digestion. Nonetheless, the absorption of iron, calcium, and sugars is reduced with increased age. In the pancreas, an alteration of the pancreatic secretion is observed with increased age (Laugier and Sarles 1984). Of all major enzymes, trypsin and lipase are markedly reduced with advancing age (Meyer and Necheles 1940). Whereas some amylase is remain unchanged (Meyer and Necheles 1940). Similarly, the stimulation of secretin-activated bicarbonate and pancreatic juice is also remained constant (Rosenberg et al. 1966). In the context of the liver changes, increased age is related to a progressive decreased in liver blood flow and liver volume (Koff et al. 1973). The enzymatic functions and hepatic structure is altered moderately (Mangoni and Jackson 2004). Several studies reported by Koff et al. (1973) and Kampmann et al. (1975) evaluated the liver function on healthy elderly individuals. The data demonstrated that the metabolism and protein synthesis was not shown any significant changes in elderly between aged 50–69 and 70–89 years old (Koff et al. 1973; Kampmann et al. 1975; Fu and Sreekumaran Nair 1998).

2.7 Changes in Neuroendocrine Response

Aging is accompanied by changes in the neuroendocrine response to physical or psychosocial stress (Heffner 2011). An alteration of the hypothalamic-pituitary-adrenal (HPA) axis is increased with age (Harris and Saltzman 2013). Hypersecretion of glucocorticoids and overexpression of HPA can cause the dendritic atrophy in neurons of the hippocampus, and thereby leading to memory and learning impairment (Myers et al. 2014). Loss or damage of hippocampal neurons impairs glucocorticoid secretion and the feedback suppression of the HPA axis and further destroys the consequence caused by increased glucocorticoid levels (Sapolsky 1987). The positive feedback on hippocampal neuronal loss is also known as the glucocorticoid cascade hypothesis (Sapolski et al. 1985). In this regard, glucocorticoids may sensitize hippocampal neurons to functional impairment and cell death (Horchar and Wohleb 2019). Under the chronic stress state, it was insufficient modulating of HPA axis activity in response to the challenge of sustained glucocorticoid concentrations. This finding could be attributed to the impaired feedback mediation of the HPA axis activity (Gust et al. 2000).

References

Abdulkader RCRM, Burdmann EA, Lebrão ML et al (2017) Aging and decreased glomerular filtration rate: an elderly population-based study. PLoS One 12:e0189935

Acosta C, Anderson HD, Anderson CM (2017) Astrocyte dysfunction in Alzheimer disease. J Neurosci Res 95:2430–2447

Ali A, Khan H, Bahadar R et al (2020) The impact of airborne pollution and exposure to solar ultraviolet radiation on skin: mechanistic and physiological insight. Environ Sci Pollut Res Int 27:28730–28736

Amaro-Ortiz A, Yan B, D'Orazio JA (2014) Ultraviolet radiation, aging and the skin: prevention of damage by topical cAMP manipulation. Molecules 19:6202–6219

Amić A, Lučić B, Stepanić V et al (2017) Free radical scavenging potency of quercetin catecholic colonic metabolites: thermodynamics of 2H+/2e− processes. Food Chem 218:144–151

An L, Sun Y, Zhang W et al (2018) Walnut diets up-regulate the decreased hippocampal neurogenesis and age-related cognitive dysfunction in d-galactose induced aged rats. Food Funct 9:4755–4762

Appel L, Appel E, Bogler O et al (2020) Older adults with cognitive and/or physical impairments can benefit from immersive virtual reality experiences: a feasibility study. Front Med 6:329

Arias-Cavieres A, Adasme T, Sánchez G et al (2017) Aging impairs hippocampal- dependent recognition memory and LTP and prevents the associated RyR up-regulation. Front Aging Neurosci 9:111

Armstrong RA (2016) Survival in the pre-senile dementia frontotemporal lobar degeneration with TDP-43 proteinopathy: effects of genetic, demographic and neuropathological variables. Folia Neuropathol 54:137–148

Ashapkin VV, Kutueva LI, Vanyushin BF (2019) Epigenetic clock: just a convenient marker or an active driver of aging? Adv Exp Med Biol 1178:175–206

Avramovski P, Avramovska M, Sotiroski K et al (2018) Ageing process and stiffening of arteries shown by increased pulse wave velocity. JSM Atheroscler 3:1040

Bacellar IOL, Tsubone TM, Pavani C et al (2015) Photodynamic efficiency: from molecular photochemistry to cell death. Int J Mol Sci 16:20523–20559

Baehr LM, West DWD, Marshall AG et al (2017) Muscle-specific and age-related changes in protein synthesis and protein degradation in response to hindlimb unloading in rats. J Appl Physiol 122:1336–1350

Bailey CH, Bartsch D, Kandel ER (1996) Toward a molecular definition of long-term memory storage. Proc Natl Acad Sci U S A 93:13445–13452

Barnes CA, McNaughton BL (1985) An age comparison of the rates of acquisition and forgetting of spatial information in relation to long-term enhancement of hippocampal synapses. Behav Neurosci 99:1040–1048

Batlle D, Ghanekar H, Jain S et al (2001) Hereditary distal renal tubular acidosis: new understandings. Annu Rev Med 52:471–484

Bell V, McCabe EL, Larson MG et al (2017) Relations between aortic stiffness and left ventricular mechanical function in the community. J Am Heart Assoc 6:e004903

Berneburg M, Grether-Beck S, Kurten V et al (1999) Singlet oxygen mediates the UVA-induced generation of the photoaging associated mitochondrial common deletion. J Biol Chem 274:15345–15349

Berneburg M, Plettenberg H, Medve-König K et al (2004) Induction of the photoaging-associated mitochondrial common deletion *in vivo* in normal human skin. J Invest Dermatol 122:1277–1283

Bessa JM, Oliveira M, Cerqueira JJ et al (2005) Age-related qualitative shift in emotional behaviour: paradoxical findings after re-exposure of rats in the elevated-plus maze. Behav Brain Res 162:135–142

Bettio LEB, Rajendran L, Gil-Mohapel J (2017) The effects of aging in the hippocampus and cognitive decline. Neurosci Biobehav Rev 79:66–86

Birch-Machin MA, Tindall M, Turner R et al (1998) Mitochondrial DNA deletions in human skin reflect photo- rather than chronologic aging. J Invest Dermatol 110:149–152

Blechman MB, Gelb AM (1999) Aging and gastrointestinal physiology. Clin Geriatr Med 15:429–438

Brem R, Guven M, Karran P (2017) Oxidatively-generated damage to DNA and proteins mediated by photosensitized UVA. Free Radic Biol Med 107:101–109

Brenner M, Hearing VJ (2008) The protective role of melanin against UV damage in human skin. Photochem Photobiol 84:539–549

Brightwell JJ, Smith CA, Countryman RA et al (2005) Hippocampal overexpression of mutant creb blocks long-term, but not short-term memory for a socially transmitted food preference. Learn Mem 12:12–17

Browning KS, Bailey-Serres J (2015) Mechanism of cytoplasmic mRNA translation. Arabidopsis Book 13:e0176

Burke SN, Barnes CA (2006) Neural plasticity in the ageing brain. Nat Rev Neurosci 7:30–40

Cai Y, Ford CP (2018) Dopamine cells differentially regulate striatal cholinergic transmission across regions through corelease of dopamine and glutamate. Cell Rep 25:3148–3157

Castellano JF, Fletcher BR, Kelley-Bell B et al (2012) Age-related memory impairment is associated with disrupted multivariate epigenetic coordination in the hippocampus. PLoS One 7:e33249

Chen JJ (2019) Functional MRI of brain physiology in aging and neurodegenerative diseases. Neuroimage 187:209–225

Darvas M, Palmiter RD (2011) Contributions of striatal dopamine signaling to the modulation of cognitive flexibility. Biol Psychiatry 69:704–707

de Jager TL, Cockrell AE, Du Plessis SS (2017) Ultraviolet light induced generation of reactive oxygen species. In: Ahmad S (ed) Ultraviolet light in human health, diseases and environment, Adv Exp Med Biol, vol 996. Springer, Cham, pp 15–23

Denic A, Glassock RJ, Rule AD (2016) Structural and functional changes with the aging kidney. Adv Chronic Kidney Dis 23:19–28

Denic A, Lieske JC, Chakkera HA et al (2017) The substantial loss of nephrons in healthy human kidneys with aging. J Am Soc Nephrol 28:313–320

Dunlosky J, Salthouse TA (1996) A decomposition of age-related differences in multitrial free recall. Aging Neuropsychol Cog 3:2–14

Dwyer BE, Fando JL, Wasterlain CG (1980) Rat brain protein synthesis declines during postdevelopmental aging. J Neurochem 35:746–749

Ekstrom R, Liu DS, Richardson A (1980) Changes in brain protein synthesis during the life span of male Fischer rats. Gerontology 26:121–128

Fando JL, Salinas M, Wasterlain CG (1980) Age-dependent changes in brain protein synthesis in the rat. Neurochem Res 5:373–383

Farage MA, Miller KW, Elsner P et al (2008) Intrinsic and extrinsic factors in skin ageing: a review. Int J Cosmetic Sci 30:87–95

Fhayli W, Boëté Q, Harki O et al (2019) Rise and fall of elastic fibers from development to aging. Consequences on arterial structure-function and therapeutical perspectives. Matrix Biol 84:41–56

Filion AM, Laughrea M (1985) Translation fidelity in the brain, liver, and hippocampus of the aging Fischer 344 rat. In: Sohal RS, Birnbaum LS, Cutler RG (eds) Molecular biology of aging: gene stability and gene expression. Raven Press, New York, pp 257–261

Fleg JL, Strait J (2012) Age-associated changes in cardiovascular structure and function: a fertile milieu for future disease. Heart Fail Rev 17:545–554

Fleg JL, Tzankoff SP, Lakatta EG (1985) Age-related augmentation of plasma catecholamines during dynamic exercise in healthy males. J Appl Physiol 59:1033–1039

Flexner JB, Flexner LB, Stellar E et al (1962) Inhibition of protein synthesis in brain and learning and memory following puromycin. J Neurochem 9:595–605

Flexner JB, Flexner LB, Stellar E (1963) Memory in mice as affected by intracerebral puromycin. Science 141:57–59

Flexner LB, Flexner JB, Roberts RB (1967) Memory in mice analyzed with antibiotics. Antibiotics are useful to study stages of memory and to indicate molecular events which sustain memory. Science 155:1377–1383

Foster TC (1999) Involvement of hippocampal synaptic plasticity in age-related memory decline. Brain Res Brain Res Rev 30:236–249

Frankland PW, Josselyn SA, Anagnostaras SG et al (2004) Consolidation of CS and US representations in associative fear conditioning. Hippocampus 14:557–569

Fu A, Sreekumaran Nair K (1998) Age effect on fibrinogen and albumin synthesis in humans. Am J Phys 275:E1023–E1030

Fukuzumi S, Shimoosako K, Suenobu T et al (2003) Mechanisms of hydrogen-, oxygen-, and electron-transfer reactions of cumylperoxyl radical. J Am Chem Soc 125:9074–9082

Gainsborough N, Maskrey VL, Nelson ML et al (1993) The association of age with gastric emptying. Age Ageing 22:37–40

Gelfo F, Mandolesi L, Serra L et al (2018) The neuroprotective effects of experience on cognitive functions: evidence from animal studies on the neurobiological bases of brain reserve. Neuroscience 370:218–235

Gonzalez GA, Montminy MR (1989) Cyclic AMP stimulates somatostatin gene transcription by phosphorylation of CREB at serine 133. Cell 59:675–680

Gust DA, Wilson ME, Stocker T et al (2000) Activity of the hypothalamic-pituitary-adrenal axis is altered by aging and exposure to social stress in female rhesus monkeys. J Clin Endocrinol Metab 85:2556–2563

Hahn A, Zuryn S (2019) Mitochondrial genome (mtDNA) mutations that generate reactive oxygen species. Antioxidants (Basel) 8:392

Halper J (2018) Basic components of vascular connective tissue and extracellular matrix. Adv Pharmacol 81:95–127

Harman MF, Martín MG (2020) Epigenetic mechanisms related to cognitive decline during aging. J Neurosci Res 98:234–246

Harris BN, Saltzman W (2013) Effects of aging on hypothalamic-pituitary-adrenal (HPA) axis activity and reactivity in virgin male and female California mice (*Peromyscus californicus*). Gen Comp Endocrinol 186:41–49

Harrison TM, La Joie R, Maass A et al (2019) Longitudinal tau accumulation and atrophy in aging and alzheimer disease. Ann Neurol 85:229–240

Hawley WR, Witty CF, Daniel JM et al (2015) Choline acetyltransferase in the hippocampus is associated with learning strategy preference in adult male rats. Behav Brain Res 289:118–124

Heffner KL (2011) Neuroendocrine effects of stress on immunity in the elderly: implications for inflammatory disease. Immunol Allergy Clin N Am 31:95–108

Hepel M, Andreescu S (2015) Chapter 1: Oxidative stress and human health. In: Hepel M, Andreescu S (eds) Oxidative stress: diagnostics, prevention, and therapy, vol 2. ACS Publications, Washington, DC, pp 1–33

Herrling T, Jung K, Fuchs J (2006) Measurements of UV-generated free radicals/reactive oxygen species (ROS) in skin. Spectrochimica Acta Part A: Mol Biomol Spectrosc 63:840–845

Hertelendy P, Varga DP, Menyhárt A et al (2019) Susceptibility of the cerebral cortex to spreading depolarization in neurological disease states: the impact of aging. Neurochem Int 127:125–136

Hipp MS, Kasturi P, Hartl U (2019) The proteostasis network and its decline in ageing. Nat Rev Mol Cell Biol 20:421–435

Höhn A, Weber D, Jung T et al (2017) Happily (n)ever after: aging in the context of oxidative stress, proteostasis loss and cellular senescence. Redox Biol 11:482–501

Horchar MJ, Wohleb ES (2019) Glucocorticoid receptor antagonism prevents microglia-mediated neuronal remodeling and behavioral despair following chronic unpredictable stress. Brain Behav Immun 81:329–340

Hughes TM, Kuller LH, Barinas-Mitchell EJM et al (2014) Arterial stiffness and β-amyloid progression in nondemented elderly adults. JAMA Neurol 71:562–568

Husebye E, Engedal K (1992) The patterns of motility are maintained in the human small intestine throughout the process of aging. Scand J Gastroenterol 27:397–404

Hwang IK, Yoo KY, Jung BK et al (2006) Correlations between neuronal loss, decrease of memory, and decrease expression of brain-derived neurotrophic factor in the gerbil hippocampus during normal aging. Exp Neurol 201:75–83

Ibarz R, Garvín A, Azuara E et al (2015) Modelling of ochratoxin A photo-degradation by a UV multi-wavelength emitting lamp. LWT-Food Sci Technol 61:385–392

Impey S, Mark M, Villacres EC et al (1996) Induction of CRE-mediated gene expression by stimuli that generate long-lasting LTP in area CA1 of the hippocampus. Neuron 16:973–982

Ingvar MC, Maeder P, Sokoloff L et al (1985) Effects of ageing on local rates of cerebral protein synthesis in Sprague-Dawley rats. Brain 108:155–170

Jacobson L, Zhang R, Elliffe D et al (2008) Correlation of cellular changes and spatial memory during aging in rats. Exp Gerontol 43:929–938

Jang S, Kumar N, Beckwitt EC et al (2019) Damage sensor role of UV-DDB during base excision repair. Nat Struc Mol Biol 26:695–703

Jeffries AR, Maroofian R, Salter CG et al (2019) Growth disrupting mutations in epigenetic regulatory molecules are associated with abnormalities of epigenetic aging. Genome Res 29:1057–1066

Jonathan S, Daniel JM, John M et al (2015) The mechanical cause of age-related dementia (Alzheimer's disease): the brain is destroyed by the pulse. J Alzheimers Dis 44:355–373

Kampmann JP, Sinding J, Moller-Jorgensen I (1975) Effect of age on liver function. Geriatrics 30:91–95

Karisetty BC, Bhatnagar A, Armour EM et al (2020) Amyloid-β peptide impact on synaptic function and neuroepigenetic gene control reveal new therapeutic strategies for Alzheimer's disease. Front Mol Neurosci 13:577622

Kedlian VR, Donertas HM, Thornton JM (2019) The widespread increase in inter-individual variability of gene expression in the human brain with age. Aging 11:2253–2280

Kim H-L, Kim S-H (2019) Pulse wave velocity in atherosclerosis. Front Cardiovasc Med 6:41

Kim I, Wilson RE, Wellman CL et al (2005) Aging and cholinergic deafferentation alter GluR1 expression in rat frontal cortex; differential preservation of AMPA receptor subunits in the hippocampi of Alzheimer's disease patients according to Braak stage. Neurobiol Aging 26:1073–1081

Kitzman D, Scholz D, Hagen P et al (1988) Age-related changes in normal human hearts during the first ten decades of life. Part II (maturity): a quantitative anatomic study of 765 specimens from subjects 20 to 99 years old. Mayo Clin Proc 63:137–146

Koff RS, Garvey AJ, Burney SW et al (1973) Absence of an age effect on sulfobromophtalein retention in healthy men. Gastroenterology 65:300–302

Kosik KS, Rapp PR, Raz N et al (2012) Mechanisms of age-related cognitive change and targets for intervention: epigenetics. J Gerontol Ser A 67:741–746

Kowluru RA, Mishra M (2018) Therapeutic targets for altering mitochondrial dysfunction associated with diabetic retinopathy. Expert Opin Ther Targets 22:233–245

Kwapis JL, Alaghband Y, Kramár EA et al (2018) Epigenetic regulation of the circadian gene Per1 contributes to age-related changes in hippocampal memory. Nat Commun 9:3323

Lacolley P, Regnault V, Segers P et al (2017) Vascular smooth muscle cells and arterial stiffening: relevance in development, aging, and disease. Physiol Rev 97:1555–1617

Langton AK, Alessi S, Hann M et al (2019) Aging in skin of color: disruption to elastic fiber organization is detrimental to skin's biomechanical function. J Invest Dermatol 139:779–788

Larsson L, Degens H, Li M et al (2019) Sarcopenia: aging-related loss of muscle mass and function. Physiol Rev 99:427–511

Laugier R, Sarles H (1984) Pancreatic function and diseases. In: Hellemans J, Vantrappen G (eds) Gastrointestinal tract disorders in the elderly. Churchill Livingstone, London, p 243

Lee SD, Ong B, Pike KE et al (2018) Prospective memory and subjective memory decline: a neuropsychological indicator of memory difficulties in community-dwelling older people. J Clin Exp Neuropsychol 40:183–197

Li Y, Xie L, Huang T et al (2019) Aging neurovascular unit and potential role of DNA damage and repair in combating vascular and neurodegenerative disorders. Front Neurosci 13:778

Lin T-Y, Wu P-Y, Hou C-W et al (2019) Protective effects of sesamin against UVB-induced skin inflammation and photodamage *in vitro* and *in vivo*. Biomol Ther 9:479

Ljungqvist A, Laggergren C (1962) Normal intrarenal arterial pattern in adult and aging human kidney. A microangiographical and histological study. J Anat 96:285–300

London GM, Pannier B, Safar ME (2019) Arterial stiffness gradient, systemic reflection coefficient, and pulsatile pressure wave transmission in essential hypertension. Hypertension 74:1366–1372

Luo L, Craik FIM (2008) Aging and memory: a cognitive approach. Can J Psychiatr 53:346–353

Malm IÅ, Alehagen U, Blomstrand P et al (2020) Higher blood pressure in elderly hypertensive females, with increased arterial stiffness and blood pressure in females with the Fibrillin-1 2/3 genotype. BMC Cardiovasc Disord 20:180

Mancusi C, Gerdts E, de Simone G et al (2017) Higher pulse pressure/stroke volume index is associated with impaired outcome in hypertensive patients with left ventricular hypertrophy the LIFE study. Blood Press 26:150–155

Mangoni AA, Jackson SHD (2004) Age-related changes in pharmacokinetics and pharmacodynamics: basic principles and practical applications. Br J Clin Pharmacol 57:6–14

Mazarakis NK, Cybulska-Klosowicz A, Grote H et al (2005) Deficits in experience-dependent cortical plasticity and sensory-discrimination learning in presymptomatic Huntington's disease mice. J Neurosci 25:3059–3066

Meyer J, Necheles H (1940) Studies in old age. IV. The clinical significance of salivary, gastric, and pancreatic secretion in the aged. JAMA 115:2050

Mielke MM, Hagen CE, Wennberg AMV et al (2017) Association of plasma total tau level with cognitive decline and risk of mild cognitive impairment or dementia in the Mayo Clinic Study on Aging. JAMA Neurol 74:1073–1080

Milani RV, Drazner MH, Lavie CJ et al (2011) Progression from concentric left ventricular hypertrophy and normal ejection fraction to left ventricular dysfunction. Am J Cardiol 108:992–996
Moghtadaei M, Jansen HJ, Mackasey M et al (2016) The impacts of age and frailty on heart rate and sinoatrial node function. J Physiol 594:7105–7126
Mohebi A, Pettibone JR, Hamid AA et al (2019) Dissociable dopamine dynamics for learning and motivation. Nature 570:65–70
Monteiro TS, Adab HZ, Chalavi S et al (2020) Reduced modulation of task-related connectivity mediates age-related declines in bimanual performance. Cereb Cortex 30:4346–4360
Morris KA, Gold PE (2012) Age-related impairments in memory and in CREB and pCREB expression in hippocampus and amygdala following inhibitory avoidance training. Mech Ageing Dev 133:291–299
Muñoz-Moreno E, Tudela R, López-Gil X et al (2018) Early brain connectivity alterations and cognitive impairment in a rat model of Alzheimer's disease. Alzheimers Res Ther 10:16
Murman DL (2015) The impact of age on cognition. Semin Hear 36:111–121
Murray EA, Wise SP, Graham KS (2018) Representational specializations of the hippocampus in phylogenetic perspective. Neurosci Lett 680:4–12
Myers B, McKlveen JM, Herman JP (2014) Glucocorticoid actions on synapses, circuits, and behavior: implications for the energetics of stress. Front Neuroendocrinol 35:180–196
Naidoo K, Birch-Machin MA (2017) Oxidative stress and ageing: the influence of environmental pollution, sunlight and diet on skin. Cosmetics 4:4
Nawkar GM, Maibam P, Park JH et al (2013) UV-induced cell death in plants. Int J Mol Sci 14:1608–1628
Nelson AR, Sweeney MD, Sagare AP et al (2016) Neurovascular dysfunction and neurodegeneration in dementia and Alzheimer's disease. Biochim Biophys Acta 1862:887–900
Nichols WW (2005) Clinical measurement of arterial stiffness obtained from noninvasive pressure waveforms. Am J Hypertens 18:3S–10S
O'Neill PA, McLean KA (1992) Water homeostasis and ageing. Med Lab Sci 49:291–298
Olivetti G, Melessari M, Capasso JM et al (1991) Cardiomyopathy of the aging human heart: myocyte loss and reactive cellular hypertrophy. Circ Res 68:1560–1568
Oxenham H, Sharpe N (2003) Cardiovascular aging and heart failure. Eur J Heart Fail 5:427–434
Özcan MF, Altınova S, Atan A (2018) Treatment approaches to small renal masses in patients of advanced age (≥75 years). Turk J Urol 44:281–286
Pagoulatou S, Stergiopulos N (2017) Evolution of aortic pressure during normal ageing: a model-based study. PLoS One 12:e0182173
Panich U, Sittithumcharee G, Rathviboon N et al (2016) Ultraviolet radiation-induced skin aging: the role of DNA damage and oxidative stress in epidermal stem cell damage mediated skin aging. Stem Cells Int 2016, Article ID 7370642, 14 pages
Papaioannou TG, Oikonomou E, Lazaros G et al (2019) The influence of resting heart rate on pulse wave velocity measurement is mediated by blood pressure and depends on aortic stiffness levels: insights from the Corinthia study. Physiol Meas 40:055005
Penner MR, Roth TL, Chawla MK et al (2011) Age-related changes in *Arc* transcription and DNA methylation within the hippocampus. Neurobiol Aging 32:2198–2210
Phillips PA, Rolls BJ, Ledingham JG et al (1984) Reduced thirst after water deprivation in healthy elderly men. N Engl J Med 311:753–759
Piccirillo AR, Cattley RT, D'Cruz LM et al (2019) Histone acetyltransferase CBP is critical for conventional effector and memory T-cell differentiation in mice. J Biol Chem 294:2397–2406
Poddar J, Pradhan M, Ganguly G et al (2019) Biochemical deficits and cognitive decline in brain aging: intervention by dietary supplements. J Chem Neuroanat 95:70–80
Poljšak B, Dahmane R (2012) Free radicals and extrinsic skin aging. Dermatol Res Pract 2012, Article ID 135206, 4 pages
Poljšak B, Fink R (2014) The protective role of antioxidants in the defence against ROS/RNS-mediated environmental pollution. Oxidative Med Cell Longev 2014, Article ID 671539, 22 pages

Pospíšil P, Prasad A, Rác M (2019) Mechanism of the formation of electronically excited species by oxidative metabolic processes: role of reactive oxygen species. Biomol Ther 9:258

Prange S, Danaila T, Laurencin C et al (2019) Age and time course of long-term motor and nonmotor complications in Parkinson disease. Neurology 92:e148–e160

Prasad A, Pospišil P (2011) Two-dimensional imaging of spontaneous ultra-weak photon emission from the human skin: role of reactive oxygen species. J Biophotonics 4:840–849

Qiu C, Kivipelto M, von Strauss E (2009) Epidemiology of Alzheimer's disease: occurrence, determinants, and strategies toward intervention. Dialogues Clin Neurosci 11:111–128

Rastogi A, Pospísil P (2011) Spontaneous ultraweak photon emission imaging of oxidative metabolic processes in human skin: effect of molecular oxygen and antioxidant defense system. J Biomed Opt 16:096005

Ravichandran VA, Kim M, Han SK et al (2018) *Stachys sieboldii* extract supplementation attenuates memory deficits by modulating BDNF-CREB and its downstream molecules, in animal models of memory impairment. Nutrients 10:917

Richard IH (2005) Anxiety disorders in Parkinson's disease. Adv Neurol 96:42–55

Richardson A (1981) The relationship between aging and protein synthesis. In: Florini JR (ed) CRC handbook of biochemistry in aging. CRC Press, Boca Raton, pp 79–101

Richardson A, Birchenall-Sparks MC (1983) Age-related changes in protein synthesis. Rev Biol Res Aging 1:255–273

Richardson A, Birchenall-Sparks MC, Staecker JL (1983) Aging and transcription. Rev Biol Res Aging 1:275–294

Rosenberg IR, Friedland N, Janowitz HD et al (1966) The effect of age and sex upon human pancreatic secretion of fluid and bicarbonate. Gastroenterology 50:191

Rosenzweig ES, Barnes CA (2003) Impact of aging on hippocampal function: plasticity, network dynamics, and cognition. Prog Neurobiol 69:143–179

Rudolf E, Cervinka M (2011) Stress responses of human dermal fibroblasts exposed to zinc pyrithione. Toxicol Lett 204:164–173

Sands JM (2012) Urine concentrating and diluting ability during aging. J Gerontol A Biol Sci Med Sci 67:1352–1357

Sapolski RM, Krey LC, McEwen BS (1985) Prolonged glucocorticoid exposure reduces hippocampal neuron number: implications for aging. J Neurosci 5:1222–1227

Sapolsky RM (1987) Glucocorticoids and hippocampal damage. Trends Neurosci 10:346–349

Sarbacher CA, Halper JT (2019) Chapter 11: Connective tissue and age-related diseases. In: Harris JR, Korolchuk VI (eds) Biochemistry and cell biology of ageing: part II Clinical science, Subcellular biochemistry, vol 91. Springer, Singapore, pp 281–310

Schimanski LA, Barnes CA (2010) Neural protein synthesis during aging: effects on plasticity and memory. Front Aging Neurosci 2:26

Schuch AP, Moreno NC, Schuch NJ et al (2017) Sunlight damage to cellular DNA: focus on oxidatively generated lesions. Free Radic Biol Med 107:110–124

Schulz S, Siemer H, Krug M et al (1999) Direct evidence for biphasic cAMP responsive element-binding protein phosphorylation during long-term potentiation in the rat dentate gyrus *in vivo*. J Neurosci 19:5683–5692

Scott CB, Clay OJ, Epps F et al (2020) Associations of knowledge of Alzheimer's disease and memory loss and employment status with burden in African American and Caucasian family caregivers. Dementia 19:847–860

Segovia G, Del Arco A, Mora F (1999) Effects of aging on the interaction between glutamate, dopamine, and GABA in striatum and nucleus accumbens of the awake rat. J Neurochem 73:2063–2072

Sgarbieri VC, Pacheco MTB (2017) Healthy human aging: intrinsic and environmental factors. Braz J Food Technol 20:e2017007

Shim H, Kim M, Won CW (2020) Motoric cognitive risk syndrome is associated with processing speed and executive function, but not delayed free recall memory: the Korean frailty and aging cohort study (KFACS). Arch Gerontol Geriatr 87:103990

Smith CB, Sun Y, Sokoloff L (1995) Effects of aging on regional rates of cerebral protein synthesis in the Sprague-Dawley rat: examination of the influence of recycling of amino acids derived from protein degradation into the precursor pool. Neurochem Int 27:407–416
Srivastava S (2017) The mitochondrial basis of aging and age-related disorders. Genes 8:398
Stout R, Birch-Machin M (2019) Mitochondria's role in skin ageing. Biology 8:29
Strait JB, Lakatta EG (2012) Aging-associated cardiovascular changes and their relationship to heart failure. Heat Fail Clin 8:143–164
Subramaniapillai S, Rajagopal S, Elshiekh A et al (2019) Sex differences in the neural correlates of spatial context memory decline in healthy aging. J Cogn Neurosci 31:1895–1916
Svobodová AR, Galandáková A, Sianská J et al (2011) Acute exposure to solar simulated ultraviolet radiation affects oxidative stress-related biomarkers in skin, liver and blood of hairless mice. Biol Pharm Bull 34:471–479
Takeda S (2019) Progression of Alzheimer's disease, tau propagation, and its modifiable risk factors. Neurosci Res 141:36–42
Tan BL, Norhaizan ME, Liew W-P-P et al (2018) Antioxidant and oxidative stress: a mutual interplay in age-related diseases. Front Pharmacol 9:1162
Tan L, Register TC, Yammani RR (2020) Age-related decline in expression of molecular chaperones induces endoplasmic reticulum stress and chondrocyte apoptosis in articular cartilage. Aging Dis 11:1091–1102
Theofilas P, Ehrenberg AJ, Dunlop S et al (2017) Locus coeruleus volume and cell population changes during Alzheimer's disease progression: a stereological study in human postmortem brains with potential implication for early-stage biomarker discovery. Alzheimers Dement 13:236–246
Tong J, Williams B, Rusjan PM et al (2020) Concentration, distribution, and influence of aging on the 18 kDa translocator protein in human brain: implications for brain imaging studies. J Cereb Blood Flow Metab 40:1061–1076
Trist BG, Hare DJ, Double KL (2019) Oxidative stress in the aging substantia nigra and the etiology of Parkinson's disease. Aging Cell 18:e13031
Trouba KJ, Hamadeh HK, Amin RP et al (2002) Oxidative stress and its role in skin disease. Antioxid Redox Signal 4:665–673
Vaitevicius P, Fleg J, Engel J et al (1993) Effects of age and aerobic capacity on arterial stiffness in healthy adults. Circulation 88:1456–1462
Van Houten B, Hunter SE, Meyer JN (2016) Mitochondrial DNA damage induced autophagy, cell death, and disease. Front Biosci (Landmark Ed) 21:42–54
Vanzi V, Toma E (2018) Recognising and managing age-related dermatoporosis and skin tears. Nurs Older People 30:26–31
Vila M (2019) Neuromelanin, aging, and neuronal vulnerability in Parkinson's disease. Mov Disord 34:1440–1451
Wang J, Sun B-L, Xiang Y et al (2020) Capsaicin consumption reduces brain amyloid-beta generation and attenuates Alzheimer's disease-type pathology and cognitive deficits in APP/PS1 mice. Transl Psychiatry 10:230
Watson M, Holman DM, Maguire-Eisen M (2016) Ultraviolet radiation exposure and its impact on skin cancer risk. Semin Oncol Nurs 32:241–254
Webster GC (1985) Protein synthesis in aging organisms. In: Sohal RS, Birnbaum LS, Cutler RG (eds) Molecular biology of aging: gene stability and gene expression. Raven Press, New York, pp 263–289
Webster SG, Leeming JT (1975) The appearance of the small bowel mucosa in old age. Age Ageing 4:168–174
Wegmann S, Bennett RE, Delorme L et al (2019) Experimental evidence for the age dependence of tau protein spread in the brain. Sci Adv 5:eaaw6404
Weinstein JR, Anderson S (2010) The aging kidney: physiological changes. Adv Chronic Kidney Dis 17:302–307

Weisfeldt M (1998) Aging, changes in the cardiovascular system, and responses to stress. Am J Hypertens 11:41S–45S

Weiss C, Disterhoft JF (2015) The impact of hippocampal lesions on trace eyeblink conditioning and forebrain-cerebellar interactions. Behav Neurosci 129:512–522

Winocur G (1988a) A neuropsychological analysis of memory loss with age. Neurobiol Aging 9:487–494

Winocur G (1988b) Long-term memory loss in senescent rats: neuropsychological analysis of interference and context effects. Psychol Aging 3:273–279

Wong VW, Akaishi S, Longaker MT et al (2011) Pushing back: wound mechanotransduction in repair and regeneration. J Investig Dermatol 131:2186–2196

Wu Y, Wu X, Wei Q et al (2020) Differences in cerebral structure associated with depressive symptoms in the elderly with Alzheimer's disease. Front Aging Neurosci 12:107

Yamamoto N, Fox MJ, Boys E et al (2019) Effects of orientation change during environmental learning on age-related difference in spatial memory. Behav Brain Res 365:125–132

Yang JW, Liu DS, Richardson A (1977) Biochemical studies of chick brain development and maturation: I. Alterations in the macromolecule content and cell-free protein synthesis. Mech Ageing Dev 6:77–93

Yin JC, Tully T (1996) CREB and the formation of long-term memory. Curr Opin Neurobiol 6:264–268

Yousufuddin M, Young N (2019) Aging and ischemic stroke. Aging (Albany NY) 11:2542–2544

Zaheer MR, Gupta A, Iqbal J et al (2016) Molecular mechanisms of drug photodegradation and photosensitization. Curr Pharm Design 22:768–782

Zeng X-S, Geng W-S, Jia J-J et al (2018) Cellular and molecular basis of neurodegeneration in Parkinson disease. Front Aging Neurosci 10:109

Zhan J-Q, Zheng L-L, Chen H-B et al (2018) Hydrogen sulfide reverses aging-associated amygdalar synaptic plasticity and fear memory deficits in rats. Front Neurosci 12:390

Zhao C, Gammie SC (2014) Glutamate, GABA, and glutamine are synchronously upregulated in the mouse lateral septum during the postpartum period. Brain Res 1591:53–62

Zhao TT, Kim KS, Shin KS et al (2017) Gypenosides ameliorate memory deficits in MPTP-lesioned mouse model of Parkinson's disease treated with L-DOPA. BMC Complement Altern Med 17:449

Chapter 3
Age-Related Oxidative Stress-Induced Redox Imbalance

3.1 Aging and Oxidative Stress

Aging is a biological process characterized by physiological, progressive, and time-dependent declines concomitantly with increased incidence of age-related diseases (Morrison and Newell 2012; Tan et al. 2018a). Several theories have been proposed to describe the underlying molecular mechanism of aging and the causality (Dice 1993). The widely accepted theory is the "oxidative stress hypothesis" (Ghezzi et al. 2017) that modified and advanced the free radical theory of aging (Harman 1956). Based on the oxidative stress hypothesis, oxidative damage is not only shown by the uncontrolled generation of ROS as described in the original free radical theory, other oxidants such as reactive lipid species and RNS are also contributed to this phenomenon. Notably, the oxidative stress hypothesis suggested the essential role of antioxidant defenses as a critical component of the overall redox balance of the organism, which was not regarded in the original theory of free radical (Sohal and Allen 1990; Sohal et al. 2002).

ROS production is an inevitable consequence of life occurred in an aerobic environment (Taverne et al. 2018). In general, ROS are characterized by an increase of chemical reactivity that includes both non-radical and free radicals species (Arman et al. 2019). Under a healthy state, there is a balance between the activities of non-enzymatic and enzymatic antioxidant systems that decreased or scavenge the ROS levels (Lobo et al. 2010). Redox imbalance is caused by a reduction in antioxidant reserve and increased ROS generation. Subsequently, this process may produce oxidative stress (Songbo et al. 2019).

Inappropriate modulating of the immune system and changing the redox status during aging may result in systemic inflammation (Weyand and Goronzy 2016). This process increased the stimulation of inflammatory mediators through oxidative stress to trigger redox imbalance (Charlotte 2019). In particular, the age-related redox imbalance could be attributed to the induction of the incessantly generate of

B. L. Tan, M. E. Norhaizan, *The Role of Antioxidants in Longevity and Age-Related Diseases*, https://doi.org/10.1007/978-3-030-82859-2_3

reactive species such as reactive lipid aldehydes, peroxynitrite (ONOO−), superoxide (O_2^-), reactive nitric oxide (NO), H_2O_2, and hydroxyl radical (•OH), and the net effect of low antioxidative defense systems (Tan et al. 2018a). Indeed, unresolved chronic inflammation during aging may act as a pathophysiologic link that shifts the normal functional changes to age-related metabolic disorders (Natarajan et al. 2020). Oxidative stress is reinforced by several reactive species, for instance, singlet oxygen, H_2O_2, O_2^- , other radicals as well as non-radicals, which are produced continuously in the body due to the aerobic metabolism, and thus potentially changing the basic structural components, for instance, lipids, proteins, and nucleic acids and cellular activities (Krumova and Cosa 2016). The biological sources of reactive species varied based on the cellular activities that linked to the plasma membrane-related NADPH oxidase, cyclooxygenase (COX), and lipoxygenase; NADH dehydrogenase, ubiquinone, and mitochondrial electron transport system; microsomal electron transport, cytochrome b_5, and cytochrome P_{450}; oxidases and flavoproteins in peroxisome; and xanthine oxidase (XO) in the cytosol (Bodamyali et al. 2000).

Oxidative stress is regarded as an imbalance between anti- and prooxidant species and thereby leading to the cellular and molecular damage (Miletić et al. 2018). Mitochondria are primary organelles that generate energy via oxidative phosphorylation to produce adenosine triphosphate (ATP), a molecule that is pivotal for cells (Nolfi-Donegan et al. 2020). Nearly 90% of total oxygen (O_2) consumed by electron transport chain (ETC) is taken up by the cells (Wallace 2013). Under these circumstances, ROS are produced as by-products for the partial four-electron reduction of O_2 to generate a water molecule, a final electron acceptor during the ATP generation process (Zhao et al. 2019). During the normal physiological phenomena, approximately 0.1–0.5% of inhaled O_2 is shifted to O_2 (Servais et al. 2009). Under a normal healthy state, the oxidation and production of ROS are occurred in a controlled condition (Tan et al. 2018a). On the other hand, the generation of ROS is stimulated under disease conditions or high-stress states (Pickering 2021). Indeed, the ROS produced from aerobic respiration can increase the oxidative damage in macromolecules, such as proteins, DNA, and lipids, and thereby resulted in cell deaths (Caliri et al. 2021), and influence the healthspan of a variety of crucial organ systems (Bornstein et al. 2020).

The interaction between inflammation and oxidative stress is tightly linked to the biosynthetic pathway of prostaglandins (PGs) that generates reactive species (Marqués et al. 2020). PGs are arachidonic acid-derived lipid metabolites that shown a potent proinflammatory and potentially pathogenic activity (Ricciotti and FitzGerald 2011). Production of reactive species from PGs metabolism aggravates inflammatory conditions and enhances tissue damage (Ricciotti and FitzGerald 2011). COX is another key enzyme that modulates PG synthetic pathway which converts arachidonic acid to prostaglandin H_2 (PGH_2) (Park et al. 2006). The reactive species generated by the PG synthesis pathway further intensify the overall reactive species pool under pathological and normal conditions, especially during aging (Tan et al. 2018a).

3.2 Redox Imbalance in Diabetes

Diabetes is the outcome of dysregulation of glucose metabolism, which is due to the stimulation of protein kinase C, activation of the poly ADP ribose polymerase (PARP) and polyol pathway, and increased protein glycation (Brownlee 2001, 2005; Luo et al. 2016; Zheng et al. 2016, 2017). These hyperglycemia activation pathways may lead to ROS production, which can trigger cell death, mitochondrial dysfunction, and oxidative stress (Robertson 2004; Kassab and Piwowar 2012).

Glucose is one of the pivotal sources of NADH. Excessive in this source can result in NAD^+ deficiency and NADH production, and thereby leading to the NAD^+/NADH redox imbalance (Wu et al. 2016). Redox imbalance between NAD^+ and NADH is the main contributor to oxidative stress and ROS production (Wu et al. 2016; Boesten et al. 2015). The predominant source of redox imbalance is believed to be activated from the stimulation of the PARP and polyol pathway (Masutani et al. 1999; Pieper et al. 1999; Tang et al. 2012; Boesten et al. 2015). In general, the reaction of the polyol pathway involves two enzymes, namely sorbitol dehydrogenase and aldose reductase (Ng et al. 1998). The polyol pathway shifts the NADPH to NADH when it converts to fructose from glucose by modulating the two-reaction mechanism (Kador and Kinoshita 1985), and thereby leading to the NADH overproduction (Yabe-Nishimura 1998; Hodgkinson et al. 2001). Another pathway is PARP that utilized NAD^+ as its substrate (Pacher and Szabo 2005). This enzyme is overexpressed by hyperglycemia triggered by DNA oxidative damage, which results in a potential depletion of NAD^+ (Szabo 2005; Obrosova et al. 2005). Subsequently, these two activated pathways, which are NAD^+/NADH redox imbalance with increased NADH expression and reduced NAD^+ levels, and thereby leading to reductive stress that may progress to oxidative stress (Yan 2014).

3.3 Redox Imbalance in Neurodegenerative Disease

Substantial evidence revealed that a positive association between neuroinflammation and oxidative damage in the neurodegenerative process (Kwon and Koh 2020; Tsukahara et al. 2020; Simpson and Oliver 2020). Neuroinflammation plays a crucial role in a common neurodegenerative disease (Song et al. 2020), which is accompanied by an elevation of $O_2^{\bullet-}$ and NO with the production of H_2O_2 (Brown 2007). The previous study has demonstrated that the brain tissues in patients with Alzheimer's disease had high oxidative stress markers, for instance, glycoxidation, protein oxidative damage, and lipid peroxidation (Barbagallo et al. 2015). The intraneuronal levels of glutathione (GSH) were markedly reduced in the cortex and hippocampus in Alzheimer's disease patients (Aquilano et al. 2014; Barbagallo et al. 2015). Therefore, the losses of ROS balance produced a chronic oxidative environment, which may trigger diminish of antioxidant activity and expression, and thereby increased the risk of developing neurodegenerative disease (García-Sánchez et al. 2020). An alteration of redox homeostasis increases the peroxidation of fatty acids, modulates the protein unfolding, insulin resistance, and oxidation of cholesterol, and activates the advanced glycosylation products formation (McGrath et al. 2001; Uttara et al. 2009; Elfrink et al. 2012; Talbot et al. 2012; Gamba et al. 2015; Kuhla et al. 2015). Further, brains from patients with Alzheimer's disease show increased levels of 8-hydroxyguanosine (8-OHG) and heme oxygenase-1 (HO-1) compared with that of the control (Moreira et al. 2010). Although the cause of redox imbalance in Alzheimer's disease pathogenesis requires further elucidation, most of the experimental studies suggest that changes of redox transition metal balance such as copper and iron could play a vital role in neurodegeneration (Lovell et al. 1998; Molina et al. 1998; Hane and Leonenko 2014). Intriguingly, the brain of patients with Alzheimer's disease had high levels of copper and iron. The previous study revealed that the concentrations of iron, copper, and zinc in senile plaques cores and rims were significantly increased in patients with Alzheimer's disease (Lovell et al. 1998). Indeed, the activities of ceruloplasmin and ferritin are crucial for the mediation of metal homeostasis, which is altered in Alzheimer's disease (Torsdottir et al. 2011). Crouch et al. (2007) further revealed that several biometals are markedly reduced in the Alzheimer's disease brain, which may increase the development of senile plaques. In support of this, the brain from Alzheimer's disease patients and cortical neurons derived from Alzheimer's disease transgenic mice showed reduced levels of intracellular copper (Schrag et al. 2011). An alteration of these levels may partially contribute to Alzheimer's disease pathogenesis. In this regard, dysregulation of biometal homeostasis in Alzheimer's disease may provide a useful approach to restore neuronal functions (Barthelson et al. 2020).

3.4 Redox Imbalance in Cancer

Redox homeostasis is regulated by the net physiologic balance between oxidizing and reducing equivalents within subcellular components, including antioxidant enzymes and ROS (Hussain et al. 2003). RNS/ROS are thought to be involved in the mediation of redox-regulated signaling pathways and DNA damage (Tan et al. 2018b). For instance, ABL/BCR oncogenes trigger ROS and thereby leading to the modulation of nuclear factor erythroid 2-related factor 2 (Nrf2), a primary regulator of cellular redox homeostasis (Irwin et al. 2013).

Posttranslational modifications involved deletion/addition of low-molecular-weight metabolites to macromolecules or the redox-sensing Cys residues, triggered by redox imbalance and/or ROS (Cuello and Eaton 2019). Most of the intracellular protein possesses cysteine residues, which is placed in the activity center of the proteins (Liu et al. 2017). ROS can reversibly oxidize the active thiol group of cysteine residues into protein S-glutathione disulfide, intra/intermolecular disulfide bridge, or sulfenic acid (Wang et al. 2008). Posttranslational modifications controlled by redox reaction, accompanied by the intramolecular protein disulfide bonds between the Cys residues, comprising of two paired sulfhydryl (–SH) groups, in which the reaction may reverse by reduction (Klomsiri et al. 2011). In general, oxygen is needed for cellular metabolism and produced ROS (Lee et al. 2013). Indeed, the physiological levels between antioxidants and oxidants are required to modulate the cellular process (Rahal et al. 2014). An alteration of these levels may lead to the progression of cancers (Harris and DeNicola 2020). Elevation of oxidants levels in a certain period may oxidize the intramolecular spaced –SH groups to disulfides, which is potentially influence the protein function (Wani et al. 2014). For instance, TRX is an antioxidant protein that provides a reducing equivalent to protein expression involved in cancer such as ribonucleotide reductase (Mohammadi et al. 2019). TRX may undergo reversible redox reactions such as Cys73, Cys69, and Cys62 located outside the active site, or other alterations such as Cys35 and Cys32 located within the active site (Lee et al. 2013).

The previous study has demonstrated that moderate ROS levels as signaling messengers, which could induce invasion and survival in cancer cells (Wang et al. 2020). Redox protein may inhibit the excessive ROS that acts as a tumor promoter (Chen et al. 2016). In particular, redox modification of proteins involved in colorectal cancer is mediated by several transcriptional factors and signaling pathways (Yang et al. 2013). In this regard, targeting redox-sensitive signaling pathways may provide great promise in the treatment of cancer including colorectal cancer.

3.5 Redox Imbalance in Cardiovascular Disease

Redox signaling has been identified as a predominant regulator in numerous disorders such as fibrosis, cardiac hypertrophy, angiogenesis, atherosclerosis, and vascular smooth muscle proliferation (Tavakoli and Asmis 2012; Sag et al. 2014; Prieto-Bermejo and Hernández-Hernández 2017; Durgin and Straub 2018; Veith et al. 2019). ROS may trigger acute stimulations in cellular functions through a covalent alteration of the target molecule (Rea et al. 2018). For instance, several critical proteins are involved in myocardial excitation-contraction coupling, including contractile proteins, sarcoplasmic reticulum calcium release channels, and sarcolemmal ion channels; all these may undergo redox-sensitive modulations in specific activity (Eisner et al. 2017). Further, ROS have also exerted a crucial acute effect on cellular energetics (Pizzino et al. 2017). An alteration of cell phenotype may lead to the mediation of intracellular signaling pathways, for instance, redox-sensitive transcriptional activity such as activator protein-1 (AP-1), hypoxia-inducible factors (HIF-1), and NF-κB levels and subsequently modify the protein and gene expression (Tan et al. 2015; Komatsu et al. 2018; Kobayashi et al. 2021). During endothelial cell stimulation, a broad spectrum of redox-sensitive genes mediated several signaling pathways including platelet-derived growth factor (PDGF), plasminogen activator inhibitor-I (PAI-1), monocyte chemoattractant protein-1 (MCP-1), and vascular cell adhesion molecule-1 (VCAM-1) (Sprague and Khalil 2009; Suganya et al. 2016; Gregg et al. 2018; Puy et al. 2019).

XO-derived ROS is implicated in the development and progression of ischemia-reperfusion (Zhou et al. 2018). Once XO is released into the blood circulation and bound to the luminal surface of endothelial cells, XO exhibits its crucial effects even in the tissues that are not usually expressed (Abooali et al. 2014). In this regard, XO appears to play a crucial role in endothelial dysfunction in certain circumstances (Dopp et al. 2011).

In addition to the different sources mentioned above, most of the experiment studies indicate that modulating NADPH oxidases could play a critical role in CVD (Touyz et al. 2019). NADPH oxidases are essential for the microbicidal activity of neutrophils, and this enzyme is extensively found in non-phagocytic tissues (Zeng et al. 2019). NADPH oxidase has been identified to contribute to the development of cardiac hypertrophy, endothelial dysfunction, atherosclerosis, and hypertension (Langbein et al. 2016; Touyz et al. 2019; Poznyak et al. 2020; Zhao et al. 2020). However, different NADPH oxidase isoforms may have distinct activity and function (Delaney et al. 2016; Fulton and Barman 2016). Overall, maintaining a precise redox balance status is important in the mediation of the physiological acid-base buffer system in the body for optimal homeostatic cellular activities. Indeed, an alteration in redox balance would provide a great impact on the cellular signaling pathways and transcriptional activity as the majority of the reactions and activation are dependent on the oxidation/reduction processes (Locato et al. 2018).

References

Abooali M, Lall GS, Coughlan K et al (2014) Crucial involvement of xanthine oxidase in the intracellular signalling networks associated with human myeloid cell function. Sci Rep 4:6307

Aquilano K, Baldelli S, Ciriolo MR (2014) Glutathione: new roles in redox signalling for an old antioxidant. Front Pharmacol 5:196

Arman MSI, Mahmud AA, Mahmud HR et al (2019) Free radical, oxidative stress and diabetes mellitus: a mini review. Discov Phytomed 6:99–101

Barbagallo M, Marotta F, Dominguez LJ (2015) Oxidative stress in patients with Alzheimer's disease: effect of extracts of fermented papaya powder. Mediat Inflamm 2015, Article ID 624801, 6 pages

Barthelson K, Pederson SM, Newman M et al (2020) Brain transcriptome analysis reveals subtle effects on mitochondrial function and iron homeostasis of mutations in the *SORL1* gene implicated in early onset familial Alzheimer's disease. Mol Brain 13:142

Bodamyali T, Stevens CR, Blake DR et al (2000) Reactive oxygen/nitrogen species and acute inflammation: a physiological process. In: Winyard PG, Blake DR, Evans CH (eds) Free radicals and inflammation. Birkha user Verlag, Basel, pp 11–16

Boesten DM, von Ungern-Sternberg SN, den Hartog GJ et al (2015) Protective pleiotropic effect of flavonoids on NAD^+ levels in endothelial cells exposed to high glucose. Oxidative Med Cell Longev 2015, Article ID 894597, 7 pages

Bornstein R, Gonzalez B, Johnson SC (2020) Mitochondrial pathways in human health and aging. Mitochondrion 54:72–84

Brown GC (2007) Mechanisms of inflammatory neurodegeneration: INOS and NADPH oxidase. Biochem Soc Trans 35:1119–1121

Brownlee M (2001) Biochemistry and molecular cell biology of diabetic complications. Nature 414:813–820

Brownlee M (2005) The pathobiology of diabetic complications: a unifying mechanism. Diabetes 54:1615–1625

Caliri AW, Tommasi S, Besaratinia A (2021) Relationships among smoking, oxidative stress, inflammation, macromolecular damage, and cancer. Mutat Res/Rev Mutat Res 787:108365

Charlotte L (2019) From oxidative stress to inflammation: redox balance and immune system. Poult Sci 98:4240–4246

Chen Y, Zhang H, Zhou HJ et al (2016) Mitochondrial redox signaling and tumor progression. Cancers 8:40

Crouch PJ, White AR, Bush AI (2007) The modulation of metal bio-availability as a therapeutic strategy for the treatment of Alzheimer's disease. FEBS J 274:3775–3783

Cuello F, Eaton P (2019) Cysteine-based redox sensing and its role in signaling by cyclic nucleotide-dependent kinases in the cardiovascular system. Annu Rev Physiol 81:63–87

Delaney MK, Kim K, Estevez B et al (2016) Differential roles of the NADPH-oxidase 1 and 2 in platelet activation and thrombosis. Arterioscler Thromb Vasc Biol 36:846–854

Dice JF (1993) Cellular and molecular mechanisms of aging. Physiol Rev 73:149–159

Dopp JM, Philippi NR, Marcus NJ et al (2011) Xanthine oxidase inhibition attenuates endothelial dysfunction caused by chronic intermittent hypoxia in rats. Respiration 82:458–467

Durgin BG, Straub AC (2018) Redox control of vascular smooth muscle cell function and plasticity. Lab Investig 98:1254–1262

Eisner DA, Caldwell JL, Kistamás K et al (2017) Calcium and excitation-contraction coupling in the heart. Circ Res 121:181–195

Elfrink HL, Zwart R, Cavanillas ML et al (2012) Rab6 is a modulator of the unfolded protein response: implications for Alzheimer's disease. J Alzheimers Dis 28:917–929

Fulton DJR, Barman SA (2016) Clarity on the isoform-specific roles of NADPH oxidases and NADPH oxidase-4 in atherosclerosis. Arterioscler Thromb Vasc Biol 36:579–581

Gamba P, Testa G, Gargiulo S et al (2015) Oxidized cholesterol as the driving force behind the development of Alzheimer's disease. Front Aging Neurosci 7:119

García-Sánchez A, Miranda-Díaz AG, Cardona-Muñoz EG (2020, 2020) The role of oxidative stress in physiopathology and pharmacological treatment with pro- and antioxidant properties in chronic diseases. Oxidative Med Cell Longev, Article ID 2082145, 16 pages

Ghezzi P, Jaquet V, Marcucci F et al (2017) The oxidative stress theory of disease: levels of evidence and epistemological aspects. Br J Pharmacol 174:1784–1796

Gregg LP, Tio MC, Adams-Huet B et al (2018) Association of monocyte chemoattractant protein-1 with death and atherosclerotic events in chronic kidney disease. Am J Nephrol 47:395–405

Hane F, Leonenko Z (2014) Effect of metals on kinetic pathways of amyloid-β aggregation. Biomol Ther 4:101–116

Harman D (1956) Aging: a theory based on free radical and radiation chemistry. J Gerontol 11:298–300

Harris IS, DeNicola GM (2020) The complex interplay between antioxidants and ROS in cancer. Trends Cell Biol 30:440–451

Hodgkinson AD, Sondergaard KL, Yang B et al (2001) Aldose reductase expression is induced by hyperglycemia in diabetic nephropathy. Kidney Int 60:211–218

Hussain SP, Hofseth LJ, Harris CC (2003) Radical causes of cancer. Nat Rev Cancer 3:276–285

Irwin ME, Rivera-Del Valle N, Chandra J (2013) Redox control of leukemia: from molecular mechanisms to therapeutic opportunities. Antioxid Redox Signal 18:1349–1383

Kador PF, Kinoshita JH (1985) Role of aldose reductase in the development of diabetes-associated complications. Am J Med 79:8–12

Kassab A, Piwowar A (2012) Cell oxidant stress delivery and cell dysfunction onset in type 2 diabetes. Biochimie 94:1837–1848

Klomsiri C, Karplus PA, Poole LB (2011) Cysteine-based redox switches in enzymes. Antioxid Redox Signal 14:1065–1077

Kobayashi Y, Oguro A, Imaoka S (2021) Feedback of hypoxia-inducible factor-1alpha (HIF-1alpha) transcriptional activity via redox factor-1 (Ref-1) induction by reactive oxygen species (ROS). Free Radic Res:1–11

Komatsu W, Kishi H, Yagasaki K et al (2018) Urolithin A attenuates pro-inflammatory mediator production by suppressing PI3-K/Akt/NF-κB and JNK/AP-1 signaling pathways in lipopolysaccharide-stimulated RAW264 macrophages: possible involvement of NADPH oxidase-derived reactive oxygen species. Eur J Pharmacol 833:411–424

Krumova K, Cosa G (2016) Chapter 1: Overview of reactive oxygen species. In: Nonell S, Flors C (eds) Singlet oxygen: applications in biosciences and nanosciences. The Royal Society of Chemistry, London, pp 1–21

Kuhla A, Ludwig SC, Kuhla B et al (2015) Advanced glycation end products are mitogenic signals and trigger cell cycle reentry of neurons in Alzheimer's disease brain. Neurobiol Aging 36:753–761

Kwon HS, Koh S-H (2020) Neuroinflammation in neurodegenerative disorders: the roles of microglia and astrocytes. Transl Neurodegener 9:42

Langbein H, Brunssen C, Hofmann A et al (2016) NADPH oxidase 4 protects against development of endothelial dysfunction and atherosclerosis in LDL receptor deficient mice. Eur Heart J 37:1753–1761

Lee S, Kim SM, Lee RT (2013) Thioredoxin and thioredoxin target proteins: from molecular mechanisms to functional significance. Antioxid Redox Signal 18:1165–1207

Liu H, Liu X, Zhang C et al (2017) Redox imbalance in the development of colorectal cancer. J Cancer 8:1586–1597

Lobo V, Patil A, Phatak A et al (2010) Free radicals, antioxidants and functional foods: impact on human health. Pharmacogn Rev 4:118–126

Locato V, Cimini S, De Gara L (2018) ROS and redox balance as multifaceted players of cross-tolerance: epigenetic and retrograde control of gene expression. J Exp Bot 69:3373–3391

Lovell MA, Robertson JD, Teesdale WJ et al (1998) Copper, iron and zinc in Alzheimer's disease senile plaques. J Neurol Sci 158:47–52

Luo X, Wu J, Jing S et al (2016) Hyperglycemic stress and carbon stress in diabetic glucotoxicity. Aging Dis 7:90–110

Marqués J, Cortés A, Pejenaute Á (2020) Induction of cyclooxygenase-2 by overexpression of the human NADPH oxidase 5 (NOX5) gene in aortic endothelial cells. Cell 9:637

Masutani M, Suzuki H, Kamada N et al (1999) Poly(ADPribose) polymerase gene disruption conferred mice resistant to streptozotocin-induced diabetes. Proc Natl Acad Sci U S A 96:2301–2304

McGrath LT, McGleenon BM, Brennan S et al (2001) Increased oxidative stress in Alzheimer's disease as assessed with 4-hydroxynonenal but not malondialdehyde. QJM 94:485–490

Miletić J, Drakulić D, Pejić S et al (2018) Prooxidant–antioxidant balance, advanced oxidation protein products and lipid peroxidation in Serbian patients with Parkinson's disease. Int J Neurosci 128:600–607

Mohammadi F, Soltani A, Ghahremanloo A et al (2019) The thioredoxin system and cancer therapy: a review. Cancer Chemother Pharmacol 84:925–935

Molina JA, Jiménez-Jiménez FJ, Aguilar MV et al (1998) Cerebrospinal fluid levels of transition metals in patients with Alzheimer's disease. J Neural Transm 105:479–488

Moreira PI, Sayre LM, Zhu X et al (2010) Detection and localization of markers of oxidative stress by in situ methods: application in the study of Alzheimer disease. Methods Mol Biol 610:419–434

Morrison S, Newell KM (2012) Aging, neuromuscular decline, and the change in physiological and behavioral complexity of upper-limb movement dynamics. J Aging Res 2012:891218

Natarajan V, Chawla R, Mah T et al (2020) Mitochondrial dysfunction in age-related metabolic disorders. Proteomics 20:1800404

Ng TF, Lee FK, Song ZT et al (1998) Effects of sorbitol dehydrogenase deficiency on nerve conduction in experimental diabetic mice. Diabetes 47:961–966

Nolfi-Donegan D, Braganza A, Shiva S (2020) Mitochondrial electron transport chain: oxidative phosphorylation, oxidant production, and methods of measurement. Redox Biol 37:101674

Obrosova IG, Drel VR, Pacher P et al (2005) Oxidative-nitrosative stress and poly(ADP-ribose) polymerase (PARP) activation in experimental diabetic neuropathy: the relation is revisited. Diabetes 54:3435–3441

Pacher P, Szabo C (2005) Role of poly(ADP-ribose) polymerase-1 activation in the pathogenesis of diabetic complications: endothelial dysfunction, as a common underlying theme. Antioxid Redox Signal 7:1568–1580

Park JY, Pillinger MH, Abramson SB (2006) Prostaglandin E_2 synthesis and secretion: the role of PGE_2 synthases. Clin Immunol 119:229–240

Pickering RJ (2021) Oxidative stress and inflammation in cardiovascular diseases. Antioxidants 10:171

Pieper AA, Brat DJ, Krug DK et al (1999) Poly(ADP-ribose) polymerase-deficient mice are protected from streptozotocin-induced diabetes. Proc Natl Acad Sci U S A 96:3059–3064

Pizzino G, Irrera N, Cucinotta M et al (2017) Oxidative stress: harms and benefits for human health. Oxidative Med Cell Longev 2017, Article ID 8416763, 13 pages

Poznyak A, Grechko AV, Orekhova VA et al (2020) NADPH oxidases and their role in atherosclerosis. Biomed 8:206

Prieto-Bermejo R, Hernández-Hernández A (2017) The importance of NADPH oxidases and redox signaling in angiogenesis. Antioxidants 6:32

Puy C, Ngo ATP, Pang J et al (2019) Endothelial PAI-1 (plasminogen activator inhibitor-1) blocks the intrinsic pathway of coagulation, inducing the clearance and degradation of FXIa (activated factor XI). Arterioscler Thromb Vasc Biol 39:1390–1401

Rahal A, Kumar A, Singh V et al (2014) Oxidative stress, prooxidants, and antioxidants: the interplay. Biomed Res Int 2014, Article ID 761264, 19 pages

Rea IM, Gibson DS, McGilligan V et al (2018) Age and age-related diseases: role of inflammation triggers and cytokines. Front Immunol 9:586

Ricciotti E, FitzGerald GA (2011) Prostaglandins and inflammation. Arterioscler Thromb Vasc Biol 31:986–1000

Robertson RP (2004) Chronic oxidative stress as a central mechanism for glucose toxicity in pancreatic islet beta cells in diabetes. J Biol Chem 279:42351–42354

Sag CM, Santos CXC, Shah AM (2014) Redox regulation of cardiac hypertrophy. J Mol Cell Cardiol 73:103–111

Schrag M, Mueller C, Oyoyo U et al (2011) Iron, zinc and copper in the Alzheimer's disease brain: a quantitative meta-analysis. Some insight on the influence of citation bias on scientific opinion. Prog Neurobiol 94:296–306

Servais S, Boussouar A, Molnar A et al (2009) Age-related sensitivity to lung oxidative stress during ozone exposure. Free Radic Res 39:305–316

Simpson DSA, Oliver PL (2020) ROS generation in microglia: understanding oxidative stress and inflammation in neurodegenerative disease. Antioxidants 9:743

Sohal RS, Allen RG (1990) Oxidative stress as a causal factor in differentiation and aging: a unifying hypothesis. Exp Gerontol 25:499–522

Sohal RS, Mockett RJ, Orr WC (2002) Mechanisms of aging: an appraisal of the oxidative stress hypothesis. Free Radic Biol Med 33:575–586

Song X, Theeng MAW, Ma F et al (2020) DNA damage induced neuroinflammation in neurodegenerative disease. Alzheimers Dement 16:e047445

Songbo M, Lang H, Xinyong C et al (2019) Oxidative stress injury in doxorubicin-induced cardiotoxicity. Toxicol Lett 307:41–48

Sprague AH, Khalil RA (2009) Inflammatory cytokines in vascular dysfunction and vascular disease. Biochem Pharmacol 78:539–552

Suganya N, Bhakkiyalakshmi E, Sarada DVL et al (2016) Reversibility of endothelial dysfunction in diabetes: role of polyphenols. Br J Nutr 116:223–246

Szabo C (2005) Roles of poly(ADP-ribose) polymerase activation in the pathogenesis of diabetes mellitus and its complications. Pharm Res 52:60–71

Talbot K, Wang H-Y, Kazi H et al (2012) Demonstrated brain insulin resistance in Alzheimer's disease patients is associated with IGF-1 resistance, IRS-1 dysregulation, and cognitive decline. J Clin Investig 122:1316–1338

Tan BL, Norhaizan ME, Huynh K et al (2015) Brewers' rice modulates oxidative stress in azoxymethane-mediated colon carcinogenesis in rats. World J Gastroenterol 21:8826–8835

Tan BL, Norhaizan ME, Liew W-P-P et al (2018a) Antioxidant and oxidative stress: a mutual interplay in age-related diseases. Front Pharmacol 9:1162

Tan BL, Norhaizan ME, Liew W-P-P (2018b) Nutrients and oxidative stress: friend or foe? Oxidative Med Cell Longev 2018, Article ID 9719584, 24 pages

Tang WH, Martin KA, Hwa J (2012) Aldose reductase, oxidative stress, and diabetic mellitus. Front Pharm 3:87

Tavakoli S, Asmis R (2012) Reactive oxygen species and thiol redox signaling in the macrophage biology of atherosclerosis. Antioxid Redox Signal 17:1785–1795

Taverne YJ, Merkus D, Bogers AJ et al (2018) Reactive oxygen species: radical factors in the evolution of animal life: a molecular timescale from Earth's earliest history to the rise of complex life. BioEssays 40:1700158

Torsdottir G, Kristinsson J, Snaedal J et al (2011) Ceruloplasmin and iron proteins in the serum of patients with Alzheimer's disease. Dement Geriatr Cogn Disord Extra 1:366–371

Touyz RM, Anagnostopoulou A, Camargo LL et al (2019) Vascular biology of superoxide-generating NADPH oxidase 5—implications in hypertension and cardiovascular disease. Antioxid Redox Signal 30:1027–1040

Tsukahara T, Haniu H, Uemura T et al (2020) Therapeutic potential of porcine liver decomposition product: new insights and perspectives for microglia-mediated neuroinflammation in neurodegenerative diseases. Biomedicines 8:446

Uttara B, Singh AV, Zamboni P et al (2009) Oxidative stress and neurodegenerative diseases: a review of upstream and downstream antioxidant therapeutic options. Curr Neuropharmacol 7:65–74

Veith C, Boots AW, Idris M et al (2019) Redox imbalance in idiopathic pulmonary fibrosis: a role for oxidant cross-talk between NADPH oxidase enzymes and mitochondria. Antioxid Redox Signal 31:1092–1115

Wallace DC (2013) A mitochondrial bioenergetic etiology of disease. J Clin Invest 123:1405–1412

Wang XJ, Sun Z, Villeneuve NF et al (2008) Nrf2 enhances resistance of cancer cells to chemotherapeutic drugs, the dark side of Nrf2. Carcinogenesis 29:1235–1243

Wang Z, Tang T, Wang S et al (2020) Aloin inhibits the proliferation and migration of gastric cancer cells by regulating NOX2–ROS-mediated pro-survival signal pathways. Drug Des Dev Ther 14:145–155

Wani R, Nagata A, Murray BW (2014) Protein redox chemistry: post-translational cysteine modifications that regulate signal transduction and drug pharmacology. Front Pharmacol 5:224

Weyand CM, Goronzy JJ (2016) Aging of the immune system. Mechanisms and therapeutic targets. Ann Am Thorac Soc 13:S422–S428

Wu J, Jin Z, Zheng H et al (2016) Sources and implications of NADH/NAD(+) redox imbalance in diabetes and its complications. Diabetes Metab Syndr Obes 9:145–153

Yabe-Nishimura C (1998) Aldose reductase in glucose toxicity: a potential target for the prevention of diabetic complications. Pharm Rev 50:21–33

Yan LJ (2014) Pathogenesis of chronic hyperglycemia: from reductive stress to oxidative stress. J Diabetes Res 2014:137919

Yang HY, Chay KO, Kwon J et al (2013) Comparative proteomic analysis of cysteine oxidation in colorectal cancer patients. Mol Cells 35:533–542

Zeng MY, Miralda I, Armstrong CL et al (2019) The roles of NADPH oxidase in modulating neutrophil effector responses. Mol Oral Microbiol 34:27–38

Zhao R-Z, Jiang S, Zhang L et al (2019) Mitochondrial electron transport chain, ROS generation and uncoupling (Review). Int J Mol Med 44:3–15

Zhao G-J, Zhao C-L, Ouyang S et al (2020) Ca^{2+}-dependent NOX5 (NADPH oxidase 5) exaggerates cardiac hypertrophy through reactive oxygen species production. Hypertension 76:827–838
Zheng H, Wu J, Jin Z et al (2016) Protein modifications as manifestations of hyperglycemic glucotoxicity in diabetes and its complications. Biochem Insights 9:1–9
Zheng H, Wu J, Jin Z et al (2017) Potential biochemical mechanisms of lung injury in diabetes. Aging Dis 8:7–16
Zhou T, Prather ER, Garrison DE et al (2018) Interplay between ROS and antioxidants during ischemia-reperfusion injuries in cardiac and skeletal muscle. Int J Mol Sci 19:417

Chapter 4
Chronic Inflammation and Aging (Inflammaging)

Aging is an inherent mode of action present in all living cells (Campisi et al. 2019). Indeed, organ functions decrease with age. Despite the interactions between the age-related diseases and the aging process require further elucidation, most of the studies revealed that aging is implicated in the development of many age-related diseases (Franceschi et al. 2018; Tan et al. 2018a; Krisko and Radman 2019). One of the most common aspects of the inflammation hypothesis of aging is that age-related diseases undergo several pathways related to the inflammatory process, which may lead to the progression and development of a variety of age-related diseases (DeBalsi et al. 2017). For instance, many age-related diseases including arthritis, osteoporosis, dementia, CVD, cancer, metabolic syndrome, diabetes have been recognized as inflammatory disorders (Tan et al. 2015; Franceschi et al. 2018; Tan and Norhaizan 2021) (Fig. 4.1).

Mitochondrial and free radical theories are the two most common theories related to aging. These theories described that a vicious cycle is produced within the mitochondria, while the ROS is markedly generated and thus promotes the damage potential (Romano et al. 2010). Oxidative stress is existing in all living beings at the system, tissue, cellular, molecular, and genetic levels. This is often manifested as a progressive increase or accumulation of detrimental changes in tissues and cells with advancing age (Knupp and Miura 2018). It has been demonstrated that ROS levels increased with age and usually accumulates in major organ systems such as skeletal muscle, brain, heart, and liver (Olgar et al. 2018; Zhou et al. 2018; Hunt et al. 2019; Stefanatos and Sanz 2018) either due to reduced detoxification or increased production. In this regard, aging is considered as a progressive reduction in the biological function of the tissues and thereby increased the vulnerability to the diseases (Kregel and Zhang 2007). The widely accepted theory, namely "oxidative stress hypothesis", describes that increases in ROS resulted in pathological conditions and observable signs related to aging as well as functional alterations, and ultimately death (Hagen 2003). Despite ETC damage and mitochondrial DNA

B. L. Tan, M. E. Norhaizan, *The Role of Antioxidants in Longevity and Age-Related Diseases*, https://doi.org/10.1007/978-3-030-82859-2_4

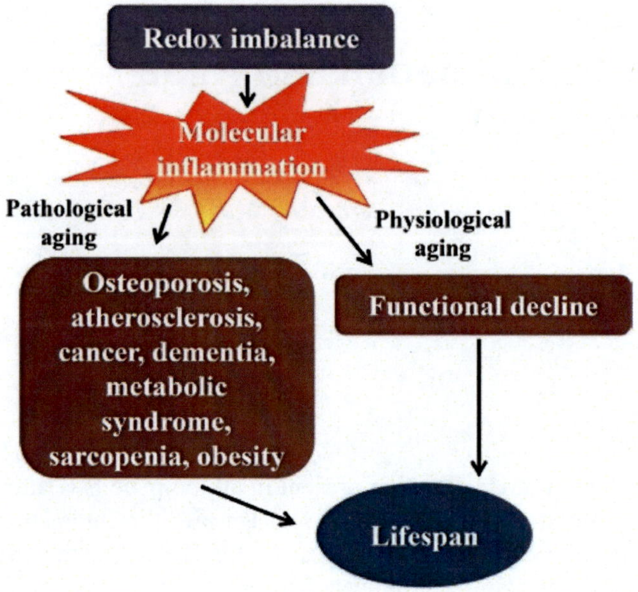

Fig. 4.1 The effects of inflammation in pathophysiological process

damage that may responsible for aging, the stimulation of redox-sensitive transcriptional factors or mediation of cellular signal response to stress by age-related oxidative stress upregulate the proinflammatory gene expression, and thereby enhances the ROS levels (Kregel and Zhang 2007).

4.1 Sources of Chronic Inflammation during Aging

Chronic, low-grade inflammation is thought to be a predominant contributor to a broad spectrum of natural processes and age-related pathologies in aging tissues, for instance, musculoskeletal and nervous systems (Libby and Kobold 2019; Cervo et al. 2020; Lin et al. 2020). In general, some tissues in the elderly are chronically inflamed (Gnani et al. 2019; Ziegler et al. 2019). Inflammatory cytokines including tumor necrosis factor-α (TNF-α), interleukin-1beta (IL-1β), and interleukin-6 (IL-6) has been recognized as one of the contributor to diminish the anabolic signaling cascade such as erythropoietin and insulin signaling pathway, and thereby leading to an increased risk of developing sarcopenia (Beyer et al. 2012). Figure 4.2 summarizes the sources of chronic inflammation in aging.

Aging promotes the production of COX-derived reactive species and enhances the release of inducible nitric oxide synthase (iNOS), COX-2, TNF-α, IL-6, and IL-1β (Wojdasiewicz et al. 2014; Begg et al. 2020). Other proinflammatory proteins, for instance, P-, E-selectin, intercellular cell adhesion molecule-1 (ICAM-1), and VCAM-1 are also upregulated during the aging process (Zou et al. 2006). The

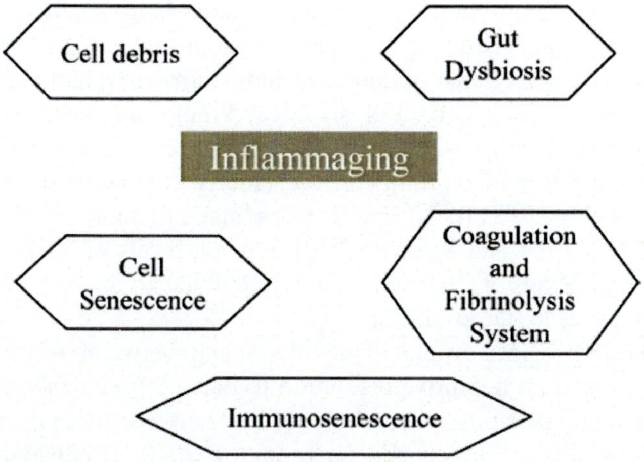

Fig. 4.2 Sources of chronic inflammation in aging

nuclear factor-kappa B (*NF-κB*) transcriptional activity is regarded as the master regulator of the inflammatory process and can be stimulated by oxidative stimuli (Park and Hong 2016; Liu et al. 2017). The stimulation of *NF-κB*-dependent genes is a key transcriptional factor for the systemic inflammatory process (Jakkampudi et al. 2016). During activation, proinflammatory genes encode proinflammatory proteins, for instance, chemokines, growth factors, and cytokines (Drago et al. 2015). The NF-κB activity is mediated by upstream signaling, for instance, mitogen-activated protein kinase (MAPK) and IκB kinase (IKK). The IκB subunits of NF-κB/IκB are phosphorylated by activated IKK complexes and thus stimulating the degradation of IκB, which subsequently lead to the activation of NF-κB. IKK activity is activated by NF-κB during aging (Tilstra et al. 2011), and subsequently promotes the activation of p38 MAPK, c-Jun N-terminal kinase (JNK), and extracellular signal-regulated kinase (ERK) activities that control *NF-κB*-dependent gene expression during an inflammatory response (Jnawali et al. 2014). A previous study revealed that aging promotes p38 MAPK, JNK, and ERK signaling pathways with an increase in ROS production (Ito et al. 2010).

Under normal circumstances, stimulation of NF-κB in response to oxidative stimuli is short-lived, and the reaction is halted with resolution. Nonetheless, when the input signal is not well-maintained during aging, chronic proinflammatory conditions may create a conducive environment for many chronic diseases (Rea et al. 2018). Several NF-κB-induced proteins such as COX-2, IL-6, and TNF-α are potent NF-κB activators that form an auto-activating loop (Oeckinghaus and Ghosh 2009). Substantial studies evaluated the changes of redox-sensitive transcriptional factors such as NF-κB in rodent models (Hansen et al. 2002; Kim et al. 2002; George et al. 2009). The study revealed that old rodents have consistently expressed high NF-κB activities in a variety of tissues, for instance, brain, kidney, liver, and heart compared to the young rodents (Korhonen et al. 1997; Radák et al. 2004; Ungvari et al. 2007; Lim et al. 2012). Data from the human study have also demonstrated that circulating levels of proinflammatory cytokines such as IL-1ra, IL-6, and TNF-α are increased during aging (Bruunsgaard 2006). In addition, aging is also linked to the high inflammatory cell counts (monocytes and neutrophils) and increased levels of C-reactive protein (CRP) (Ritzel et al. 2018; Wong and Wagner 2018; Álvarez-Sánchez et al. 2020). High IL-6 plasma levels were shown a greater likelihood of morbidity, disability, and mortality in the elderly (Puzianowska-Kuznicka et al. 2016). High levels of CRP, IL-1β, and IL-6 are linked to many diseases in the elderly (Ng et al. 2018; Poole and Steptoe 2020). Plasma levels of TNF-α are positively linked to the high levels of CRP and IL-6, implied that an interrelated stimulation of inflammatory cascade (Oe et al. 2015).

Numerous studies have evaluated the relationship between insulin resistance and obesity, but there is no definitive resolution to date. A previous study showed that inflammation could be a possible underlying link between these metabolic ailments (Fiordelisi et al. 2019; Greten and Grivennikov 2019; Diedisheim et al. 2020). Excessive caloric consumption increased adiposity and thus leads to macrophage infiltration into adipose tissues that promote local chronic inflammation which potentiates insulin resistance (Poli et al. 2017; Tan et al. 2018b). Overexpression of *Mcp1* promotes insulin resistance, inflammation, and macrophage infiltration (Kanda et al. 2006; Patsouris et al. 2014; Gogh et al. 2016). Furthermore, knockout of *Mcp1* and its receptor (*Ccr2*) impairs migration of macrophages, and thus increases insulin sensitivity and reduces inflammation (Tamura et al. 2008; Sawyer et al. 2014).

4.1.1 Immunoglobulin or Cell Debris Production

Immunoglobulin or debris accumulation caused by an inappropriate cell elimination system in aging, which induces innate immune activity stimulation and thereby leading to inflammation (Sanada et al. 2018). In particular, glycosylation is the most often posttranslational modification of protein (Carnino et al. 2020). The protein-linked sugar chain plays a crucial role in the "fine-tuning" of molecules and cells (Ohtsubo and Marth 2006; Dall'Olio et al. 2013). High-throughput studies of the N-glycome, a sugar chain N-linked to asparagine, demonstrated a potential biomarker for natural aging, for instance, N-glycans devoid of galactose residues on the branch, in human studies (Parekh et al. 1988; Vanhooren et al. 2007; Ruhaak et al. 2011). This agalactosylated biantennary structure primarily decorates Asn297 of the Fc portion of IgG (IgG-G0) and is present in patients with inflammatory/autoimmune diseases or progeria syndromes (Dall'Olio et al. 2013). Accelerated aging syndromes or progerias are partially recapitulated normal aging (Dreesen and Stewart 2011). Progerias are predominantly triggered by defects in DNA repair systems or an alteration of the nuclear envelope (Burla et al. 2018). Indeed, IgG-G0 shows a proinflammatory effect via a few mechanisms including formation of autoantibody aggregates, binding to Fcγ receptors, and lectin pathway of complement

(Gudelj et al. 2018). Further, the age-related production of IgG-G0 can stimulate the immune system and hence result in inflammaging (Barrientos et al. 2020). By contrast, mitochondrial dysfunction has also drawn attention among scientists (Manolis et al. 2021). Mitochondria-derived damage-associated molecular patterns (DAMPs) such as cell-free circulating mitochondrial DNA have been extensively studied due to the involvement in chronic diseases and aging (Zhang et al. 2010; Dall'Olio et al. 2013). Through their bacterial ancestry, these molecules may promote the inflammatory response via interaction with receptors similar to those involved in pathogen-related response (Sanada et al. 2018).

4.1.2 The Microbiota and Gut Mucosa in Elderly

The ability of gut mucosa to sequester bacteria deteriorates with age (Shoemark and Allen 2015). Periodontal disease has been reported to cause chronic low-grade inflammation (Loos and Van Dyke 2020). The study found that the diversity of gut microbiota is reduced in older people (Claesson et al. 2011; Kinross and Nicholson 2012). In particular, the anti-inflammatory microbiota, for example, *F. prausnitzii*, *Bifidobacterium* spp., and *Clostridium* cluster XIVa, are reduced in the elderly (Toward et al. 2012). A study by Okada et al. (2009) further supported that the *Bifidobacterium* species is negatively linked to the serum IL-1β and TNF-α levels. By contrast, pathogenic and inflammatory microbiota, such as *Enterobacter* spp., *Enterococcus* spp., *Staphylococcus* spp., and *Streptococcus* spp., are increased with age (Toward et al. 2012). Alteration in gut microbiota diversity may increase the susceptibility to infectious agents by pathobionts colonization (Mosca et al. 2016).

4.1.3 Cell Senescence

Cellular senescence is an irreversible cell cycle arrest mediated by a few mechanisms such as inflammatory cytokines, mitogen stimuli, genotoxic stress, and telomere shortening, which can lead to the stimulation of the cyclin-dependent kinase inhibitor p16 and/or p53 tumor suppressor (de Magalhães and Passos 2018).

Senescence is a cellular response to damage and stress (Franceschi and Campisi 2014). It was evident that the number of senescent cells is increased with age, in which these organs secrete many inflammatory cytokines and produce low-grade inflammation. Senescent cells are linked to age-related diseases or aging through the secretion of proinflammatory cytokines that alter the function of normal cells or the tissue microenvironment (Baker et al. 2011). The phenotype of senescent cells is known as senescence-associated secretory phenotype (SASP), which is suggested as the primary origin of inflammaging in age-related diseases and aging (Sanada et al. 2009; Tchkonia et al. 2013; He and Sharpless 2017). The previous study showed that the elimination of senescent cells in prematurely aged mice ameliorates the progression of age-related diseases (Coppé et al. 2010). Such findings indicate that the mediation of proinflammatory pathways linked to the acquisition of SASP, reprogramming of senescent cells, and elimination of senescent cells could be used as a potential anti-aging approach for extending healthspan and ameliorating the metabolic ailments (van Deursen 2014).

4.1.4 Immunosenescence

Immunosenescence is characterized by the chronic inflammatory response, due to the age-related dysregulation of an innate immune system (Shaw et al. 2013). Immunosenescence impairs wound healing, reduces the response to vaccinations, and increases the susceptibility to malignancy (Aw et al. 2007; Gruver et al. 2007). Aging modifies the immune system and thus contributes to inflammaging (Fulop et al. 2018). Indeed, the immunosenescence process can be accelerated by chronic inflammatory disease (Barbé-Tuana et al. 2020). The mechanisms underlying the persistent aging-associated basal inflammation are not fully understood, but it is hypothesized that the changes in functions and numbers of innate immune cells contribute to these phenomena. Most of the studies so far indicated that unusual downstream signaling pathway of pattern recognition receptors (PRRs) stimulation, activation of PRRs by endogenous ligands related to cellular damage, and changes in the PRRs levels may lead to the induction of chronic cytokine secretion (Hung and Suzuki 2017; Zhu et al. 2019). In this regard, dysregulation of immunological imprinting modulated by innate immunity as well as cell senescence may contribute to chronic low-grade inflammation. In addition, adaptive immunity declines with age; while innate immunity showed minute changes in mild hyperactivity (Santoro et al. 2018). However, the innate immune response might activate when adaptive immunosenescence progresses. Collectively, the age-related changes could be attributed to the intrinsic changes in immune cells and lifelong exposure to pathogens and antigens (Stephenson et al. 2018).

4.1.5 Coagulation and Fibrinolysis System

Activation coagulation and fibrinolysis system in the elderly increased inflammation by modulating the protease-activated receptor (PAR) (Chu 2010; Hess and Grant 2011; Sanada et al. 2016), and thereby lead to an increased risk for lung fibrosis and atherosclerosis (Biagi et al. 2011). Coagulation is considered as part of the inflammation system. The inflammatory process is linked to the potentially aggravating phenomenon of obesity (Tan et al. 2018b). Age-related obesity is predominantly due to the increased adiposity, especially visceral fat deposits, during aging via redistribution of fat deposits with age (Villarroya et al. 2018). Indeed, most of the proinflammatory cytokines are generated by resident macrophages and adipocytes in adipose tissues, and thereby leading to systemic inflammation (Makki et al. 2013). Elevation of proinflammatory status is more likely to increase the susceptibility of several age-related diseases (Ackermann et al. 2020; Tu et al. 2020; Tahir et al. 2021). For instance, osteopenia and sarcopenia are characterized as the normal aging processes, which are good examples of the involvement of inflammation in the normal aging process to pathogenesis (Fig. 4.1).

Research evidence indicates that plasma concentrations of coagulation factor IX, VIII, VII, and V were increased in healthy subjects in conjunction with the physiological processes of aging (Chu 2011; Favaloro et al. 2014). The fibrinogen levels (coagulation factor I), a predominant risk factor for thrombotic disorders, are increased with age (Gligorijević et al. 2018). In particular, the coagulation factor X is overexpressed in human atherosclerotic plaques, such as inflammatory cells, smooth muscle cells, and endothelial cells (Sanada et al. 2017). Based on the evidence, increased plasma levels and local coagulation factors during physiological aging may increase the risk of CVD progression in the elderly. The previous study showed that the direct coagulation factor rivaroxaban, Xa inhibitor, decreased the risk of the composite endpoint of death from stroke, myocardial infarction, and CVD in patients with acute coronary syndrome (Mega et al. 2012). Despite the molecular mechanisms underlying the coagulation factor and reduced risk of CVD require further elucidation, most of the experimental studies indicate that stimulation of coagulation cascade following fibrinogen activation may elevate the thrombosis (Palta et al. 2014). Further, increased levels of thrombin and coagulation factor Xa may improve the inflammatory response via modulation of PAR-1/2 signaling (Spronk et al. 2014). Notably, PAR-1/2 signaling triggered by fibrinolytic factor plasmin and coagulation factor Xa (FXa) was shown to elevate the insulin-like growth factor binding protein-5 (IGFBP-5) levels (Kamio et al. 2008; Carmo et al. 2014; Sanada et al. 2016, 2017). A study by Kojima et al. (2012) found that IGFBP-5 mediates IL-6 expression to trigger ROS generation, and thereby leading to the DNA damage and senescence of fibroblast cells. Further, IGFBP-5 also stimulates a fibrotic phenotype by stimulating nuclear early growth response-1 (EGR-1) translocation and MAPK signaling that interacts with IGFBP-5 and upregulates inflammatory and fibrotic transcriptional activity (Yasuoka et al. 2009). In addition, activation of FXa in endothelial progenitor cells, endothelial cells, and smooth muscle cells promote cellular senescence by modulating the EGR-1-IGFBP-5-p53 signaling pathway (Sanada et al. 2016). This finding implies that cell senescence, hypercoagulability, and inflammaging may share a common pathway mediated by IGFBP-5 signaling. Intriguingly, some research has emerged to suggest that IGFBP-5- and FXa-positive areas were distributed in the human atherosclerotic plaques (Sanada et al. 2017). Collectively, locally produced coagulation factor Xa in atherosclerotic plaques may promote cellular senescence with SASP and trigger IGFBP-5 levels (Sparkenbaugh et al. 2014). Because aging is a complex mechanisms results from the epigenetic, genetic, and environmental factors, further studies focused on interventions that selectively damage senescent cells, for instance, "senolytic therapies" in the aging host may enhance the therapeutic approach (Roos et al. 2016; Farr et al. 2017; Lehmann et al. 2017; Collins et al. 2018).

References

Ackermann A, Lafferton B, Plotz G et al (2020) Expression and secretion of the pro-inflammatory cytokine IL-8 is increased in colorectal cancer cells following the knockdown of non-erythroid spectrin αII. Int J Oncol 56:1551–1564

Álvarez-Sánchez N, Álvarez-Ríos AI, Guerrero JM et al (2020) Homocysteine and C-reactive protein levels are associated with frailty in older Spaniards: The Toledo Study for Healthy Aging. J Gerontol Ser A 75:1488–1494

Aw D, Silva AB, Palmer DB (2007) Immunosenescence: emerging challenges for an ageing population. Immunology 120:435–446

Baker DJ, Wijshake T, Tchkonia T et al (2011) Clearance of p16Ink4a-positive senescent cells delays ageing-associated disorders. Nature 479:232–236

Barbé-Tuana F, Funchal G, Schmitz CRR et al (2020) The interplay between immunosenescence and age-related diseases. Semin Immunopathol 42:545–557

Barrientos G, Habazin S, Novokmet M et al (2020) Changes in subclass-specific IgG Fc glycosylation associated with the postnatal maturation of the murine immune system. Sci Rep 10:15243

Begg DP, Sinclair AJ, Weisinger RS (2020) Impaired fluid intake, but not sodium appetite, in aged rats is mediated by the cyclooxygenase-prostaglandin E2 pathway. Front Aging Neurosci 12:19

Beyer I, Mets T, Bautmans I (2012) Chronic low-grade inflammation and age-related sarcopenia. Curr Opin Clin Nutr Metab Care 15:12–22

Biagi E, Candela M, Franceschi C et al (2011) The aging gut microbiota: new perspectives. Ageing Res Rev 10:428–429

Bruunsgaard H (2006) The clinical impact of systemic low-level inflammation in elderly populations. With special reference to cardiovascular disease, dementia and mortality. Dan Med Bull 53:285–309

Burla R, La Torre M, Merigliano C et al (2018) Genomic instability and DNA replication defects in progeroid syndromes. Nucleus 9:368–379

Campisi J, Kapahi P, Lithgow GJ et al (2019) From discoveries in ageing research to therapeutics for healthy ageing. Nature 571:183–192

Carmo AA, Costa BR, Vago JP et al (2014) Plasmin induces in vivo monocyte recruitment through protease-activated receptor-1-, MEK/ERK-, and CCR2-mediated signaling. J Immunol 193:3654–3663

Carnino JM, Ni K, Jin Y (2020) Post-translational modification regulates formation and cargo-loading of extracellular vesicles. Front Immunol 11:948

Cervo MM, Shivappa N, Hebert JR et al (2020) Longitudinal associations between dietary inflammatory index and musculoskeletal health in community-dwelling older adults. Clin Nutr 39:516–523

Chu AJ (2010) Blood coagulation as an intrinsic pathway for proinflammation: a mini review. Inflamm Allergy Drug Targets 9:32–44

Chu AJ (2011) Tissue factor, blood coagulation, and beyond: an overview. Int J Inflam 2011, Article ID 367284, 30 pages

Claesson MJ, Cusack S, O'Sullivan O et al (2011) Composition, variability, and temporal stability of the intestinal microbiota of the elderly. Proc Natl Acad Sci U S A 108:4586–4591

Collins JA, Diekman BO, Loeser RF (2018) Targeting aging for disease modification in osteoarthritis. Curr Opin Rheumatol 30:101–107

Coppé JP, Desprez PY, Krtolica A et al (2010) The senescence associated secretory phenotype: the dark side of tumor suppression. Annu Rev Pathol 5:99–118

Dall'Olio F, Vanhooren V, Chen CC et al (2013) N-glycomic biomarkers of biological aging and longevity: a link with inflammaging. Ageing Res Rev 12:685–698

de Magalhães JP, Passos JF (2018) Stress, cell senescence and organismal ageing. Mech Ageing Dev 170:2–9

DeBalsi KL, Hoff KE, Copeland WC (2017) Role of the mitochondrial DNA replication machinery in mitochondrial DNA mutagenesis, aging and age-related diseases. Ageing Res Rev 33:89–104

Diedisheim M, Carcarino E, Vandiedonck C et al (2020) Regulation of inflammation in diabetes: from genetics to epigenomics evidence. Mol Metab 41:101041

Drago F, Ciccarese G, Broccolo F et al (2015) The role of cytokines, chemokines, and growth factors in the pathogenesis of Pityriasis Rosea. Mediat Inflamm 2015, Article ID 438963, 6 pages

Dreesen O, Stewart CL (2011) Accelerated aging syndromes, are they relevant to normal human aging? Aging 3:889–895

Farr JN, Xu M, Weivoda MM et al (2017) Targeting cellular senescence prevents age-related bone loss in mice. Nat Med 23:1072–1079

Favaloro EJ, Franchini M, Lippi G (2014) Aging hemostasis: changes to laboratory markers of hemostasis as we age – a narrative review. Semin Thromb Hemost 40:621–633

Fiordelisi A, Iaccarino G, Morisco C et al (2019) NFkappaB is a key player in the crosstalk between inflammation and cardiovascular diseases. Int J Mol Sci 20:1599

Franceschi C, Campisi J (2014) Chronic inflammation (inflammaging) and its potential contribution to age-associated diseases. J Gerontol A Biol Sci Med Sci 69:S4–S9

Franceschi C, Garagnani P, Morsiani C et al (2018) The continuum of aging and age-related diseases: common mechanisms but different rates. Front Med 5:61

Fulop T, Larbi A, Dupuis G et al (2018) Immunosenescence and inflamm-aging as two sides of the same coin: friends or foes? Front Immunol 8:1960

George L, Lokhandwala MF, Asghar M (2009) Exercise activates redox-sensitive transcription factors and restores renal D1 receptor function in old rats. Am J Physiol Renal Physiol 297:F1174–F1180

Gligorijević N, Križáková MZ, Penezić A et al (2018) Structural and functional changes of fibrinogen due to aging. Int J Biol Macromol 108:1028–1034

Gnani D, Crippa S, della Volpe L et al (2019) An early-senescence state in aged mesenchymal stromal cells contributes to hematopoietic stem and progenitor cell clonogenic impairment through the activation of a pro-inflammatory program. Aging Cell 18:e12933

Gogh IJAE, Oteng A-B, Alex S et al (2016) Muscle-specific inflammation induced by MCP-1 overexpression does not affect whole-body insulin sensitivity in mice. Diabetologia 59:624–633

Greten FR, Grivennikov SI (2019) Inflammation and cancer: triggers, mechanisms, and consequences. Immunity 51:27–41

Gruver AL, Hudson LL, Sempowski GD (2007) Immunosenescence of ageing. J Pathol 211:144–156

Gudelj I, Lauc G, Pezer M (2018) Immunoglobulin G glycosylation in aging and diseases. Cell Immunol 333:65–79

Hagen TM (2003) Oxidative stress, redox imbalance, and the aging process. Antioxid Redox Signal 5:503–506

Hansen JM, Gong S-G, Philbert M et al (2002) Misregulation of gene expression in the redox-sensitive NF-kappab-dependent limb outgrowth pathway by thalidomide. Dev Dyn 225:186–194

He S, Sharpless NE (2017) Senescence in health and disease. Cell 169:1000–1011

Hess K, Grant PJ (2011) Inflammation and thrombosis in diabetes. Thromb Haemost 105:S43–S54

Hung Y-L, Suzuki K (2017) The pattern recognition receptors and lipopolysaccharides (LPS)-induced systemic inflammation. Int J Res Stud Med Heal Sci 2:1–7

Hunt NJ, Kang SW, Lockwood GP et al (2019) Hallmarks of aging in the liver. Comput Struc Biotechnol J 17:1151–1161

Ito M, Miyado K, Nakagawa K et al (2010) Age-associated changes in the subcellular localization of phosphorylated p38 MAPK in human granulosa cells. Mol Hum Reprod 16:928–937

Jakkampudi A, Jangala R, Reddy BR et al (2016) NF-κB in acute pancreatitis: mechanisms and therapeutic potential. Pancreatology 16:477–488

Jnawali HN, Lee E, Jeong K-W et al (2014) Anti-inflammatory activity of Rhamnetin and a model of its binding to c-Jun NH2-terminal kinase 1 and p38 MAPK. J Nat Prod 77:258–263

Kamio N, Hashizume H, Nakao S et al (2008) Plasmin is involved in inflammation via protease-activated receptor-1 activation in human dental pulp. Biochem Pharmacol 75:1974–1980

Kanda H, Tateya S, Tamori Y et al (2006) MCP-1 contributes to macrophage infiltration into adipose tissue, insulin resistance, and hepatic steatosis in obesity. J Clin Invest 116:1494–1505
Kim HJ, Jung KJ, Yu BP et al (2002) Modulation of redox-sensitive transcription factors by calorie restriction during aging. Mech Ageing Dev 123:1589–1595
Kinross J, Nicholson JK (2012) Gut microbiota: dietary and social modulation of gut microbiota in the elderly. Nat Rev Gastroenterol Hepatol 9:563–564
Knupp D, Miura P (2018) CircRNA accumulation: a new hallmark of aging? Mech Ageing Dev 173:71–79
Kojima H, Kunimoto H, Inoue T et al (2012) The STAT3-IGFBP5 axis is critical for IL-6/gp130-induced premature senescence in human fibroblasts. Cell Cycle 11:730–739
Korhonen P, Helenius M, Salminen A (1997) Age-related changes in the regulation of transcription factor NF-κB in rat brain. Neurosci Lett 225:61–64
Kregel KC, Zhang HJ (2007) An integrated view of oxidative stress in aging: basic mechanisms, functional effects, and pathological considerations. Am J Phys 292:R18–R36
Krisko A, Radman M (2019) Protein damage, ageing and age-related diseases. Open Biol 9:180249
Lehmann M, Korfei M, Mutze K et al (2017) Senolytic drugs target alveolar epithelial cell function and attenuate experimental lung fibrosis *ex vivo*. Eur Respir J 50:1602367
Libby P, Kobold S (2019) Inflammation: a common contributor to cancer, aging, and cardiovascular diseases—expanding the concept of cardio-oncology. Cardiovasc Res 115:824–829
Lim HA, Lee EK, Kim JM (2012) PPARγ activation by baicalin suppresses NF-κB-mediated inflammation in aged rat kidney. Biogerontology 13:133–145
Lin J-Y, Kuo W-W, Baskaran R et al (2020) Swimming exercise stimulates IGF1/PI3K/Akt and AMPK/SIRT1/PGC1α survival signaling to suppress apoptosis and inflammation in aging hippocampus. Aging 12:6852–6864
Liu T, Zhang L, Joo D et al (2017) NF-κB signaling in inflammation. Signal Transduct Target Ther 2:17023
Loos BG, Van Dyke TE (2020) The role of inflammation and genetics in periodontal disease. Periodontol 83:26–39
Makki K, Froguel P, Wolowczuk I (2013) Adipose tissue in obesity-related inflammation and insulin resistance: cells, cytokines, and chemokines. ISRN Inflamm 2013:139239
Manolis AS, Manolis AA, Manolis TA et al (2021) Mitochondrial dysfunction in cardiovascular disease: current status of translational research/clinical and therapeutic implications. Med Res Rev 41:275–313
Mega JL, Braunwald E, Wiviott SD et al (2012) ATLAS ACS 2–TIMI 51 investigators. Rivaroxaban in patients with a recent acute coronary syndrome. N Engl J Med 366:9–19
Mosca A, Leclerc M, Hugot JP (2016) Gut microbiota diversity and human diseases: should we reintroduce key predators in our ecosystem? Front Microbiol 7:455
Ng A, Tam WW, Zhang MW et al (2018) IL-1β, IL-6, TNF-α and CRP in elderly patients with depression or Alzheimer's disease: systematic review and meta-analysis. Sci Rep 8:12050
Oe Y, Mochizuki K, Mitauchi R et al (2015) Plasma TNF-α is associated with inflammation and nutrition status in community-dwelling Japanese elderly. J Nutr Sci Vitaminol 61:263–269
Oeckinghaus A, Ghosh S (2009) The NF-κB family of transcription factors and its regulation. Cold Spring Harb Perspect Biol 1:a000034
Ohtsubo K, Marth JD (2006) Glycosylation in cellular mechanisms of health and disease. Cell 126:855–867
Okada Y, Tsuzuki Y, Hokari R et al (2009) Anti-inflammatory effects of the genus *Bifidobacterium* on macrophages by modification of phospho-IκB and SOCS gene expression. Int J Exp Pathol 90:131–140
Olgar Y, Degirmenci S, Durak A et al (2018) Aging related functional and structural changes in the heart and aorta: MitoTEMPO improves aged-cardiovascular performance. Exp Gerontol 110:172–181
Palta S, Saroa R, Palta A (2014) Overview of the coagulation system. Indian J Anaesth 58:515–523

Parekh RB, Roitt IM, Isenberg DA et al (1988) Galactosylation of IgG associated oligosaccharides: reduction in patients with adult and juvenile onset rheumatoid arthritis and relation to disease activity. Lancet 1:966–969

Park MH, Hong JT (2016) Roles of NF-κB in cancer and inflammatory diseases and their therapeutic approaches. Cell 5:15

Patsouris D, Cao J-J, Vial G et al (2014) Insulin resistance is associated with MCP1-mediated macrophage accumulation in skeletal muscle in mice and humans. PLoS One 9:e110653

Poli VFS, Sanches RB, dos Santos MA et al (2017) The excessive caloric intake and micronutrient deficiencies related to obesity after a long-term interdisciplinary therapy. Nutrition 38:113–119

Poole L, Steptoe A (2020) The combined association of depressive symptoms and C-reactive protein for incident disease risk up to 12 years later. Findings from the English Longitudinal Study of Ageing (ELSA). Brain Behav Immun 88:908–912

Puzianowska-Kuznicka M, Owczarz M, Wieczorowska-Tobis K et al (2016) Interleukin-6 and C-reactive protein, successful aging, and mortality: the PolSenior study. Immun Ageing 13:21

Radák Z, Chung HY, Naito H et al (2004) Age-associated increases in oxidative stress and nuclear transcription factor κB activation are attenuated in rat liver by regular exercise. FASEB J 18:749–750

Rea IM, Gibson DS, McGilligan V et al (2018) Age and age-related diseases: role of inflammation triggers and cytokines. Front Immunol 9:586

Ritzel RM, Lai Y-J, Crapser JD et al (2018) Aging alters the immunological response to ischemic stroke. Acta Neuropathol 136:89–110

Romano AD, Serviddio G, De Matthaeis A et al (2010) Oxidative stress and aging. J Nephrol 23:S29–S36

Roos CM, Zhang B, Palmer AK et al (2016) Chronic senolytic treatment alleviates established vasomotor dysfunction in aged or atherosclerotic mice. Aging Cell 15:973–977

Ruhaak LR, Uh HW, Beekman M et al (2011) Plasma protein N-glycan profiles are associated with calendar age, familial longevity and health. J Proteome Res 10:1667–1674

Sanada F, Taniyama Y, Azuma J et al (2009) Hepatocyte growth factor, but not vascular endothelial growth factor, attenuates angiotensin II-induced endothelial progenitor cell senescence. Hypertension 53:77–82

Sanada F, Taniyama Y, Muratsu J et al (2016) Activated factor X induces endothelial cell senescence through IGFBP-5. Sci Rep 6:35580

Sanada F, Muratsu J, Otsu R et al (2017) Local production of activated factor X in atherosclerotic plaque induced vascular smooth muscle cell senescence. Sci Rep 7:17172

Sanada F, Taniyama Y, Muratsu J et al (2018) Source of chronic inflammation in aging. Front Cardiovasc Med 5:12

Santoro A, Spinelli CC, Martucciello S et al (2018) Innate immunity and cellular senescence: the good and the bad in the developmental and aged brain. J Leukoc Biol 103:509–524

Sawyer AJ, Tian W, Saucier-Sawyer JK et al (2014) The effect of inflammatory cell-derived MCP-1 loss on neuronal survival during chronic neuroinflammation. Biomaterials 35:6698–6706

Shaw AC, Goldstein DR, Montgomery RR (2013) Age-dependent dysregulation of innate immunity. Nat Rev Immunol 13:875–887

Shoemark DK, Allen SJ (2015) The microbiome and disease: reviewing the links between the oral microbiome, aging, and Alzheimer's disease. J Alzheimers Dis 43:725–738

Sparkenbaugh EM, Chantrathammachart P, Mickelson J et al (2014) Differential contribution of FXa and thrombin to vascular inflammation in a mouse model of sickle cell disease. Blood 123:1747–1756

Spronk HM, de Jong AM, Crijns HJ et al (2014) Pleiotropic effects of factor Xa and thrombin: what to expect from novel anticoagulants. Cardiovasc Res 101:344–351

Stefanatos R, Sanz A (2018) The role of mitochondrial ROS in the aging brain. FEBS Lett 592:743–758

Stephenson J, Nutma E, van der Valk P et al (2018) Inflammation in CNS neurodegenerative diseases. Immunology 154:204–219

Tahir A, Martinez PJ, Ahmad F et al (2021) An evaluation of lipid profile and pro-inflammatory cytokines as determinants of cardiovascular disease in those with diabetes: a study on a Mexican American cohort. Sci Rep 11:2435

Tamura Y, Sugimoto M, Murayama T et al (2008) Inhibition of CCR2 ameliorates insulin resistance and hepatic steatosis in db/db mice. Arterioscler Thromb Vasc Biol 28:2195–2201

Tan BL, Norhaizan ME (2021) Oxidative stress, diet and prostate cancer. World J Mens Health 39:195–207

Tan BL, Norhaizan ME, Huynh K et al (2015) Brewers' rice modulates oxidative stress in azoxymethane-mediated colon carcinogenesis in rats. World J Gastroenterol 21:8826–8835

Tan BL, Norhaizan ME, Liew W-P-P et al (2018a) Antioxidant and oxidative stress: a mutual interplay in age-related diseases. Front Pharmacol 9:1162

Tan BL, Norhaizan ME, Liew W-P-P (2018b) Nutrients and oxidative stress: friend or foe? Oxidative Med Cell Longev 2018, Article ID 9719584, 24 pages

Tchkonia T, Zhu Y, van Deursen J et al (2013) Cellular senescence and the senescent secretory phenotype: therapeutic opportunities. J Clin Invest 123:966–972

Tilstra JS, Clauson CL, Niedernhofer LJ et al (2011) NF-κB in aging and disease. Aging Dis 2:449–465

Toward R, Montandon S, Walton G et al (2012) Effect of prebiotics on the human gut microbiota of elderly persons. Gut Microbes 3:57–60

Tu Y, Zhu M, Wang Z et al (2020) Melatonin inhibits Müller cell activation and pro-inflammatory cytokine production via upregulating the MEG3/miR-204/Sirt1 axis in experimental diabetic retinopathy. J Cell Physiol 235:8724–8735

Ungvari Z, Orosz Z, Labinskyy N et al (2007) Increased mitochondrial H2O2 production promotes endothelial NF-κB activation in aged rat arteries. Am J Physiol Heart Circ Physiol 293:H37–H47

van Deursen JM (2014) The role of senescent cells in ageing. Nature 509:439–446

Vanhooren V, Desmyter L, Liu XE et al (2007) N-glycomic changes in serum proteins during human aging. Rejuvenation Res 10:521–531

Villarroya F, Cereijo R, Gavaldà-Navarro A et al (2018) Inflammation of brown/beige adipose tissues in obesity and metabolic disease. J Intern Med 284:492–504

Wojdasiewicz P, Poniatowski ŁA, Szukiewicz D (2014) The role of inflammatory and anti-inflammatory cytokines in the pathogenesis of osteoarthritis. Mediat Inflamm 2014, Article ID 561459, 19 pages

Wong SL, Wagner DD (2018) Peptidylarginine deiminase 4: a nuclear button triggering neutrophil extracellular traps in inflammatory diseases and aging. FASEB J 32:6258–6370

Yasuoka H, Hsu E, Ruiz XD et al (2009) The fibrotic phenotype induced by IGFBP-5 is regulated by MAPK activation and egr-1-dependent and -independent mechanisms. Am J Pathol 175:605–615

Zhang Q, Raoof M, Chen Y et al (2010) Circulating mitochondrial DAMPs cause inflammatory responses to injury. Nature 464:104–107

Zhou T, Prather ER, Garrison DE et al (2018) Interplay between ROS and antioxidants during ischemia-reperfusion injuries in cardiac and skeletal muscle. Int J Mol Sci 19:417

Zhu Y, Deng J, Nan M-L et al (2019) The interplay between pattern recognition receptors and autophagy in inflammation. In: Cui J (ed) Autophagy regulation of innate immunity, Advances in experimental medicine and biology, vol 1209. Springer, Singapore, pp 79–108

Ziegler AK, Damgaard A, Mackey AL et al (2019) An anti-inflammatory phenotype in visceral adipose tissue of old lean mice, augmented by exercise. Sci Rep 9:12069

Zou Y, Yoon S, Jung KJ et al (2006) Upregulation of aortic adhesion molecules during aging. J Gerontol Ser A 61:232–244

Chapter 5
Implications of Inflammation in Aging and Age-Related Diseases

Research evidence indicates that low-grade chronic systemic inflammation plays a crucial role in mediating the aging process and age-related diseases, for instance, osteoporosis, diabetes, cancer, atherosclerosis, neurodegenerative disease, and metabolic syndrome (Tan et al. 2018a, b), which is also known as "inflammatory diseases" (McGeer and McGeer 1999). Despite the precise molecular inflammatory process involved in each disease may vary from other diseases, the fundamental mode of actions of the proinflammatory cytokines and inflammatory mediators are similar (Kany et al. 2019; Rea et al. 2018). The role of chronic inflammation in age-related diseases will be described in the following section.

5.1 Obesity

Metabolic syndrome was originally known as Syndrome X, refers to a cluster of clinical factors, such as dyslipidemia with low high-density lipoprotein (HDL) levels and increased triglycerides (TG), hypertension, high blood glucose, hyperinsulinemia, and insulin resistance (Huang 2009; Bonomini et al. 2015). Nonetheless, reduced insulin sensitivity is linked to abdominal obesity, which is the predominant feature of the syndrome (Galderisi et al. 2019).

The prevalence of obesity has doubled from 1980 to 2008 worldwide. Nearly 50% of women and men in the WHO European Region were overweight in 2008, in which 23% of women and 20% of men were obese (World Health Organization 2017). Approximately 1.5 billion people are overweight or obese worldwide. Individuals who are obese or overweight are at risk of developing coronary heart disease, CVD, type 2 diabetes, nonalcoholic fatty liver disease, and inflammatory disturbances (Wang et al. 2011; Matsuda and Shimomura 2013). The rates of obesity have increased in both sexes and all ages irrespective of socioeconomic status, ethnicity, or geographical locality. Women and elderly people were shown to have a

B. L. Tan, M. E. Norhaizan, *The Role of Antioxidants in Longevity and Age-Related Diseases*, https://doi.org/10.1007/978-3-030-82859-2_5

greater likelihood of obesity (Chooi et al. 2019). Obesity is linked to the high levels of 4-hydroxynonenal (4-HNE) adducts in skeletal muscle (Russell et al. 2003) and protein carbonyls in subcutaneous adipose tissues (Frohnert et al. 2011).

Under normal circumstances, the oxidation of fatty acids or glucose produces ATP via mitochondrial respiration, and thus results in the production of superoxide (Brownlee 2001). Fatty acid oxidation and glycolysis generate $FADH_2$ and NADH, which allow the electrons that fuel ATP and ETC production (Murphy 2009). $FADH_2$ and NADH donate their electrons to complexes II and I of the ETC, respectively. These complexes transfer electrons to ubiquinone, and thereby passed through the electrons to complex III, cytochrome c, and complex IV, and ultimately to molecular oxygen (Zhao et al. 2019a). The produced electrons are used to shuttle protons across the mitochondrial membrane to generate membrane voltage potential that drives mitochondrial respiration (Alberts et al. 2002). Superoxide can be generated when the electrons from coenzyme Q are received by molecular oxygen as they are shuttled from complexes I and II to complex III (Brownlee 2001). Excessive high-carbohydrate or high-fat diet may produce more substrates to enter into mitochondrial respiration. Subsequently, the number of electrons donated to the ETC may increase. Once the membrane potential reaches the threshold voltage, excess electrons may back up at complex III, with a subsequent donation to molecular oxygen, and thus produce high levels of superoxide (Brownlee 2001). Data from *in vitro* study showed that free fatty acids promote ROS production, suggesting that high fatty acids in obesity may provide an additional source of excess ETC substrates by increasing fatty acids oxidation (Furukawa et al. 2004). Obesity or increased caloric intake is linked to the activation of mitochondrial superoxide production. Feeding a high-fat diet significantly increased H_2O_2 emission from mitochondria isolated from skeletal muscle (Jain et al. 2014). The data further revealed that obese or high-fat diet animals promote ROS production in mitochondria of adipose tissue (Masschelin et al. 2020), liver (Vial et al. 2011), and kidney (Chang et al. 2019).

Another prominent contributor that promotes a prooxidant environment in obesity is through the mediation of proinflammatory response (Rodríguez-Cerdeira et al. 2019). Inflammation can lead to oxidative stress in *in vivo* model when there is a large increase in free radical production by immune cells (Pizzino et al. 2017). Obesity is linked to the changes of T-cell subsets related to adipose tissues (Misumi et al. 2019) and the activation of infiltration of macrophages into the adipose tissue (Surmi and Hasty 2008). These processes increase production of some proinflammatory cytokines by preadipocytes, mature adipocytes, and immune cells (Asghar and Sheikh 2017). Elevation of oxidative stress is related to chronic inflammation (Wiegman et al. 2015); other sources may also further enhance the production of proinflammatory cytokines in a "vicious cycle" (di Penta et al. 2013). ROS promotes the expression of proinflammatory monocyte chemotactic protein-1 (MCP-1) and cytokine IL-6 (Elmarakby and Sullivan 2012). In adipose tissue, this process further enhances the macrophage infiltration and thereby leading to the proinflammatory environment (Lee and Lam 2019). ROS can also induce NF-κB signaling pathway that activates and IL-6 and TNF-α (Yoshida et al. 1999; Kamata et al. 2005; Zhang et al. 2020). Furthermore, oxidative stress may promote cellular senescence, particularly adipocyte senescence, partly via the cellular oxidation damage (Minamino et al. 2009; MacKellar et al. 2010). Adipocyte senescence may increase the generation of proinflammatory cytokines and recruit macrophages (Minamino et al. 2009; Lafontan 2014).

Adipose tissues are divided into two classes, namely brown adipose tissue (BAT) and white adipose tissue (WAT) (Kaisanlahti and Glumoff 2019). WAT is one of the predominant adipose tissues found in the organisms, which is a site for energy storage. BAT is mainly used for non-shivering thermogenesis, primarily in human neonates and small mammals (Fantuzzi 2005). Among all the cell types found in WAT, adipocytes are one of the abundant cells. The previous study has revealed that adipose tissue is characterized by an elevation infiltration of macrophages, suggesting the presence of inflammation in adipose tissues (Suganami and Ogawa 2010; Sharma et al. 2020). The macrophage infiltration may lead to a dysregulation in the adipose tissue endocrine system and stimulate the production of the inflammatory cytokines, and subsequently result in insulin resistance and endothelial dysfunction (Chait and den Hartigh 2020). Intriguingly, the inflammatory response in obesity is induced by adipose tissues (Pan et al. 2019). Adipocytes constitutively increased TNF-α levels, in which the adipocytes are markedly elevated with obesity (Poret et al. 2018; Tan and Norhaizan 2019). TNF-α activates the cellular kinase complex, which in turn promotes the NF-κB. Subsequently, this transcription factor mediates the production of proinflammatory cytokines, such as IL-6 and IL-1β (Kern et al. 2019).

The previous study has revealed that increased oxidative stress in obesity can reduce antioxidant defense system and increase superoxide production by NADPH oxidase (Olusi 2002; Furukawa et al. 2004; Serpillon et al. 2009; Marseglia et al. 2015). NADPH oxidase is mainly used by neutrophils to produce superoxide to kill fungi and invading bacteria (Gazendam et al. 2016; Belambri et al. 2018). It is stimulated by advanced glycation end products (AGEs), which are increased after high glucose diet consumption (Goldin et al. 2006; El-Bassossy et al. 2018). Interestingly, obesity is inversely associated with catalase, glutathione peroxidase 1 (GPx1), and SOD in the adipose tissues, but not in the liver or skeletal muscle (Furukawa et al. 2004). It remains unknown why the reduction of antioxidant enzymes is only observed in adipose tissues. It could be due to the high oxidative stress in this tissue that caused an inhibition of the antioxidant enzyme (Furukawa et al. 2004). This phenomenon suggests that the general increase in oxidative stress related to obesity may be due to the alteration in fat (Walle et al. 2017). Taken together, systemic oxidative stress-related high-fat diet and obesity may promote inflammation and metabolic dysfunction.

5.2 Neurodegenerative Diseases

Neurodegenerative diseases are characterized by progressive deterioration of neural cells and thus lead to compromised cognitive or motor function (Dugger and Dickson 2017; Liu et al. 2017a). There are several neurodegenerative diseases including Huntington's disease, Parkinson's disease, Alzheimer's disease, spinocerebellar ataxia, and amyotrophic lateral sclerosis (Albers and Beal 2000; Silvade et al. 2000; Matilla-Duenas et al. 2014; Alzheimer's Association 2016). These diseases represent one of the greatest health threats, particularly among the elderly (Albers and Beal 2000; Hamer and Chida 2009). Dementia is a progressive or chronic syndrome, in which there is declined in their ability to conduct daily activities (World Health Organization 2020a). It affects behavior, thinking, memory, as well as orientation and judgment. Despite dementia is primarily affected the elderly, it is not an exclusive part for aging. Dementia accounts for nearly 50 million people worldwide (World Health Organization 2020a). In particular, Alzheimer's disease is the most common form of dementia and contributes to 60–70% of cases (World Health Organization 2020a). ROS are chemically reactive molecules that have been implicated in the pathogenesis of neurodegenerative diseases (Collin 2019). Research evidence suggested that ROS may play a crucial role in neurodegenerative disorders (Ballance et al. 2019). Patients with neurodegenerative diseases were shown high amounts of oxidative stress (Cenini et al. 2019). Despite ROS may not be an essential factor for neurodegenerative diseases, they are more likely to aggravate disease progression via interaction with mitochondria and oxidative damage (Dias et al. 2013). Notably, neuron cells are susceptible to oxidative damage due to the low antioxidant defense, high rates of oxygen consumption, and increased levels of polyunsaturated fatty acid in membranes (Rego and Oliveira 2003). The redox imbalance can lead to abnormal mitochondrial function and neural inflammation (Rego and Oliveira 2003; Mosley et al. 2006). The pathogenesis of several neurodegenerative diseases including Parkinson's disease and Alzheimer's disease are linked to the accumulation of misfolded proteins (Lim 2019). The aggregation of these proteins can induce an inflammatory response in the brain, and thus markedly release ROS (Wyss-Coray and Mucke 2002; Zuo et al. 2015). Mitochondrial dysfunction is accompanied by aberrant ROS generation, which is closely related to neurodegenerative diseases (Albers and Beal 2000; Lin and Beal 2006).

Alzheimer's disease represents the most common prevalent neurodegenerative condition (Magalingam et al. 2018). The Alzheimer's disease population is expected to increase to 65.69 million people by 2030 (Alzheimer's disease international 2009). The pathophysiology of Alzheimer's disease is primarily linked to the accumulation of intracellular tau neurofibrillary tangles and extracellular deposition of amyloid-beta (Aβ) plaques (Querfurth and LaFerla 2010; Butterfield 2014). Aβ plaques can diminish calcium ions (Ca^{2+}) storage in the endoplasmic reticulum and thus leading to an overload of cytosolic Ca^{2+} (Corona et al. 2011). When cytosolic Ca^{2+} is upregulated, it can cause an accumulation of ROS inside the cells and reduce endogenous GSH levels (Ferreiro et al. 2008). In particular, ROS-induced oxidative stress is a key factor in the pathogenesis of Alzheimer's disease. This finding implies that overexpression of ROS is considered critical in the deposition and accumulation of Aβ in Alzheimer's disease (Bonda et al. 2010). Mitochondrial dysfunction can cause an alteration of Ca^{2+} homeostasis, excitotoxicity, reduction of ATP, and misregulation of ROS (Gleichmann and Mattson 2011). All these modifications have been implicated in the development of Alzheimer's disease (Huang et al.

2016). Overproduction of oxidative stress in Alzheimer's disease patients could be attributed to the overactivation of N-methyl-D-aspartate-type glutamate receptors (NMDARs). Stimulation of NMDAR causes an excessive influx of Ca^{2+} by enhancing cell permeability and thus producing neurotoxic amounts of ROS/RNS (Nakamura and Lipton 2010, 2011). ROS plays a crucial role in modulating the JNK/stress-activated protein kinase signaling pathway (Benhar et al. 2002). The stimulation of these pathways triggered the Aβ-induced cell death and hyperphosphorylation of tau proteins (Patten et al. 2010). Furthermore, Aβ proteins can initiate the free radical formation through stimulation of NADPH oxidase (Shelat et al. 2008). Aβ-induced ROS accumulation alters cellular signaling pathways and initiates tau hyperphosphorylation through stimulation of p38 MAPK. Giraldo et al. (2014) and Bulat and Widmann (2009) found that dysfunction of hyperphosphorylated tau proteins can lead to the formation of intracellular neurofibrillary tangles. It has been reported that Aβ plays a crucial role in modulating cellular apoptotic cascades (Agostinho et al. 2008). In particular, Aβ can enhance calcineurin activity, and thereby induces Bcl-2-related death promoter. Ultimately, these lead to cytochrome c release from mitochondria (Awasthi et al. 2005). It may also link to the caspases, which in turn trigger neuron apoptosis (Awasthi et al. 2005).

The study has shown that environmental stress, inflammation, and aging can trigger oxidative stress that may contribute to the increased of Aβ production (Hamilton and Holscher 2012; Aseervatham et al. 2013). Elderly people are prone to oxidative stress, hence partially explaining the susceptibility of Alzheimer's disease among aging populations (Stadtman 2001; Hamilton and Holscher 2012). Inflammation promotes cellular toxicity, enhances ROS levels, and increases the amount of cytokines; all these may aggravate Alzheimer's disease progression (Blaser et al. 2016; Varvel et al. 2016; Italiani et al. 2018). In addition, Aβ accumulation leads to microglial activation (Hemonnot et al. 2019). It has been reported that prolonged stimulation of microglia may result in the release of proinflammatory cytokines, as well as initiating the proinflammatory cascade, and thereby resulting in neuronal loss and damage (Wang et al. 2015).

Oxidative stress can also be triggered by environmental factors such as radiation, chemicals, and pollutants (Nizzari et al. 2012; Aseervatham et al. 2013). For instance, excess iron deposits lead to the formation of ROS (Nizzari et al. 2012). Aβ interacts with metal ions and produces free radicals (Butterfield and Boyd-Kimball 2005; Nizzari et al. 2012). Notably, Cu^{2+}/Zn^{2+} -bound Aβ was shown a structure similar to SOD with antioxidant activity (Curtain et al. 2001). In this regard, supplementation of Zn^{2+} and Cu^{2+} could be served as a potential approach to reduce metal-catalyzed Aβ deposition and Aβ-induced ROS production (Curtain et al. 2001). Collectively, the molecular mechanisms that implicate the development of neurodegenerative diseases are complex and involved a few mechanisms. In particular, mitochondria dysfunction is linked to the oxidative stress in neurodegenerative diseases.

5.3 Atherosclerosis

CVD has become the leading cause of death worldwide, accounting for nearly 17.9 million lives annually, an estimated 31% of all deaths globally (World Health Organization 2020b). CVD is a cluster of disorders of the blood vessels and heart including rheumatic heart disease, cerebrovascular disease, coronary heart disease, and other conditions (World Health Organization 2020b). About 85% of all CVD are due to strokes and heart attacks, with one-third of this death occurred prematurely in individuals below 70 years old (World Health Organization 2020b). The major contributor to CVD is atherosclerosis, which is a multifactorial and progressive process that affects the large arteries characterized by the accumulation of fibrous particles and lipids in the walls. Inflammation has been demonstrated in all phases of the atherosclerotic process, from the initial steps of leukocyte recruitment to the rupture of a vulnerable plaque (Pant et al. 2014).

The initial stage of atherosclerosis has been implicated by the adherence or attraction of monocytes to the vascular endothelium as well as their migration into the vessel wall (Čejková et al. 2016). The adhesion molecules such as VCAM-1, ICAM-1, and selectins promotes the adhesion of leukocytes to the vascular endothelium (Pierangeli et al. 2001) and thus triggered by inflammatory factors, for instance, CRP, TNF-α, and interleukin-1 (IL-1) (Guo et al. 2018). Indeed, VCAM-1 binds to certain leukocytes was found in nascent atheroma. Both endothelial cells and macrophages generate ICAM-1 in response to inflammatory cytokines such as interferon-γ, TNF-α, and IL-1 (Ren et al. 2010; Paulsen et al. 2015; Wiesolek et al. 2020). The accumulation of macrophages and their uptake of oxidized-low density lipoprotein (ox-LDL) may result in the production of foam cells and thus promote fatty streaks (Moore et al. 2013).

The development of lesions is characterized by the migration of smooth muscle cells into the subendothelial space from the medial layer of the artery wall (Lusis 2000; Milutinović et al. 2020). Smooth muscle cells grow and migrate with the facilitation of several cytokines and growth factors produced by endothelial and inflammatory cells (Louis and Zahradka 2010). Nonetheless, the smooth muscle cell is a source of inflammatory mediators, for instance, monocyte chemoattractant protein-1 (MCP-1) (Lim and Park 2014). Mice lack of MCP-1 expression or its chemokine receptor, CCR2 was shown to develop fewer atherosclerotic lesions (Gosling et al. 1999; Wezel et al. 2015). Similar to the increase of chemoattractant proteins and adhesion molecules, initiators of the atherosclerotic cascade (ox-LDL) were found to promote the production of growth factors including macrophage colony-stimulating factor (M-CSF) (Wu et al. 2017).

The growth of smooth muscle cells further recruits the inflammatory cells and the synthesis of extracellular matrix proteins and thus leading to the formation of the atheroma, a mature atherosclerotic plaque comprised of fibrous cap separating the pro-thrombotic lipid pool from luminal blood flow (Newby and Zaltsman 1999; Lim and Park 2014; Basatemur et al. 2019). The stimulation of smooth muscle cells and inflammatory cells (T lymphocytes and macrophages) is linked to the release of mediators such as growth factors, cytokines, and adhesion molecules (Ramel et al. 2019). IL-6 has been reported to increase plasma levels of CRP, plasminogen activator inhibitor type 1, and fibrinogen, and thereby enhance and amplify both procoagulant and inflammatory responses (Yamaguchi et al. 1998; Fischer et al. 2006; Cesari et al. 2010). This proinflammatory mediator has been suggested to be involved in the modulation of atherosclerosis pathogenesis (Moriya 2019).

They are several enzyme systems that can generate ROS in the vascular wall. Among the ROS-producing enzymes, four of them seem to play a crucial role, namely, a dysfunctional endothelial NO synthase, enzymes of the mitochondrial respiratory chain, XO, and NADPH oxidase (Landmesser et al. 2002; George and Struthers 2009; Faria et al. 2018; Daiber et al. 2019; Ding et al. 2019; Mazat et al. 2020). Notably, NADPH oxidase can induce eNOS uncoupling as well as XO activation (Landmesser et al. 2007).

5.3.1 NADPH Oxidases

NADPH oxidases are key sources of ROS in the vasculature, which generates superoxide from molecular oxygen by using NADPH as the electron donor (Yousefian et al. 2019). These oxidases are multi-subunit enzyme complexes that interact with one of few homologs of the membrane-bound Nox catalytic subunit (Bedard and Krause 2007; Drummond et al. 2011). The activity of Nox4 depends on its association with p22phox. Indeed, p22phox is not only essential for Nox2 and Nox1 activity in vascular cells, other Nox catalytic subunit such as Rac1, p67phox (or NOXA1), and p47phox (or NOXO1) also plays a crucial role in this phenomenon (Brandes and Kreuzer 2005).

Nox4 (Ellmark et al. 2005) and Nox1 (Lassegue et al. 2001) are found in vascular smooth muscle cells, while Nox4 (Ago et al. 2004; Xu et al. 2008) and Nox2 are expressed predominantly in endothelial cells. Atherosclerosis is linked to the upregulation of NADPH oxidase subunits (p67phox, p47phox, p22phox, and Nox2) in the coronary artery (Guzik et al. 2006).

Despite data from animal studies of global p47phox (Hsich et al. 2000) and Nox2 (Kirk et al. 2000) knockout mice demonstrated that no difference in atherosclerosis of the aortic sinus, few studies stated that global knockout of p47phox leads to a moderate decline in aortic atherosclerosis for both high-fat and chow fed animals (Barry-Lane et al. 2001). This finding implies that global Nox2 phosphorylation results in a decline in aortic atherosclerotic burden (Judkins et al. 2010).

The cell-specific effects of Nox2 were evaluated. An animal study showed that endothelium-specific overexpressing Nox2 enhanced the macrophage recruitment in early lesions, elevated the endothelial levels of VCAM-1, and increased vascular superoxide production in ApoE-KO mice. Nevertheless, neither the native nor the angiotensin II-driven atherosclerosis is influenced by endothelial Nox2 overexpression (Douglas et al. 2012). This finding implied that despite the stimulation of endothelial Nox2 alone is sufficient for the initiation of atherosclerosis, Nox2 in other cell types could be an indispensable contribution to the development of atherosclerosis. Notably, the plaque core of atherosclerotic human arteries was shown the highest Nox2 expression, the Nox2 in macrophage is also abundant. Therefore, as long as the macrophage function is not changed and the lipid composition does not alter, the elevation of endothelial-derived ROS formation alone may not lead to atherosclerotic plaque progression (Haendeler et al. 2012).

5.3.2 *Xanthine Oxidase*

Similar to NADPH oxidases, XO has been demonstrated to be another key source of superoxide in human coronary artery disease (Guzik et al. 2006). XO was found in endothelial cells and the activity and expression of endothelial XO are increased in response to angiotensin II treatment (Landmesser et al. 2007). Furthermore, XO can be released from the circulation XO and the liver adheres to endothelial cells by interacting with the endothelial glycosaminoglycans (White et al. 1996). XO transfers the electrons to molecular oxygen and thereby generating H_2O_2 and superoxide. The activity of both plasma XO (White et al. 1996) and endothelial XO (Ohara et al. 1993) is elevated in human atherosclerotic plaque and atherosclerosis (Patetsios et al. 2001; Guzik et al. 2006), implied that XO-derived superoxide promotes atherosclerosis. Data from *in vivo* study showed that XO inhibitors improve the endothelium-dependent, NO-modulated vasodilation in aorta rings of hypercholesterolemic animals (White et al. 1996). In another study, XO inhibitor tungsten prevents endothelial dysfunction, alleviates the vascular superoxide anion formation, and decreases the development of atherosclerosis in the aorta (Schroder et al. 2006).

5.3.3 *Mitochondria*

Atherosclerosis is developed by inflammation and a mutually enhancing circuit of ROS (Marchio et al. 2019). In addition to the different sources of ROS including uncoupled eNOS, XO, and NADPH oxidases, the mitochondrial respiratory ETC is of vitally important to CVD, heart function, and atherogenesis (Madamanchi and Runge 2007; Madungwe et al. 2016; Chistiakov et al. 2018). It has been exemplified by the cardiac loss of mitochondrial Mn-SOD (SOD2), and causes perinatal lethality due to the congestive heart failure or cardiac myopathy (Nojiri et al. 2006). Furthermore, it has been demonstrated that S-nitrosation of the mitochondrial complex I, which is the site where electrons enter the ETC. This site is critically involved in the oxidative burst during reperfusion/ischemia damage (Shiva 2010). Given the systemic uptake of an antioxidant such as vitamins failed to show any benefit of atheroprotection in clinical studies (Sesso et al. 2012), and thus the mediation of local antioxidant processes is a critical option. Due to the vital role of mitochondria, some of the new compounds were recently developed by targeting antioxidative components to the mitochondria to produce direct, efficient, and local ROS scavenging substances (Battogtokh et al. 2018; Goleva et al. 2019; Zhao et al. 2019b). Mitochondria-targeted antioxidants have great potential against detrimental impacts caused by ROS production. The ability of mitochondria-targeted antioxidants confer greater protection against oxidative damage, and this is primarily attributed to the ability to cross the phospholipid bilayer of mitochondria and thereby eliminating ROS (Oyewole and Birch-Machin 2015). In particular, MitoQ is one of the promising candidates, which is a mitochondria-targeted Q10 moiety that alleviates mitochondrial ROS production. In principle, a broad spectrum of antioxidants could be targeted to mitochondria through the conjugation of triphenylphosphonium (TPP) moiety (Smith and Murphy 2011). Indeed, ubiquinol (MitoQ) is the best-characterized antioxidant targeted to mitochondria through conjugation to the TPP cation (Smith and Murphy 2011). In an animal study, MitoQ decreased metabolic syndromes (hyperglycemia, hypercholesterolemia, and adiposity) by modulating oxidative DNA damage as well as macrophage infiltration into the plaques (Bond et al. 2019). The role of MitoQ in relation to atherosclerosis will be described in the ubiquinone section.

The "free radical theory" or "oxidative response to injury" has proposed that increased formation of radicals such as oxidative stress, is one of the predominant underlying modes of action that responsible for atherosclerosis (Volobueva et al. 2019). These data implied that the importance of mitochondria in atherosclerosis, which may involve ROS. Nonetheless, the elevation of ROS production is not necessarily the only link that responsible for atherosclerosis (Li et al. 2014). The previous study showed that increased mitochondrial DNA damage, which is observed in both hematopoietic and vascular cells as well as in early lesions that promotes vascular smooth muscle cells apoptosis, hyperlipidemia, atherogenesis, and release of IL-1β and TNF-α; all of these processes occur in the absence of increased ROS levels (Yu et al. 2013). mtDNA damage was also associated with plaque vulnerability (Yu et al. 2013).

5.3.4 Dysfunctional, "Uncoupled" Endothelial NO Synthase

Endothelial nitric oxide synthase (eNOS) generates NO under physiological circumstances, which represents a crucial element in the vasoprotective function of the endothelium (Suvorava et al. 2015). Under pathological conditions linked to oxidative stress, eNOS may become dysfunctional (Daiber et al. 2019). Oxidative stress leads to endothelial dysfunction, mainly due to the rapid oxidative inactivation of NO with an accumulation of superoxide (Daiber and Chlopicki 2020). In a subsequent stage, the oxidative stress renders eNOS uncoupled, in which it is no longer generates NO, but produces superoxide (Förstermann et al. 2017). In particular, phosphorylation of (6R-)5,6,7,8-tetrahydrobiopterin (BH_4), a crucial cofactor for the eNOS enzyme, is likely to be the primary cause for endothelial dysfunction and eNOS uncoupling (Förstermann and Sessa 2012). Peroxynitrite is another direct reaction product of superoxide and NO, which is capable of oxidizing BH_4 and thus resulting in BH_4 deficiency (Forstermann and Munzel 2006; Li and Forstermann 2013). ApoE-KO mice demonstrate an eNOS uncoupling and increased oxidative phosphorylation of BH_4 in CVD (Alp et al. 2004; Wohlfart et al. 2008; Xia et al. 2010). Data from the human study showed that eNOS uncoupling was shown in patients with endothelial dysfunction due to diabetes mellitus (Heitzer et al. 2000) or hypercholesterolemia (Stroes et al. 1997).

It has been reported that eNOS-derived H_2O_2 is essential in both flow- (Drouin and Thorin 2009) and acetylcholine-induced (Drouin et al. 2007) vasodilation of cerebral arteries isolated from healthy young mice. The catalase-sensitive acetylcholine-induced dilation could be restored by eNOS cofactor BH_4 (Drouin et al. 2007). Nonetheless, it remains obscure whether the physiological role of eNOS-derived H_2O_2 is a ubiquitous pathway in some vasoactive stimuli in some specified vascular beds or the vasculature (Chen et al. 2018). Despite endothelial cells in the vasculature share several basic characteristics, they differ between species in function and structure as well as vascular beds (Yokoyama and Hirata 2007). Taken together, decreased antioxidant defense mechanisms and extramitochondrial and mitochondrial sources of ROS have occurred in the myocardium of animals and humans (Ilkun and Boudina 2013).

5.4 Cancer

Cancer is the second leading cause of death worldwide, accounting for nearly 9.6 million deaths (or one in six deaths) in 2018 (World Health Organization 2020c). Breast, colorectal, lung, cervical, and thyroid cancers are the most common types of cancer in women, while lung, prostate, colorectal, stomach, and liver cancers are the most common among men (World Health Organization 2020c). Indeed, 30–50% of cancers could be prevented (World Health Organization 2020c).

The incidence of cancers increases steeply with age. A study has found that a single cell lineage must occur before malignant tissues are formed (Quinn et al. 2021). A study reported by Peto et al. (1977) stated that cancer is not associated with aging. Nonetheless, aging has been suggested as a time-dependent process, and probably the summation of innumerable different events. In this regard, the time-dependent emergence of cancer may relate to other age-related conditions (Le et al. 2019).

ROS is a hallmark of many types of cancer. ROS and its functions with respect to cancer progression and initiation in cancer cells are prime concerns in cancer research (Tan et al. 2014). Inflammatory cells produce most of the ROS (Aminjan et al. 2019). The stimulation of the redox metabolism of the inflammatory cells creates a highly oxidative environment within an organ of aerobic organisms (Di Marzo et al. 2018). The oxygen biochemistry via the stimulation of neutrophils, macrophages, and plasma membrane NADPH oxidase, and thus release the hydroxyl radicals, H_2O_2, and superoxide anion (Nguyen et al. 2017). Furthermore, inflammation reacts via the formation of RNS and ROS, and leads to oxidative damage in the cellular components (Mittal et al. 2014; Di Meo et al. 2016). Many proinflammatory mediators such as prostaglandins, chemokines, and cytokines turn on the angiogenesis switches primarily controlled by vascular endothelial growth factors (VEGF) (Fu et al. 2020). Cancer-related inflammation is also associated with the immune suppression that allows cancer cells to evade detection by the immune system (Grabowski et al. 2021). Inflammation is a crucial component for tumor progression (Chakraborty et al. 2020). In particular, several cancers may arise from sites of inflammation, chronic irritation, and infection (Piemonte et al. 2018; Siegfried et al. 2020; Eyvazi et al. 2020). It has been demonstrated that the tumor microenvironment, which is primarily orchestrated by inflammatory cells, is an indispensable player in the promotion of migration, survival, proliferation as well as neoplastic process (Guo and Deng 2018). Pathological angiogenesis is a hallmark of inflammatory diseases, ischemic, and cancer (Aguilar-Cazares et al. 2019; Johnson et al. 2019). Chronic inflammation is linked to angiogenesis, a process that promotes the proliferation of cancer cells. Angiogenesis is essential for the expansion of tumor mass, tumor cells, fibroblasts, platelets, and macrophages, which is a major source of angiogenic factors (Huong et al. 2019; Ireland and Mielgo 2018; Wang et al. 2019; Teleanu et al. 2020). The inflammation in the tumor microenvironment is characterized by leukocyte infiltration, ranging from composition, distribution, and size, like lymphocytes, eosinophils, neutrophils, natural killer (NK) cells, dendritic cells, mast cells, and tumor-associated macrophages (TAM) (Eiró and Vizoso 2012; Comen et al. 2018). These cells generate a wide range of cytotoxic mediators, for instance, RNS and ROS, cysteine and serine proteases, interferons (IFNs), interleukins (IL-8, IL-6, and IL-1), matrix metalloproteinase (MMP), and enzymes such as phospholipase A2 (PLA2), lipooxygenase-5 (LOX-5), and cyclooxygenase-2 (COX-2) (Tan et al. 2018a, b; Andreou et al. 2020). The AP-1 may cause basal gene expression in biological systems. ROS can stimulate AP-1 via several biological mechanisms. Activation of AP-1 promotes cell growth by inhibiting p21waf protein levels and increasing growth stimulatory gene expression such as cyclin D1 (de los Fayos Alonso et al. 2018). It has been demonstrated that NF-κB and AP-1, inducible by tumor promoters of oxidative stimuli, show activation in response to tumor promoters or differential protein levels in JB6 cells (Hsu et al. 2000). Several growth factors such as fibroblast growth factor 2 or insulin-like growth factor I produce ROS in PANC-1 and MIA PaCa-2 cells, which are human pancreatic adenocarcinoma cells that promote cell proliferation (Vaquero et al. 2004). Suppressing ROS production with NADPH oxidase (Nox4) antisense or antioxidants may promote apoptosis in PANC-1 and PaCa-2 cells (Acharya et al. 2010). This finding implies that this biological mechanism may mediate the pancreatic cancer resistance for treatment and thus may represent a potential therapeutic target. It has been demonstrated that GSH antioxidant defense system and oxidative pentose pathway (OPP) play a crucial role in the regulation of colon and gastrointestinal cancer cell proliferation and apoptosis (Acharya et al. 2010). The OPP provides NADPH for the synthesis of GSH and mediates intracellular redox status, which is responsible for the inactivation of intracellular ROS that leads to cell injury and induces apoptosis

(Wang et al. 2018). Depletion of GSH enhances the sensitivity of cells toward ROS. Therefore, suppression of OPP and/or GSH defense system may increase the sensitivity of colon and gastric cancer cells to anticancer therapy (Matthews et al. 2006). The previous study has demonstrated that the enzymatic product of thymidine phosphorylase (TP) produced ROS within cancer cells that facilitate maintaining the proliferation of colon cancer cells (Allavena et al. 2008).

The MAPK signaling pathway is stimulated by a broad spectrum of receptors involved in differentiation and proliferation including ion channels, integrins, and receptor tyrosine kinases (RTKs). This signaling pathway is stimulated in more than 50% of acute lymphocytic leukemia and acute myelogenous leukemia and is often stimulated in the prostate, breast, and other cancers (McCubrey et al. 2008; Mebratu and Tesfaigzi 2009). Under optimal circumstances, transiently increased ROS levels confer a growth advantage to tumor cells (Acharya et al. 2010). Nevertheless, treatment of cancer cells with anticancer agents trigger ROS levels and thus leading to potentiating of apoptosis. In this regard, ROS mediates the ability of stress kinases to activate cell death or cell proliferation, and this depends on the signal duration and signal intensity (Gechev et al. 2006). Proline oxidase (POX) is considered as a "housekeeping enzyme", which triggers apoptosis via extrinsic and intrinsic pathways. These signalings are involved in the regulation of the ERK/MEK pathway and mediation of nuclear factor of activated T cells (NFAT) signaling in colon cancer cells (Salaroglio et al. 2019). As a nutritional factor, POX may mediate apoptosis signals triggered by p52 or other anticancer agents and thus promote apoptosis under stress conditions (Surh 2008).

5.5 Diabetes

Insulin resistance is a condition that plays a prominent role in the development and progression of vascular complications in diabetes including type 2 diabetes (Wu et al. 2020a). Insulin resistance refers to failed or impaired cell response to insulin receptor-activated signaling in insulin-sensitive tissues, for instance, brain (Talbot et al. 2012; Bomfim et al. 2012), adipose, skeletal muscle, and liver (Newsholme et al. 2014). Reduced glucose uptake by these tissues concomitantly with the increase in hepatic glucose output leading to the increased plasma glucose levels (Kubota et al. 2011). The subsequent changes in glucose homeostasis increased the burden on pancreatic β-cells to secrete and produce more insulin to restore normal blood carbohydrate levels. Despite this compensatory process may attenuate glucose levels in prediabetes, exposure of β-cells to excess lipids and blood glucose as well as persistent insulin resistance increase β-cells dysfunction, failure, and lastly death, culminating in overt diabetes (Eizirik et al. 2020).

Insulin elicits its anabolic metabolism through association with the transmembrane insulin receptor (IR) in target tissues. The interaction of insulin stimulates related downstream signaling cascades such as protein kinase B (Akt) and phosphatidylinositol 3-kinase (PI3K), recruits and phosphorylates insulin receptor substrate

(IRS) proteins, and triggers autophosphorylation of the receptor (White et al. 1988; White 2003). Akt is a crucial regulator for GLUT-4 vesicle translocation to the plasma membrane, which is important for the intracellular uptake of free glucose in insulin-sensitive tissues (Taniguchi et al. 2006; Henriksen et al. 2011).

Insulin resistance occurs when there is interference in the insulin signaling cascade due to the structural or mutation alterations in the insulin signaling pathways. Serine-related hyperphosphorylation of insulin receptor substrate (IRS) and mutation has been associated with the development of insulin resistance. This finding could be due to the decreased interaction with PI3K (Saini 2010). An animal study has shown that homozygous interruption of IRS1 level in mice results in mild insulin resistance (Araki et al. 1994), while phosphorylation of IRS2 levels in rodents led to severe insulin resistance (Kubota et al. 2000).

Apart from that, structural alteration through hyperphosphorylation of serine at residues Ser^{632}, Ser^{612}, Ser^{307}, and Ser^{602} in IRS1 has been suggested to be a crucial mechanistic element that responsible for the elevation of insulin resistance in rodent models (Saini 2010). Increased expression of signaling proteins, for instance, JNK1 and TNF-α, could be derived from the adipose expansion and can trigger serine hyperphosphorylation of IRS1 (Hotamisligil et al. 1996; Stuart et al. 2014), especially at residue Ser^{636}. Nonetheless, it remains unknown whether the combination of residues or individual residues must be hyperphosphorylated to enhance insulin-resistant phenotype.

Insulin resistance in type 2 diabetes has been characterized by hyperinsulinemia, dyslipidemia, and hyperglycemia (American Diabetes Association 2010; Czech 2017; Ormazabal et al. 2018). Oxidative stress plays a crucial role in the development of type 2 diabetes (Molehin et al. 2020). RNS and ROS, for instance, peroxynitrite ($ONOO^-$), nitric oxide (NO), hydroxyl radical ($OH^\bullet$), H_2O_2, and superoxide anion ($O_2^{\bullet-}$) are key physiologic and metabolic processes. Impairment of mitochondrial function may decrease ATP production capacity, and thereby detrimental to the β-cell glucose-stimulated insulin secretion (GSIS) (Cerf 2013).

With respect to the progression of type 2 diabetes and insulin resistance, overnutrition leads to the production of RNS and ROS, and thereby promotes oxidative stress in organs, tissues, and cells. Ultimately, NO and oxygen based free radicals destroy protein, DNA, and cell membrane structures, as well as mediating the transcriptional activity via redox chemistry, such as NF-κB, and thus leading to cell apoptosis and chronic inflammation (Newsholme et al. 2009). Despite every single cell is potentially destroyed by oxidative stress, the decreased capacity of peroxidase-based antioxidant defense may expose the β-cells to damage and thus leading to the development of type 2 diabetes. Increased cytosolic Ca^{2+} levels, AGEs, inflammation, and mitochondrial dysfunction can enhance redox dysregulation and oxidative stress in type 2 diabetes (Rosales-Corral et al. 2015).

Decreased capacity in scavenging of free radicals is the main avenue for ROS production (Poprac et al. 2017). GSH is a free radical scavenger that is produced when there is a reduction in glutathione disulphide (GSSG) by glutathione reductase (Kurutas 2016). Besides reducing enzymes, NADPH is a crucial cofactor for glutathione reductase activity (Jain and Micinski 2013). The metabolic pathways that are mediated the hyperglycemia or overnutrition utilized NADPH, and thereby decreasing the cell capacity to produce GSH. For example, during polyol pathway flux, glucose is reduced by aldose reductase to sorbitol. Subsequently, sorbitol is converted into fructose by sorbitol dehydrogenase (Zhou et al. 2006). Despite aldose reductase has a low affinity to glucose, their activities are increased under hyperglycemic circumstances and thereby more NADPH is consumed (Rosales-Corral et al. 2015). Reduced amounts of GSH can lead to type 2 diabetes and this condition could be attributed to the deficiency in essential amino acids that are required to synthesize GSH or the impairment of protein turnover (De Luca et al. 2001). This finding is in line with the previous study who reported that reduced GSSG expression in Alzheimer's disease patients are linked to the decline of cognitive function (McCaddon et al. 2003; Cristalli et al. 2012). Suppression of the polyol pathway has been demonstrated to normalize sorbitol in the brain in the presence of hyperglycemia (Malone et al. 2008), implied that impaired cognitive function linked to hyperglycemia may be ameliorated by preventing the breakdown of sorbitol. Collectively, inflammation plays a pivotal role in insulin resistance and type 2 diabetes. Excessive consumption of saturated fats and carbohydrates affects insulin secretion. Intracellular signaling pathways in type 2 diabetes such as active inflammatory process, oxidative stress, and aberrant redox regulation may detrimental insulin secretion and its signaling.

5.6 Osteoporosis

Osteoporosis has been characterized by microarchitectural deterioration and low bone mass of bone tissues, and thus increased fracture risk and enlarged bone fragility (Sözen et al. 2017). Fractures caused by either disease-induced skeletal fragility (Hernlund et al. 2013) or traumatic injury (Mathew and Hanson 2009) leading a significant global health burden (Odén et al. 2015). It is due to the changed in balance between activities of osteoclasts and osteoblasts (Kim et al. 2020). These specialized cells are responsible for bone resorption and formation (Kim et al. 2018). The negative balance between osteoclastic and osteoblastic activity may lead to osteoporosis (Demontiero et al. 2012). The activity of bone cells can be affected by several cellular and nutritional factors, such as free radicals, growth factors, cytokines, endocrines, nutrients, and supply of oxygen (Abdollahi et al. 2005). The differentiation of osteoclasts and osteoblasts are believed to contribute to the pathogenesis of osteoporosis (Li et al. 2020; Wu et al. 2020b).

Osteoblasts are derived from osteoprogenitors that reside in the bone marrow. Osteoblasts are present throughout the lifespan, in which the highest activity was demonstrated during embryonic skeletal growth and formation (Boskey and Coleman 2010). In an adult organism, osteoblast is stimulated when the bone matrix has been depleted or when there is a need to regenerate a defect (Rutkovskiy et al. 2016). Osteoblasts secrete several matrix protein expressions, such as alkaline phosphatase (ALP), osteocalcin (OC), and collagen type 1 alpha 1 (Col1α1) (Rutkovskiy et al. 2016). It's worth taking notes that osteoblasts are not fast responders, in which the osteoblast takes 4 months to mature until the synthesis of bone matrix by the cells is detected (Canalis 2008). Osteoblasts are post-mitotic cells, but they are not terminally differentiated (Rutkovskiy et al. 2016). The osteoblasts that encircled with bone matrix may subsequently differentiate into osteocytes, which are interconnected stellar cells that modulate the turnover of bone matters (Florencio-Silva et al. 2015). When the mature osteoblast cells are reduced, new osteoblasts are differentiated from mesenchymal progenitor cells (Long 2012).

Despite research studies revealed that several signaling pathways and transcriptional factors may be involved in osteoblast differentiation, the molecular mechanism that diminishes the osteoblastic differentiation in osteoporosis are remained obscure (Carina et al. 2020). Oxidative stress has been described as a disruption balance of ROS and antioxidant defense levels (Birben et al. 2012). Oxidative stress is involved in the mediation of cadmium-induced osteoblast apoptosis (Branca et al. 2020). Increased production of intracellular ROS, including hydroxyl radicals, H_2O_2, and superoxide anions may subsequently result in the oxidation of proteins, DNA, and lipids (Schieber and Chandel 2014). The previous study showed that oxidative stress contributes to several age-related diseases including osteoporosis and that antioxidants can alleviate the damaging effect of oxidative stress (Zhu et al. 2018). Cadmium is a toxic heavy metal released from agricultural and industrial activities into water and soil, in which it can be accumulated in and absorbed by aquatic organisms and plants destined for the food supply (Agency for Toxic Substances and Disease Registry 2012). Indeed, diet is the key source of cadmium exposure to most individuals (Kim et al. 2019). Cadmium can trigger oxidative stress via suppression of antioxidant enzyme activity or depletion of the antioxidant GSH (Unsal et al. 2020). Bone is recognized as one of the target organs for cadmium toxicity (Chen et al. 2019). Exposure to cadmium has been recognized as a risk factor for osteoporosis (Buha et al. 2019). A study found that high levels of cumulative intake of cadmium are linked to an increased rate of fractures and osteoporosis in women (Chen et al. 2019). Data from *in vivo* study revealed that environmental cadmium disrupts bone mineralization in male rats (Buha et al. 2019). Hydrogen peroxide can induce oxidative stress in osteoblastic cells (MC3T3-E1), and thereby promotes apoptosis and suppresses osteoblast differentiation (Bai et al. 2004; Arai et al. 2007).

A study found that treatment osteoblast with H_2O_2 downregulated the expression of Runt-related transcription factor 2 (RUNX2) (Arai et al. 2007). The runt-associated transcriptional activity, for instance, RUNX2 is a crucial mediator of osteoblast phenotype that is involved in the osteoblast function and differentiation by regulating bone sialoprotein, osteopontin, and osteocalcin (Ducy et al. 1997). The primary role of RUNX2 in osteoblast differentiation was described by Komori et al. (1997). The data from the human study further revealed that patients with mutations in RUNX2 have been linked to the skeletal disorder of cleidocranial dysplasia (Komori 2002).

In addition, RUNX2 protects against osteoporosis in postmenopausal women by modulating bone mineral density (Vaughan et al. 2004). Declining of estrogen in women after menopause leads to derangement of osteoblasts and osteoclasts, with an elevation of bone turnover rate and a phenomenon in which resorption exceeds formation (Ji and Yu 2015). This metabolic change underlies the onset of postmenopausal osteoporosis, a progressive disease characterized by low bone mass density (White et al. 2010; Faienza et al. 2013). The molecular mechanisms underlying the deficiency of estrogen in postmenopausal osteoporosis are complex and multifaceted (Cervellati et al. 2014). Several studies from animal and *in vitro* have demonstrated that estrogen removal alters the ROS and the antioxidant defense capacity (Li et al. 2017), and subsequently resulting in an accumulation of these oxidant species. Ultimately, these lead to stimulation of resorption activity and osteoclast formation (Hodge et al. 2011). In support of this, research evidence suggests that a relationship between oxidative stress and postmenopausal osteoporosis onset, implied that a potential role of reactive species in uncoupling bone turnover (Bonaccorsi et al. 2018).

ROS, for instance, H_2O_2 and superoxide have been identified as regulatory factors that are involved in osteoclastic bone resorption (Agidigbi and Kim 2019). Superoxide produced from osteoclasts leads to bone degradation (Darden et al. 1996). Key et al. (1990, 1994) exploring the impact of superoxide on osteoclasts. The data showed that the accumulation of superoxide may lead to osteoclastic bone resorption. Similarly, suppression of osteoclastic superoxide production caused a decrease in bone resorption (Ries et al. 1992; Darden et al. 1996). Nicotinamide adenine dinucleotide phosphate, a reduced form (NADPH)-oxidase is the enzyme involved in the generation of ROS by osteoclasts (Wegner and Haudenschild 2020). A study found that interferon-γ, a stimulator of NADPH oxidase activity, enhanced the defective osteoclastic function in osteoporotic in calvaria cultured from animals or microphthalmic mice (Rodriguiz et al. 1993). The enzymatic production of superoxide by XO and xanthine was increased concomitantly with bone resorption in mouse calvarial organ cultures (Fraser et al. 1996). Some studies suggest that H_2O_2 may promote bone resorption by stimulating osteoclast formation in marrow cultures (Bax et al. 1992; Baek et al. 2010). Despite the study has reported a positive association between H_2O_2 and osteoclast formation, not all data showed such a link. Garrett et al. (1990) showed that H_2O_2 level does not promote bone resorption. Furthermore, calvarial bone treated with H_2O_2 significantly increased the osteoclast numbers, implied that H_2O_2 may improve mature osteoclasts and activate the osteoclast formation (Fraser et al. 1996).

The nuclear factor-kappa B (NF-κB) transcriptional factor has been demonstrated as a crucial factor for inflammation, which can be activated by oxidative stimuli (Liu et al. 2017b). In general, the activation of NF-κB-dependent gene expression is a key culprit involved in the systemic inflammatory process (Lin et al. 2017). In the context of osteoporosis, the major role of NF-κB was demonstrated by the phenotype of NF-κB knockout mice, which is primarily due to the impairment of osteoclastic and osteoclastogenesis function (Herrington et al. 2016). The molecular mechanisms underlying the increased of oxidative stress and bone loss are not fully understood, but it is hypothesized that NF-κB (Zhang et al. 2011), augmented by TNF-α (de Araújo et al. 2019), and increased oxidative stress production (Xu et al. 2020) contribute to these phenomena. In addition, oxidative stress also destroys fibronectin (Abdollahi et al. 2005). Fibronectin is the key component in the extracellular matrix of bone (Abdollahi et al. 2005). The glycoprotein serves as a substratum of the osteoblast that is involved in several cellular activities, for instance, differentiation, cell shape, migration, proliferation, and adhesion (Miller et al. 2020). Because the metabolic turnover of fibronectin is slightly slower than that of some cellular components; it will be influenced by several nonenzymatic modifications, such as the production of free radicals during the aging process. ROS can lead to the partial modification and degradation of fibronectin molecules (Eble and de Rezende 2014). Subsequently, the damaged fibronectin molecule may lose its function in bone nodule formation.

Cytokines are crucial regulators for bone cell activity (Zhou et al. 2019). Among all the cytokines, interleukin (IL-1) is one of the most potent stimulators for bone resorption through suppression of collagen synthesis and elevation of osteoclasts number (Weitzmann 2013). IL-6 appears to be a potent osteotropic factor that may play a crucial role in disease characterized by promoting osteoclast fusion and increasing bone resorption (Hienz et al. 2015). TNF-α improves bone resorption by enhancing the activity of mature cells and activating the development of osteoclast progenitors (Hienz et al. 2015). A study found that oxidative stress not only a powerful stimulant for proinflammatory cytokines in the pancreas (Yu and Kim 2014), inflammatory bowel disease (Tian et al. 2017), and heart (Voigt et al. 2013); in fact, it has also been reported to increase cytokine levels in bone, suggesting that oxidative stress may induce osteoporosis (Bonaccorsi et al. 2018).

References

Abdollahi M, Larijani B, Rahimi R et al (2005) Role of oxidative stress in osteoporosis. Therapy 2:787–796

Acharya A, Das I, Chandhok D et al (2010) Redox regulation in cancer: a double-edged sword with therapeutic potential. Oxidative Med Cell Longev 3:23–34

Agency for Toxic Substances and Disease Registry (2012) Toxicological profile for cadmium; Agency for Toxic Substances and Disease Registry, Public Health Service. U.S. Department Of Health and Human Services, Atlanta, pp 1–487

Agidigbi TS, Kim C (2019) Reactive oxygen species in osteoclast differentiation and possible pharmaceutical targets of ROS-mediated osteoclast diseases. Int J Mol Sci 20:3576

Ago T, Kitazono T, Ooboshi H et al (2004) Nox4 as the major catalytic component of an endothelial NAD(P)H oxidase. Circulation 109:227–233

Agostinho P, Lopes JP, Velez Z et al (2008) Overactivation of calcineurin induced by amyloid-beta and prion proteins. Neurochem Int 52:1226–1233

Aguilar-Cazares D, Chavez-Dominguez R, Carlos-Reyes A et al (2019) Contribution of angiogenesis to inflammation and cancer. Front Oncol 9:1399

Albers DS, Beal MF (2000) Mitochondrial dysfunction and oxidative stress in aging and neurodegenerative disease. J Neural Transm Suppl 59:133–154

Alberts B, Johnson A, Lewis J et al (2002) The mitochondrion. Molecular biology of the cell, 4th edn. Garland Science, New York

Allavena P, Sica A, Solinas G et al (2008) The inflammatory microenvironment in tumor progression: the role of tumor-associated macrophages. Crit Rev Oncol Hematol 66:1–9

Alp NJ, McAteer MA, Khoo J et al (2004) Increased endothelial tetrahydrobiopterin synthesis by targeted transgenic GTP-cyclohydrolase I overexpression reduces endothelial dysfunction and atherosclerosis in ApoE-knockout mice. Arterioscler Thromb Vasc Biol 24:445–450

Alzheimer's Association (2016) 2016 Alzheimer's disease facts and figures. Alzheimers Dement 12:459–509

Alzheimer's disease international (2009) World Alzheimer Report, 2009, Executive Summary [R]. London

American Diabetes Association (2010) Diagnosis and classification of diabetes mellitus. Diabetes Care 33:S62–S69

Aminjan HH, Abtahi SR, Hazrati E et al (2019) Targeting of oxidative stress and inflammation through ROS/NF-kappaB pathway in phosphine-induced hepatotoxicity mitigation. Life Sci 232:116607

Andreou N-P, Legaki E, Gazouli M (2020) Inflammatory bowel disease pathobiology: the role of the interferon signature. Ann Gastroenterol 33:125–133

Arai M, Shibata Y, Pugdee K et al (2007) Effects of reactive oxygen species (ROS) on antioxidant system and osteoblastic differentiation in MC3T3-E1 cells. IUBMB Life 59:27–33

Araki E, Lipes MA, Patti M-E et al (1994) Alternative pathway of insulin signalling in mice with targeted disruption of the IRS-1 gene. Nature 372:186–190

Aseervatham GS, Sivasudha T, Jeyadevi R et al (2013) Environmental factors and unhealthy lifestyle influence oxidative stress in humans–an overview. Environ Sci Pollut Res Int 20:4356–4369

Asghar A, Sheikh N (2017) Role of immune cells in obesity induced low grade inflammation and insulin resistance. Cell Immunol 315:18–26

Awasthi A, Matsunaga Y, Yamada T (2005) Amyloid-beta causes apoptosis of neuronal cells via caspase cascade, which can be prevented by amyloid-beta-derived short peptides. Exp Neurol 196:282–289

Baek KH, Oh KW, Lee WY et al (2010) Association of oxidative stress with postmenopausal osteoporosis and the effects of hydrogen peroxide on osteoclast formation in human bone marrow cell cultures. Calcif Tissue Int 87:226–235

Bai X, Lu D, Bai J et al (2004) Oxidative stress inhibits osteoblastic differentiation of bone cells by ERK and NF-kB. Biochem Biophys Res Commun 314:197–207

Ballance WC, Qin EC, Chung HJ et al (2019) Reactive oxygen species-responsive drug delivery systems for the treatment of neurodegenerative diseases. Biomaterials 217:119292

Barry-Lane PA, Patterson C, van der Merwe M et al (2001) p47phox is required for atherosclerotic lesion progression in ApoE(-/-) mice. J Clin Invest 108:1513–1522

Basatemur GL, Jørgensen HF, Clarke MCH et al (2019) Vascular smooth muscle cells in atherosclerosis. Nat Rev Cardiol 16:727–744

Battogtokh G, Choi YS, Kang DS et al (2018) Mitochondria-targeting drug conjugates for cytotoxic, anti-oxidizing and sensing purposes: current strategies and future perspectives. Acta Pharm Sin B 8:862–880

Bax BE, Alam AS, Banerji B et al (1992) Stimulation of osteoclastic bone resorption by hydrogen peroxide. Biochem Biophys Res Commun 183:1153–1158

Bedard K, Krause KH (2007) The NOX family of ROS-generating NADPH oxidases: physiology and pathophysiology. Physiol Rev 87:245–313

Belambri SA, Rolas L, Raad H et al (2018) NADPH oxidase activation in neutrophils: role of the phosphorylation of its subunits. Eur J Clin Investig 48:e12951

Benhar M, Engelberg D, Levitzki A (2002) ROS, stress-activated kinases and stress signaling in cancer. EMBO Rep 3:420–425

Birben E, Sahiner UM, Sackesen C et al (2012) Oxidative stress and antioxidant defense. World Allergy Organ J 5:9–19

Blaser H, Dostert C, Mak TW et al (2016) TNF and ROS crosstalk in inflammation. Trends Cell Biol 26:249–261

Bomfim TR, Forny-Germano L, Sathler LB et al (2012) An anti-diabetes agent protects the mouse brain from defective insulin signaling caused by Alzheimer's disease-associated Aβ oligomers. J Clin Investig 122:1339–1353
Bonaccorsi G, Piva I, Greco P et al (2018) Oxidative stress as a possible pathogenic cofactor of post-menopausal osteoporosis: existing evidence in support of the axis oestrogen deficiency-redox imbalance-bone loss. Indian J Med Res 147:341–351
Bond ST, Kim J, Calkin AC et al (2019) The antioxidant moiety of MitoQ imparts minimal metabolic effects in adipose tissue of high fat fed mice. Front Physiol 10:543
Bonda DJ, Wang X, Perry G et al (2010) Oxidative stress in Alzheimer disease: a possibility for prevention. Neuropharmacology 59:290–294
Bonomini F, Rodella LF, Rezzani R (2015) Metabolic syndrome, aging and involvement of oxidative stress. Aging Dis 6:109–120
Boskey AL, Coleman R (2010) Aging and bone. J Dent Res 89:1333–1348
Branca JJV, Fiorillo C, Carrino D et al (2020) Cadmium-induced oxidative stress: focus on the central nervous system. Antioxidants 9:492
Brandes RP, Kreuzer J (2005) Vascular NADPH oxidases: molecular mechanisms of activation. Cardiovasc Res 65:16–27
Brownlee M (2001) Biochemistry and molecular cell biology of diabetic complications. Nature 414:813–820
Buha A, Jugdaohsingh R, Matovic V et al (2019) Bone mineral health is sensitively related to environmental cadmium exposure-experimental and human data. Environ Res 176:108539
Bulat N, Widmann C (2009) Caspase substrates and neurodegenerative diseases. Brain Res Bull 80:251–267
Butterfield DA (2014) The 2013 SFRBM discovery award: selected discoveries from the butterfield laboratory of oxidative stress and its sequela in brain in cognitive disorders exemplified by Alzheimer disease and chemotherapy induced cognitive impairment. Free Radic Biol Med 74:157–174
Butterfield DA, Boyd-Kimball D (2005) The critical role of methionine 35 in Alzheimer's amyloid beta-peptide (1-42)-induced oxidative stress and neurotoxicity. Biochim Biophys Acta 1703:149–156
Canalis E (2008) Notch signaling in osteoblasts. Sci Signal 1:pe17
Carina V, Bella ED, Costa V et al (2020) Bone's response to mechanical loading in aging and osteoporosis: molecular mechanisms. Calcif Tissue Int 107:301–318
Čejková S, Králová-Lesná I, Poledne R (2016) Monocyte adhesion to the endothelium is an initial stage of atherosclerosis development. Cor Vasa 58:e419–e425
Cenini G, Lloret A, Cascella R (2019) Oxidative stress and mitochondrial damage in neurodegenerative diseases: from molecular mechanisms to targeted therapies. Oxidative Med Cell Longev 2019, Article ID 2105607, 18 pages
Cerf ME (2013) Beta cell dysfunction and insulin resistance. Front Endocrinol 4:37
Cervellati C, Bonaccorsi G, Cremonini E et al (2014) Oxidative stress and bone resorption interplay as a possible trigger for postmenopausal osteoporosis. Biomed Res Int 2014, Article ID 569563, 8 pages
Cesari M, Pahor M, Incalzi RA (2010) Plasminogen activator inhibitor-1 (PAI-1): a key factor linking fibrinolysis and age-related subclinical and clinical conditions. Cardiovasc Ther 28:e72–e91
Chait A, den Hartigh LJ (2020) Adipose tissue distribution, inflammation and its metabolic consequences, including diabetes and cardiovascular disease. Front Cardiovasc Med 7:22
Chakraborty C, Sharma AR, Sharma G et al (2020) The interplay among miRNAs, major cytokines, and cancer-related inflammation. Mol Ther Nucleic Acids 20:606–620
Chang J-F, Yeh Y-C, Ho C-T et al (2019) Targeting ROS and cPLA2/COX2 expressions ameliorated renal damage in obese mice with endotoxemia. Int J Mol Sci 20:4393
Chen Q, Wang Q, Zhu J et al (2018) Reactive oxygen species: key regulators in vascular health and diseases. Br J Pharmacol 175:1279–1292

Chen X, Wang Z, Zhu G et al (2019) The association between cumulative cadmium intake and osteoporosis and risk of fracture in a Chinese population. J Expo Sci Environ Epidemiol 29:435–443
Chistiakov DA, Shkurat TP, Melnichenko AA et al (2018) The role of mitochondrial dysfunction in cardiovascular disease: a brief review. Ann Med 50:121–127
Chooi YC, Ding C, Magkos F (2019) The epidemiology of obesity. Metabolism 92:6–10
Collin F (2019) Chemical basis of reactive oxygen species reactivity and involvement in neurodegenerative diseases. Int J Mol Sci 20:2407
Comen EA, Bowman RL, Kleppe M (2018) Underlying causes and therapeutic targeting of the inflammatory tumor microenvironment. Front Cell Dev Biol 6:56
Corona C, Pensalfini A, Frazzini V et al (2011) New therapeutic targets in Alzheimer's disease: brain deregulation of calcium and zinc. Cell Death Dis 2:e176
Cristalli DO, Arnal N, Marra FA et al (2012) Peripheral markers in neurodegenerative patients and their first-degree relatives. J Neurol Sci 314:48–56
Curtain CC, Ali F, Volitakis I et al (2001) Alzheimer's disease amyloid-beta binds copper and zinc to generate an allosterically ordered membrane-penetrating structure containing superoxide dismutase-like subunits. J Biol Chem 276:20466–20473
Czech MP (2017) Insulin action and resistance in obesity and type 2 diabetes. Nat Med 23:804–814
Daiber A, Chlopicki S (2020) Revisiting pharmacology of oxidative stress and endothelial dysfunction in cardiovascular disease: evidence for redox-based therapies. Free Radic Biol Med 157:15–37
Daiber A, Xia N, Steven S et al (2019) New therapeutic implications of endothelial nitric oxide synthase (eNOS) function/dysfunction in cardiovascular disease. Int J Mol Sci 20:187
Darden AG, Ries WL, Wolf WC et al (1996) Osteoclastic superoxide production and bone resorption: stimulation and inhibition by modulators of NADPH oxidase. J Bone Miner Res 11:671–675
de Araújo AA, de Morais HB, de Medeiros CACX et al (2019) Gliclazide reduced oxidative stress, inflammation, and bone loss in an experimental periodontal disease model. J Appl Oral Sci 27:e20180211
de los Fayos Alonso IG, Liang H-C, Turner SD et al (2018) The role of activator protein-1 (AP-1) family members in CD30-positive lymphomas. Cancers 10:93
De Luca G, Calpona PR, Caponetti A et al (2001) Preliminary report: amino acid profile in platelets of diabetic patients. Metabolism 50:739–741
Demontiero O, Vidal C, Duque G (2012) Aging and bone loss: new insights for the clinician. Ther Adv Musculoskelet Dis 4:61–76
Di Marzo N, Chisci E, Giovannoni R (2018) The role of hydrogen peroxide in redox-dependent signaling: homeostatic and pathological responses in mammalian cells. Cell 7:156
Di Meo S, Reed TT, Venditti P et al (2016) Role of ROS and RNS sources in physiological and pathological conditions. Oxidative Med Cell Longev 2016, Article ID 1245049, 44 pages
di Penta A, Moreno B, Reix S et al (2013) Oxidative stress and proinflammatory cytokines contribute to demyelination and axonal damage in a cerebellar culture model of neuroinflammation. PLoS One 8:e54722
Dias V, Junn E, Mouradian MM (2013) The role of oxidative stress in Parkinson's disease. J Parkinsons Dis 3:461–491
Ding L, Kang Y, Dai H-B et al (2019) Adipose afferent reflex is enhanced by TNFα in paraventricular nucleus through NADPH oxidase-dependent ROS generation in obesity-related hypertensive rats. J Transl Med 17:256
Douglas G, Bendall JK, Crabtree MJ et al (2012) Endothelial-specific Nox2 overexpression increases vascular superoxide and macrophage recruitment in ApoE(-)/(-) mice. Cardiovasc Res 94:20–29
Drouin A, Thorin E (2009) Flow-induced dilation is mediated by Akt-dependent activation of endothelial nitric oxide synthase-derived hydrogen peroxide in mouse cerebral arteries. Stroke 40:1827–1833

Drouin A, Thorin-Trescases N, Hamel E et al (2007) Endothelial nitric oxide synthase activation leads to dilatory H2O2 production in mouse cerebral arteries. Cardiovasc Res 73:73–81
Drummond GR, Selemidis S, Griendling KK et al (2011) Combating oxidative stress in vascular disease: NADPH oxidases as therapeutic targets. Nat Rev Drug Discov 10:453–471
Ducy P, Zhang R, Geoffroy V et al (1997) Osf2/Cbfa1: transcriptional activator of osteoblast differentiation. Cell 89:747–754
Dugger BN, Dickson DW (2017) Pathology of neurodegenerative diseases. Cold Spring Harb Perspect Biol 9:a028035
Eble J, de Rezende FF (2014) Redox-relevant aspects of the extracellular matrix and its cellular contacts via integrins. Antioxid Redox Signal 20:1977–1993
Eiró N, Vizoso FJ (2012) Inflammation and cancer. World J Gastrointest Surg 4:62–72
Eizirik DL, Pasquali L, Cnop M (2020) Pancreatic β-cells in type 1 and type 2 diabetes mellitus: different pathways to failure. Nat Rev Endocrinol 16:349–362
El-Bassossy HM, Neamatallah T, Balamash KS et al (2018) Arginase overexpression and NADPH oxidase stimulation underlie impaired vasodilation induced by advanced glycation end products. Biochem Biophys Res Commun 499:992–997
Ellmark SH, Dusting GJ, Fui MN et al (2005) The contribution of Nox4 to NADPH oxidase activity in mouse vascular smooth muscle. Cardiovasc Res 65:495–504
Elmarakby AA, Sullivan JC (2012) Relationship between oxidative stress and inflammatory cytokines in diabetic nephropathy. Cardiovasc Ther 30:49–59
Eyvazi S, Vostakolaei MA, Dilmaghani A et al (2020) The oncogenic roles of bacterial infections in development of cancer. Microb Pathog 141:104019
Faienza MF, Ventura A, Marzano F et al (2013) Postmenopausal osteoporosis: the role of immune system cells. Clin Dev Immunol 2013, Article ID 575936, 6 pages
Fantuzzi G (2005) Adipose tissue, adipokines, and inflammation. J Allergy Clin Immunol 115:911–919
Faria TO, Simões MR, Vassallo DV et al (2018) Xanthine oxidase activation modulates the endothelial (vascular) dysfunction related to $HgCl_2$ exposure plus myocardial infarction in rats. Cardiovasc Toxicol 18:161–174
Ferreiro E, Oliveira CR, Pereira CM (2008) The release of calcium from the endoplasmic reticulum induced by amyloid-beta and prion peptides activates the mitochondrial apoptotic pathway. Neurobiol Dis 30:331–342
Fischer CP, Berntsen A, Perstrup LB et al (2006) Plasma levels of interleukin-6 and C-reactive protein are associated with physical inactivity independent of obesity. Scand J Med Sci Sports 17:580–587
Florencio-Silva R, da Silva Sasso GR, Sasso-Cerri E et al (2015) Biology of bone tissue: structure, function, and factors that influence bone cells. Biomed Res Int 2015, Article ID 421746, 17 pages
Forstermann U, Munzel T (2006) Endothelial nitric oxide synthase in vascular disease: from marvel to menace. Circulation 113:1708–1714
Förstermann U, Sessa WC (2012) Nitric oxide synthases: regulation and function. Eur Heart J 33:829–837
Förstermann U, Xia N, Li H (2017) Roles of vascular oxidative stress and nitric oxide in the pathogenesis of atherosclerosis. Circ Res 120:713–735
Fraser JH, Helfrich MH, Wallace HM et al (1996) Hydrogen peroxide, but not superoxide, stimulates bone resorption in mouse calvariae. Bone 19:223–226
Frohnert BI, Sinaiko AR, Serrot FJ et al (2011) Increased adipose protein carbonylation in human obesity. Obesity 19:1735–1741
Fu L-Q, Du W-L, Cai M-H et al (2020) The roles of tumor-associated macrophages in tumor angiogenesis and metastasis. Cell Immunol 353:104119
Furukawa S, Fujita T, Shimabukuro M et al (2004) Increased oxidative stress in obesity and its impact on metabolic syndrome. J Clin Invest 114:1752–1761

Galderisi A, Polidori D, Weiss R et al (2019) Lower insulin clearance parallels a reduced insulin sensitivity in obese youths and is associated with a decline in β-cell function over time. Diabetes 68:2074–2084

Garrett IR, Boyce BF, Oreffo RO et al (1990) Oxygen-derived free radicals stimulate osteoclastic bone resorption in rodent bone *in vitro* and *in vivo*. J Clin Invest 85:632–639

Gazendam RP, van de Geer A, Roos D et al (2016) How neutrophils kill fungi. Immunol Rev 273:299–311

Gechev TS, Van Breusegem F, Stone JM et al (2006) Reactive oxygen species as signals that modulate plant stress responses and programmed cell death. BioEssays 28:1091–1101

George J, Struthers AD (2009) Role of urate, xanthine oxidase and the effects of allopurinol in vascular oxidative stress. Vasc Health Risk Manag 5:265–272

Giraldo E, Lloret A, Fuchsberger T et al (2014) Aβ and tau toxicities in Alzheimer's are linked via oxidative stress-induced p38 activation: protective role of vitamin E. Redox Biol 2:873–877

Gleichmann M, Mattson MP (2011) Neuronal calcium homeostasis and dysregulation. Antioxid Redox Signal 14:1261–1273

Goldin A, Beckman JA, Schmidt AM et al (2006) Advanced glycation end products sparking the development of diabetic vascular injury. Circulation 114:597–605

Goleva TN, Rogov AG, Korshunova GA et al (2019) SkQThy, a novel and promising mitochondria-targeted antioxidant. Mitochondrion 49:206–216

Gosling J, Slaymaker S, Gu L et al (1999) MCP-1 deficiency reduces susceptibility to atherosclerosis in mice that overexpress human apolipoprotein B. J Clin Investig 103:773–778

Grabowski MM, Sankey EW, Ryan KJ et al (2021) Immune suppression in gliomas. J Neuro-Oncol 151:3–12

Guo S, Deng C-X (2018) Effect of stromal cells in tumor microenvironment on metastasis initiation. Int J Biol Sci 14:2083–2093

Guo M, Hao M, Tang Y (2018) Relationships among serum IL-6, TNF-α, CRP and coronary artery lesion in patients with acute myocardial infarction. Int J Clin Exp Med 11:6987–6992

Guzik TJ, Sadowski J, Guzik B et al (2006) Coronary artery superoxide production and nox isoform expression in human coronary artery disease. Arterioscler Thromb Vasc Biol 26:333–339

Haendeler J, Eckers A, Lukosz M et al (2012) Endothelial NADPH oxidase 2: when does it matter in atherosclerosis? Cardiovasc Res 94:1–2

Hamer M, Chida Y (2009) Physical activity and risk of neurodegenerative disease: a systematic review of prospective evidence. Psychol Med 39:3–11

Hamilton A, Holscher C (2012) The effect of ageing on neurogenesis and oxidative stress in the APPswe/PS1deltaE9 mouse model of Alzheimer's disease. Brain Res 1449:83–93

Heitzer T, Krohn K, Albers S et al (2000) Tetrahydrobiopterin improves endothelium-dependent vasodilation by increasing nitric oxide activity in patients with type II diabetes mellitus. Diabetologia 43:1435–1438

Hemonnot A-L, Hua J, Ulmann L et al (2019) Microglia in Alzheimer disease: well-known targets and new opportunities. Front Aging Neurosci 11:233

Henriksen EJ, Diamond-Stanic MK, Marchionne EM (2011) Oxidative stress and the etiology of insulin resistance and type 2 diabetes. Free Radic Biol Med 51:993–999

Hernlund E, Svedbom A, Ivergard M et al (2013) Osteoporosis in the European Union: medical management, epidemiology and economic burden. A report prepared in collaboration with the International Osteoporosis Foundation (IOF) and the European Federation of Pharmaceutical Industry Associations (EFPIA). Arch Osteoporos 8:136

Herrington FD, Carmody RJ, Goodyear CS (2016) Modulation of NF-κB signaling as a therapeutic target in autoimmunity. J Biomol Screen 21:223–242

Hienz SA, Paliwal S, Ivanovski S (2015) Mechanisms of bone resorption in periodontitis. J Immunol Res 2015, Article ID 615486, 10 pages

Hodge JM, Collier FM, Pavlos NJ et al (2011) M-csf potently augments rankl-induced resorption activation in mature human osteoclasts. PLoS One 6:e21462

Hotamisligil GS, Peraldi P, Budavari A et al (1996) IRS-1-mediated inhibition of insulin receptor tyrosine kinase activity in TNF-alpha- and obesity-induced insulin resistance. Science 271:665–668

Hsich E, Segal BH, Pagano PJ et al (2000) Vascular effects following homozygous disruption of p47(phox): an essential component of NADPH oxidase. Circulation 101:1234–1236

Hsu TC, Young MR, Cmarik J et al (2000) Activator protein 1 (AP-1)- and nuclear factor kappaB (NF-kappaB)-dependent transcriptional events in carcinogenesis. Free Radic Biol Med 28:1338–1348

Huang PL (2009) A comprehensive definition for metabolic syndrome. Dis Model Mech 2:231–237

Huang WJ, Zhang X, Chen WW (2016) Role of oxidative stress in Alzheimer's disease. Biomed Rep 4:519–522

Huong PT, Nguyen LT, Nguyen X-B et al (2019) The role of platelets in the tumor-microenvironment and the drug resistance of cancer cells. Cancers 11:240

Ilkun O, Boudina S (2013) Cardiac dysfunction and oxidative stress in the metabolic syndrome: an update on antioxidant therapies. Curr Pharm Des 19:4806–4817

Ireland LV, Mielgo A (2018) Macrophages and fibroblasts, key players in cancer chemoresistance. Front Cell Dev Biol 6:131

Italiani P, Puxeddu I, Napoletano S et al (2018) Circulating levels of IL-1 family cytokines and receptors in Alzheimer's disease: new markers of disease progression? J Neuroinflammation 15:342

Jain SK, Micinski D (2013) Vitamin D upregulates glutamate cysteine ligase and glutathione reductase, and GSH formation, and decreases ROS and MCP-1 and IL-8 secretion in high-glucose exposed U937 monocytes. Biochem Biophys Res Commun 437:7–11

Jain SS, Paglialunga S, Vigna C et al (2014) High-fat diet–induced mitochondrial biogenesis is regulated by mitochondrial-derived reactive oxygen species activation of CaMKII. Diabetes 63:1907–1913

Ji M-X, Yu Q (2015) Primary osteoporosis in postmenopausal women. Chronic Dis Transl Med 1:9–13

Johnson T, Zhao L, Manuel G et al (2019) Approaches to therapeutic angiogenesis for ischemic heart disease. J Mol Med 97:141–151

Judkins CP, Diep H, Broughton BR et al (2010) Direct evidence of a role for Nox2 in superoxide production, reduced nitric oxide bioavailability, and early atherosclerotic plaque formation in ApoE-/- mice. Am J Physiol Heart Circ Physiol 298:H24–H32

Kaisanlahti A, Glumoff T (2019) Browning of white fat: agents and implications for beige adipose tissue to type 2 diabetes. J Physiol Biochem 75:1–10

Kamata H, Honda S-I, Maeda S et al (2005) Reactive oxygen species promote TNFalpha-induced death and sustained JNK activation by inhibiting MAP kinase phosphatases. Cell 120:649–661

Kany S, Vollrath JT, Relja B (2019) Cytokines in inflammatory disease. Int J Mol Sci 20:6008

Kern L, Mittenbühler MJ, Vesting AJ et al (2019) Obesity-induced TNFα and IL-6 signaling: the missing link between obesity and inflammation—driven liver and colorectal cancers. Cancers 11:24

Key LL, Ries WL, Taylor RG et al (1990) Oxygen-derived free radicals in osteoclasts: the specificity and location of the nitroblue tetrazolium reaction. Bone 11:115–119

Key LL, Wolf WC, Guadberg CM et al (1994) Superoxide and bone resorption. Bone 5:431–436

Kim B-J, Lee Y-S, Lee S-Y et al (2018) Osteoclast-secreted SLIT3 coordinates bone resorption and formation. J Clin Invest 128:1429–1441

Kim K, Melough MM, Vance TM et al (2019) Dietary cadmium intake and sources in the US. Nutrients 11:2

Kim J-M, Lin C, Stavre Z et al (2020) Osteoblast-osteoclast communication and bone homeostasis. Cell 9:2073

Kirk EA, Dinauer MC, Rosen H et al (2000) Impaired superoxide production due to a deficiency in phagocyte NADPH oxidase fails to inhibit atherosclerosis in mice. Arterioscler Thromb Vasc Biol 20:1529–1535

Komori T (2002) Runx2, a multifunctional transcription factor in skeletal development. J Cell Biochem 87:1–8
Komori T, Yagi H, Nomura S et al (1997) Targeted disruption of Cbfa1 results in a complete lack of bone formation owing to maturational arrest of osteoblasts. Cell 89:755–764
Kubota N, Tobe K, Terauchi Y et al (2000) Disruption of insulin receptor substrate 2 causes type 2 diabetes because of liver insulin resistance and lack of compensatory β-cell hyperplasia. Diabetes 49:1880–1889
Kubota T, Kubota N, Kumagai H et al (2011) Impaired insulin signaling in endothelial cells reduces insulin-induced glucose uptake by skeletal muscle. Cell Metab 13:294–307
Kurutas EB (2016) The importance of antioxidants which play the role in cellular response against oxidative/nitrosative stress: current state. Nutr J 15:71
Lafontan M (2014) Adipose tissue and adipocyte dysregulation. Diabetes Metab 40:16–28
Landmesser U, Spiekermann S, Dikalov S et al (2002) Vascular oxidative stress and endothelial dysfunction in patients with chronic heart failure: role of xanthine-oxidase and extracellular superoxide dismutase. Circulation 106:3073–3078
Landmesser U, Spiekermann S, Preuss C et al (2007) Angiotensin II induces endothelial xanthine oxidase activation: role for endothelial dysfunction in patients with coronary disease. Arterioscler Thromb Vasc Biol 27:943–948
Lassegue B, Sorescu D, Szocs K et al (2001) Novel gp91(phox) homologues in vascular smooth muscle cells: nox1 mediates angiotensin II-induced superoxide formation and redox-sensitive signaling pathways. Circ Res 88:888–894
Le A, Udupa S, Zhang C (2019) The metabolic interplay between cancer and other diseases. Trends Cancer 5:809–821
Lee CH, Lam KSL (2019) Obesity-induced insulin resistance and macrophage infiltration of the adipose tissue: a vicious cycle. J Diabetes Investig 10:29–31
Li H, Forstermann U (2013) Uncoupling of endothelial NO synthase in atherosclerosis and vascular disease. Curr Opin Pharmacol 13:161–167
Li H, Horke S, Förstermann U (2014) Vascular oxidative stress, nitric oxide and atherosclerosis. Atherosclerosis 237:208–219
Li J, Wang Q, Yang R et al (2017) BMI-1 mediates estrogen-deficiency–induced bone loss by inhibiting reactive oxygen species accumulation and T cell activation. J Bone Miner Res 32:962–973
Li K, Chen S, Cai P et al (2020) MiRNA-483–5p is involved in the pathogenesis of osteoporosis by promoting osteoclast differentiation. Mol Cell Probes 49:101479
Lim KH (2019) Diverse misfolded conformational strains and cross-seeding of misfolded proteins implicated in neurodegenerative diseases. Front Mol Neurosci 12:158
Lim S, Park S (2014) Role of vascular smooth muscle cell in the inflammation of atherosclerosis. BMB Rep 47:1–7
Lin MT, Beal MF (2006) Mitochondrial dysfunction and oxidative stress in neurodegenerative diseases. Nature 443:787–795
Lin T-h, Pajarinen J, Lu L et al (2017) NF-κB as a therapeutic target in inflammatory-associated bone diseases. Adv Protein Chem Struct Biol 107:117–154
Liu Z, Zhou T, Ziegler AC et al (2017a) Oxidative stress in neurodegenerative diseases: from molecular mechanisms to clinical applications. Oxidative Med Cell Longev 2017, Article ID 2525967, 11 pages
Liu T, Zhang L, Joo D et al (2017b) NF-κB signaling in inflammation. Signal Transduct Target Ther 2:e17023
Long F (2012) Building strong bones: molecular regulation of the osteoblast lineage. Nat Rev Mol Cell Biol 13:27–38
Louis SF, Zahradka P (2010) Vascular smooth muscle cell motility: from migration to invasion. Exp Clin Cardiol 15:e75–e85
Lusis AJ (2000) Atherosclerosis. Nature 407:233–241

MacKellar J, Cushman SW, Periwal V (2010) Waves of adipose tissue growth in the genetically obese Zucker fatty rat. PLoS One 5:e8197
Madamanchi NR, Runge MS (2007) Mitochondrial dysfunction in atherosclerosis. Circ Res 100:460–473
Madungwe NB, Zilberstein NF, Feng Y et al (2016) Critical role of mitochondrial ROS is dependent on their site of production on the electron transport chain in ischemic heart. Am J Cardiovasc Dis 6:93–108
Magalingam KB, Radhakrishnan A, Ping NS et al (2018) Current concepts of neurodegenerative mechanisms in Alzheimer's disease. Biomed Res Int 2018, Article ID 3740461, 12 pages
Malone JI, Hanna S, Saporta S et al (2008) Hyperglycemia not hypoglycemia alters neuronal dendrites and impairs spatial memory. Pediatr Diabetes 9:531–539
Marchio P, Guerra-Ojeda S, Vila JM et al (2019) Targeting early atherosclerosis: a focus on oxidative stress and inflammation. Oxidative Med Cell Longev 2019, Article ID 8563845, 32 pages
Marseglia L, Manti S, D'Angelo G et al (2015) Oxidative stress in obesity: a critical component in human diseases. Int J Mol Sci 16:378–400
Masschelin PM, Cox AR, Chernis N et al (2020) The impact of oxidative stress on adipose tissue energy balance. Front Physiol 10:1638
Mathew G, Hanson B (2009) Global burden of trauma: need for effective fracture therapies. Indian J Orthop 43:111–116
Matilla-Duenas A, Ashizawa T, Brice A et al (2014) Consensus paper: pathological mechanisms underlying neurodegeneration in spinocerebellar ataxias. Cerebellum 13:269–302
Matsuda M, Shimomura I (2013) Increased oxidative stress in obesity: implications for metabolic syndrome, diabetes, hypertension, dyslipidemia, atherosclerosis, and cancer. Obes Res Clin Pract 7:e330–e341
Matthews GM, Howarth GS, Butler RN (2006) Nutrient and antioxidant modulation of apoptosis in gastric and colon cancer cells. Cancer Biol Ther 5:569–572
Mazat J-P, Devin A, Ransac S (2020) Modelling mitochondrial ROS production by the respiratory chain. Cell Mol Life Sci 77:455–465
McCaddon A, Hudson P, Hill D et al (2003) Alzheimer's disease and total plasma aminothiols. Biol Psychiatry 53:254–260
McCubrey JA, Steelman LS, Abrams SL et al (2008) Targeting survival cascades induced by activation of Ras/Raf/MEK/ERK, PI3K/PTEN/Akt/mTOR and Jak/STAT pathways for effective leukemia therapy. Leukemia 22:708–722
McGeer PL, McGeer EG (1999) Inflammation of the brain in Alzheimer's disease: implications for therapy. J Leukoc Biol 65:409–415
Mebratu Y, Tesfaigzi Y (2009) How ERK1/2 activation controls cell proliferation and cell death is subcellular localization the answer? Cell Cycle 8:1168–1175
Miller AE, Hu P, Barker TH (2020) Feeling things out: bidirectional signaling of the Cell–ECM interface, implications in the mechanobiology of cell spreading, migration, proliferation, and differentiation. Adv Healthc Mater 9:1901445
Milutinović A, Šuput D, Zorc-Pleskovič R (2020) Pathogenesis of atherosclerosis in the tunica intima, media, and adventitia of coronary arteries: an updated review. Bosn J Basic Med Sci 20:21–30
Minamino T, Orimo M, Shimizu I et al (2009) A crucial role for adipose tissue p53 in the regulation of insulin resistance. Nat Med 15:1082–1087
Misumi I, Starmer J, Uchimura T et al (2019) Obesity expands a distinct population of T cells in adipose tissue and increases vulnerability to infection. Cell Rep 27:514–524
Mittal M, Siddiqui MR, Tran K et al (2014) Reactive oxygen species in inflammation and tissue injury. Antioxid Redox Signal 20:1126–1167
Molehin OR, Adefegha SA, Adeyanju AA (2020) Role of oxidative stress in the pathophysiology of type 2 diabetes and cardiovascular diseases. In: Maurya P, Dua K (eds) Role of oxidative stress in pathophysiology of diseases. Springer, Singapore, pp 277–297

Moore K, Sheedy F, Fisher E (2013) Macrophages in atherosclerosis: a dynamic balance. Nat Rev Immunol 13:709–721

Moriya J (2019) Critical roles of inflammation in atherosclerosis. J Cardiol 73:22–27

Mosley RL, Benner EJ, Kadiu I et al (2006) Neuroinflammation, oxidative stress and the pathogenesis of Parkinson's disease. Clin Neurosci Res 6:261–281

Murphy MP (2009) How mitochondria produce reactive oxygen species. Biochem J 417:1–13

Nakamura T, Lipton SA (2010) Preventing Ca^{2+}-mediated nitrosative stress in neurodegenerative diseases: possible pharmacological strategies. Cell Calcium 47:190–197

Nakamura T, Lipton SA (2011) Redox modulation by S-nitrosylation contributes to protein misfolding, mitochondrial dynamics, and neuronal synaptic damage in neurodegenerative diseases. Cell Death Differ 18:1478–1486

Newby AC, Zaltsman AB (1999) Fibrous cap formation or destruction—the critical importance of vascular smooth muscle cell proliferation, migration and matrix formation. Cardiovasc Res 41:345–360

Newsholme P, Morgan D, Rebelato E et al (2009) Insights into the critical role of NADPH oxidase(s) in the normal and dysregulated pancreatic beta cell. Diabetologia 52:2489–2498

Newsholme P, Cruzat V, Arfuso F et al (2014) Nutrient regulation of insulin secretion and action. J Endocrinol 221:R105–R120

Nguyen GT, Green ER, Mecsas J (2017) Neutrophils to the ROScue: mechanisms of NADPH oxidase activation and bacterial resistance. Front Cell Insect Microbiol 7:373

Nizzari M, Thellung S, Corsaro A et al (2012) Neurodegeneration in Alzheimer disease: role of amyloid precursor protein and presenilin 1 intracellular signaling. J Toxicol 2012, Article ID 187297, 13 pages

Nojiri H, Shimizu T, Funakoshi M et al (2006) Oxidative stress causes heart failure with impaired mitochondrial respiration. J Biol Chem 281:33789–33801

Odén A, McCloskey EV, Kanis JA et al (2015) Burden of high fracture probability worldwide: secular increases 2010–2040. Osteoporos Int 26:2243–2248

Ohara Y, Peterson TE, Harrison DG (1993) Hypercholesterolemia increases endothelial superoxide anion production. J Clin Invest 91:2546–2551

Olusi S (2002) Obesity is an independent risk factor for plasma lipid peroxidation and depletion of erythrocyte cytoprotectic enzymes in humans. Int J Obes Relat Metab Disord 26:1159–1164

Ormazabal V, Nair S, Elfeky O et al (2018) Association between insulin resistance and the development of cardiovascular disease. Cardiovasc Diabetol 17:122

Oyewole AO, Birch-Machin MA (2015) Mitochondria-targeted antioxidants. FASEB J 29:4766–4771

Pan Y, Hui X, Hoo RLC et al (2019) Adipocyte-secreted exosomal microRNA-34a inhibits M2 macrophage polarization to promote obesity-induced adipose inflammation. J Clin Invest 129:834–849

Pant S, Deshmukh A, GuruMurthy GS et al (2014) Inflammation and atherosclerosis-revisited. J Cardiovasc Pharmacol Ther 19:170–178

Patetsios P, Song M, Shutze WP et al (2001) Identification of uric acid and xanthine oxidase in atherosclerotic plaque. Am J Cardiol 88:188–191

Patten DA, Germain M, Kelly MA et al (2010) Reactive oxygen species: stuck in the middle of neurodegeneration. J Alzheimers Dis 20:S357–S367

Paulsen K, Tauber S, Dumrese C et al (2015) Regulation of ICAM-1 in cells of the monocyte/macrophage system in microgravity. Biomed Res Int 2015, Article ID 538786, 18 pages

Peto R, Parish SE, Gray RG (1977) There is no such thing as ageing, and cancer is not related to it. IARC Publ 58:23–53

Piemonte E, Lazos J, Belardinelli P et al (2018) Oral cancer associated with chronic mechanical irritation of the oral mucosa. Med Oral Patol Oral Cir Bucal 23:e151–e160

Pierangeli SS, Espinola RG, Liu X et al (2001) Thrombogenic effects of antiphospholipid antibodies are mediated by intercellular cell adhesion molecule-1, vascular cell adhesion molecule-1, and P-selectin. Circ Res 88:245–250

Pizzino G, Irrera N, Cucinotta M et al (2017) Oxidative stress: harms and benefits for human health. Oxidative Med Cell Longev 2017, Article ID 8416763, 13 pages

Poprac P, Jomova K, Simunkova M et al (2017) Targeting free radicals in oxidative stress-related human diseases. Trends Pharmacol Sci 38:592–607

Poret JM, Souza-Smith F, Marcell SJ et al (2018) High fat diet consumption differentially affects adipose tissue inflammation and adipocyte size in obesity-prone and obesity-resistant rats. Int J Obes 42:535–541

Querfurth HW, LaFerla FM (2010) Alzheimer's disease. New Engl J Med 362:329–344

Quinn JJ, Jones MG, Okimoto RA et al (2021) Single-cell lineages reveal the rates, routes, and drivers of metastasis in cancer xenografts. Science 371:eabc1944

Ramel D, Gayral S, Sarthou M-K et al (2019) Immune and smooth muscle cells interactions in atherosclerosis: how to target a breaking bad dialogue? Front Pharmacol 10:1276

Rea IM, Gibson DS, McGilligan V et al (2018) Age and age-related diseases: role of inflammation triggers and cytokines. Front Immunol 9:586

Rego AC, Oliveira CR (2003) Mitochondrial dysfunction and reactive oxygen species in excitotoxicity and apoptosis: implications for the pathogenesis of neurodegenerative diseases. Neurochem Res 28:1563–1574

Ren G, Zhao X, Zhang L et al (2010) Inflammatory cytokine-induced intercellular adhesion molecule-1 and vascular cell adhesion molecule-1 in mesenchymal stem cells are critical for immunosuppression. J Immunol 184:2321–2328

Ries WL, Key LL, Rodriguiz R (1992) Nitroblue tetrazolium reduction and bone resorption by osteoclasts *in vitro* inhibited by a manganese-based superoxide dismutase mimic. J Bone Miner Res 7:931–939

Rodríguez-Cerdeira C, Cordeiro-Rodríguez M, Carnero-Gregorio M et al (2019) Biomarkers of inflammation in obesity-psoriatic patients. Mediat Inflamm 2019, Article ID 7353420, 14 pages

Rodriguiz RM, Key LL, Ries WL (1993) Combination macrophage-colony stimulating factor and interferon-γ administration ameliorates the osteopetrotic condition in microphthalmic (mi/mi) mice. Pediatr Res 33:382–389

Rosales-Corral S, Tan DX, Manchester L et al (2015) Diabetes and Alzheimer disease, two overlapping pathologies with the same background: oxidative stress. Oxidative Med Cell Longev 2015, Article ID 985845, 14 pages

Russell AP, Gastaldi G, Bobbioni-Harsch E et al (2003) Lipid peroxidation in skeletal muscle of obese as compared to endurance-trained humans: a case of good vs. bad lipids? FEBS Lett 551:104–106

Rutkovskiy A, Stensløkken K-O, Vaage IJ (2016) Osteoblast differentiation at a glance. Med Sci Monit Basic Res 22:95–106

Saini V (2010) Molecular mechanisms of insulin resistance in type 2 diabetes mellitus. World J Diabetes 1:68–75

Salaroglio IC, Mungo E, Gazzano E et al (2019) ERK is a pivotal player of chemo-immune-resistance in cancer. Int J Mol Sci 20:2505

Schieber M, Chandel NS (2014) ROS function in redox signaling and oxidative stress. Curr Biol 24:R453–R462

Schroder K, Vecchione C, Jung O et al (2006) Xanthine oxidase inhibitor tungsten prevents the development of atherosclerosis in ApoE knockout mice fed a Western-type diet. Free Radic Biol Med 41:1353–1360

Serpillon S, Floyd BC, Gupte RS et al (2009) Superoxide production by NAD(P)H oxidase and mitochondria is increased in genetically obese and hyperglycemic rat heart and aorta before the development of cardiac dysfunction. The role of glucose-6-phosphate dehydrogenase-derived NADPH. Am J Physiol Heart Circ Physiol 297:H153–H162

Sesso HD, Christen WG, Bubes V et al (2012) Multivitamins in the prevention of cardiovascular disease in men: The Physicians' Health Study II randomized controlled trial. JAMA 308:1751–1760

Sharma M, Boytard L, Hadi T et al (2020) Enhanced glycolysis and HIF-1α activation in adipose tissue macrophages sustains local and systemic interleukin-1β production in obesity. Sci Rep 10:5555

Shelat PB, Chalimoniuk M, Wang JH et al (2008) Amyloid beta peptide and NMDA induce ROS from NADPH oxidase and AA release from cytosolic phospholipase A2 in cortical neurons. J Neurochem 106:45–55

Shiva S (2010) Mitochondria as metabolizers and targets of nitrite. Nitric Oxide 22:64

Siegfried G, Descarpentrie J, Evrard S et al (2020) Proprotein convertases: key players in inflammation-related malignancies and metastasis. Cancer Lett 473:50–61

Silvade HR, Khan NL, Wood NW (2000) The genetics of Parkinson's disease. Curr Opin Genet Dev 10:292–298

Smith RAJ, Murphy MP (2011) Mitochondria-targeted antioxidants as therapies. Discov Med 11:106–114

Sözen T, Özışık L, Çalık Başaran N (2017) An overview and management of osteoporosis. Eur J Rheumatol 4:46–56

Stadtman ER (2001) Protein oxidation in aging and age-related diseases. Ann N Y Acad Sci 928:22–38

Stroes E, Kastelein J, Cosentino F et al (1997) Tetrahydrobiopterin restores endothelial function in hypercholesterolemia. J Clin Invest 99:41–46

Stuart CA, Howell ME, Cartwright BM et al (2014) Insulin resistance and muscle insulin receptor substrate-1 serine hyperphosphorylation. Physiol Rep 2:e12236

Suganami T, Ogawa Y (2010) Adipose tissue macrophages: their role in adipose tissue remodeling. J Leukoc Biol 88:33–39

Surh Y-J (2008) NF-κB and Nrf2 as potential chemopreventive targets of some anti-inflammatory and antioxidative phytonutrients with anti-inflammatory and antioxidative activities. Asia Pac J Clin Nutr 17:269–272

Surmi BK, Hasty AH (2008) Macrophage infiltration into adipose tissue. Future Lipidol 3:545–556

Suvorava T, Nagy N, Pick S et al (2015) Impact of eNOS-dependent oxidative stress on endothelial function and neointima formation. Antioxid Redox Signal 23:711–723

Talbot K, Wang H-Y, Kazi H et al (2012) Demonstrated brain insulin resistance in Alzheimer's disease patients is associated with IGF-1 resistance, IRS-1 dysregulation, and cognitive decline. J Clin Investig 122:1316–1338

Tan BL, Norhaizan ME (2019) Effect of high-fat diets on oxidative stress, cellular inflammatory response and cognitive function. Nutrients 11:2579

Tan BL, Esa NM, Rahman HS et al (2014) Brewers' rice induces apoptosis in azoxymethane-induced colon carcinogenesis in rats via suppression of cell proliferation and the Wnt signaling pathway. BMC Complement Altern Med 14:304

Tan BL, Norhaizan ME, Liew W-P-P (2018a) Nutrients and oxidative stress: friend or foe? Oxidative Med Cell Longev 2018, Article ID 9719584, 24 pages

Tan BL, Norhaizan ME, Liew W-P-P et al (2018b) Antioxidant and oxidative stress: a mutual interplay in age-related diseases. Front Pharmacol 9:1162

Taniguchi CM, Emanuelli B, Kahn CR (2006) Critical nodes in signalling pathways: insights into insulin action. Nat Rev Mol Cell Biol 7:85–96

Teleanu RI, Chircov C, Grumezescu AM et al (2020) Tumor angiogenesis and anti-angiogenic strategies for cancer treatment. J Clin Med 9:84

Tian T, Wang Z, Zhang J (2017) Pathomechanisms of oxidative stress in inflammatory bowel disease and potential antioxidant therapies. Oxidative Med Cell Longev 2017, Article ID 4535194, 18 pages

Unsal V, Dalkıran T, Çiçek M et al (2020) The role of natural antioxidants against reactive oxygen species produced by cadmium toxicity: a review. Adv Pharm Bull 10:184–202

Vaquero EC, Edderkaoui M, Pandol SJ et al (2004) Reactive oxygen species produced by NAD(P) H oxidase inhibit apoptosis in pancreatic cancer cells. J Biol Chem 279:34643–34654

Varvel NH, Neher JJ, Bosch A et al (2016) Infiltrating monocytes promote brain inflammation and exacerbate neuronal damage after status epilepticus. Proc Natl Acad Sci U S A 113:E5665–E5674

Vaughan T, Reid DM, Morrison NA et al (2004) RUNX2 alleles associated with BMD in Scottish women; interaction of RUNX2 alleles with menopausal status and body mass index. Bone 34:1029–1036

Vial G, Dubouchaud H, Couturier K et al (2011) Effects of a high-fat diet on energy metabolism and ROS production in rat liver. J Hepatol 54:348–356

Voigt A, Rahnefeld A, Kloetzel PM et al (2013) Cytokine-induced oxidative stress in cardiac inflammation and heart failure – how the ubiquitin proteasome system targets this vicious cycle. Front Physiol 4:42

Volobueva A, Grechko A, Yet S-F et al (2019) Changes in mitochondrial genome associated with predisposition to atherosclerosis and related disease. Biomol Ther 9:377

Walle P, Takkunen M, Männistö V et al (2017) Alterations in fatty acid metabolism in response to obesity surgery combined with dietary counseling. Nutr Diabetes 7:e285

Wang YC, McPherson K, Marsh T et al (2011) Health and economic burden of the projected obesity trends in the USA and the UK. Lancet 378:815–825

Wang WY, Tan MS, Yu JT et al (2015) Role of pro-inflammatory cytokines released from microglia in Alzheimer's disease. Ann Transl Med 3:136

Wang W, Cai Q, Zhou F et al (2018) Impaired pentose phosphate pathway in the development of 3D MCF-7 cells mediated intracellular redox disturbance and multi-cellular resistance without drug induction. Redox Biol 15:253–265

Wang F-T, Sun W, Zhang J-T et al (2019) Cancer-associated fibroblast regulation of tumor neoangiogenesis as a therapeutic target in cancer (Review). Oncol Lett 17:3055–3065

Wegner AM, Haudenschild DR (2020) NADPH oxidases in bone and cartilage homeostasis and disease: a promising therapeutic target. J Orthop Res 38:2104–2112

Weitzmann MN (2013) The role of inflammatory cytokines, the RANKL/OPG axis, and the immunoskeletal interface in physiological bone turnover and osteoporosis. Scientifica 2013, Article ID 125705, 29 pages

Wezel A, van der Velden D, Maassen JM et al (2015) RP105 deficiency attenuates early atherosclerosis via decreased monocyte influx in a CCR2 dependent manner. Atherosclerosis 238:132–139

White MF (2003) Insulin signaling in health and disease. Science 302:1710–1711

White MF, Shoelson SE, Keutmann H et al (1988) A cascade of tyrosine autophosphorylation in the β-subunit activates the phosphotransferase of the insulin receptor. J Biol Chem 263:2969–2980

White CR, Darley-Usmar V, Berrington WR et al (1996) Circulating plasma xanthine oxidase contributes to vascular dysfunction in hypercholesterolemic rabbits. Proc Natl Acad Sci U S A 93:8745–8749

White RE, Gerrity R, Barman SA et al (2010) Estrogen and oxidative stress: a novel mechanism that may increase the risk for cardiovascular disease in women. Steroids 75:788–793

Wiegman CH, Michaeloudes C, Haji G et al (2015) Oxidative stress–induced mitochondrial dysfunction drives inflammation and airway smooth muscle remodeling in patients with chronic obstructive pulmonary disease. J Allergy Clin Immunol 136:769–780

Wiesolek HL, Bui TM, Lee JJ et al (2020) Intercellular adhesion molecule 1 functions as an efferocytosis receptor in inflammatory macrophages. Am J Pathol 190:P874–P885

Wohlfart P, Xu H, Endlich A et al (2008) Antiatherosclerotic effects of small-molecular-weight compounds enhancing endothelial nitric-oxide synthase (eNOS) expression and preventing eNOS uncoupling. J Pharmacol Exp Ther 325:370–379

World Health Organization (2017) Data and statistics. The challenge of obesity-quick statistics, 2017. http://www.euro.who.int/en/health-topics/noncommunicable-diseases/obesity/data-and-statistics

World Health Organization (2020a) Dementia. https://www.who.int/news-room/fact-sheets/detail/dementia. Accessed 22 Feb 2020

World Health Organization (2020b) Cardiovascular diseases. https://www.who.int/health-topics/cardiovascular-diseases/#tab=tab_1. Accessed 27 Feb 2020
World Health Organization (2020c) Cancer. https://www.who.int/health-topics/cancer#tab=tab_1. Accessed 28 Feb 2020
Wu M-Y, Li C-J, Hou M-F et al (2017) New insights into the role of inflammation in the pathogenesis of atherosclerosis. Int J Mol Sci 18:2034
Wu W-C, Wei J-N, Chen S-C et al (2020a) Progression of insulin resistance: a link between risk factors and the incidence of diabetes. Diabetes Res Clin Pract 161:108050
Wu W, Xiao Z, Chen Y et al (2020b) CD39 produced from human GMSCs regulates the balance of osteoclasts and osteoblasts through the Wnt/β-Catenin pathway in osteoporosis. Mol Ther 28:1518–1532
Wyss-Coray T, Mucke L (2002) Inflammation in neurodegenerative disease–a double-edged sword. Neuron 35:419–432
Xia N, Daiber A, Habermeier A et al (2010) Resveratrol reverses endothelial nitric-oxide synthase uncoupling in apolipoprotein E knockout mice. J Pharmacol Exp Ther 335:149–154
Xu H, Goettsch C, Xia N et al (2008) Differential roles of PKCalpha and PKCepsilon in controlling the gene expression of Nox4 in human endothelial cells. Free Radic Biol Med 44:1656–1667
Xu W, Liu X, He X et al (2020) Bajitianwan attenuates D-galactose-induced memory impairment and bone loss through suppression of oxidative stress in aging rat model. J Ethnopharmacol 261:112992
Yamaguchi T, Yamamoto Y, Yokota S et al (1998) Involvement of interleukin-6 in the elevation of plasma fibrinogen levels in lung cancer patients. Jpn J Clin Oncol 28:740–744
Yokoyama M, Hirata K (2007) Endothelial nitric oxide synthase uncoupling: is it a physiological mechanism of endothelium-dependent relaxation in cerebral artery? Cardiovasc Res 73:8–9
Yoshida Y, Maruyama M, Fujita T et al (1999) Reactive oxygen intermediates stimulate interleukin-6 production in human bronchial epithelial cells. Am J Phys 276:L900–L908
Yousefian M, Shakour N, Hosseinzadeh H et al (2019) The natural phenolic compounds as modulators of NADPH oxidases in hypertension. Phytomedicine 55:200–213
Yu JH, Kim H (2014) Oxidative stress and cytokines in the pathogenesis of pancreatic cancer. J Cancer Prev 19:97–102
Yu E, Calvert PA, Mercer JR et al (2013) Mitochondrial DNA damage can promote atherosclerosis independently of reactive oxygen species through effects on smooth muscle cells and monocytes and correlates with higher-risk plaques in humans. Circulation 128:702–712
Zhang Y-B, Zhong Z-M, Hou G et al (2011) Involvement of oxidative stress in age-related bone loss. J Surg Res 169:e37–e42
Zhang P, Yin Y, Wang T et al (2020) Maresin 1 mitigates concanavalin A-induced acute liver injury in mice by inhibiting ROS-mediated activation of NF-κB signaling. Free Radic Biol Med 147:23–36
Zhao R-Z, Jiang S, Zhang L et al (2019a) Mitochondrial electron transport chain, ROS generation and uncoupling (Review). Int J Mol Med 44:3–15
Zhao Y-P, Wang F, Jiang W et al (2019b) A mitochondrion-targeting tanshinone IIA derivative attenuates myocardial hypoxia reoxygenation injury through a SDH-dependent antioxidant mechanism. J Drug Target 27:896–902
Zhou R, Cheng L, Dandekar AM (2006) Down-regulation of sorbitol dehydrogenase and up-regulation of sucrose synthase in shoot tips of the transgenic apple trees with decreased sorbitol synthesis. J Exp Bot 57:3647–3657
Zhou M, Li S, Pathak JL (2019) Pro-inflammatory cytokines and osteocytes. Curr Osteoporos Rep 17:97–104
Zhu S, Wei W, Liu Z et al (2018) Tanshinone-IIA attenuates the deleterious effects of oxidative stress in osteoporosis through the NF-κB signaling pathway. Mol Med Rep 17:6969–6976
Zuo L, Zhou T, Pannell BK et al (2015) Biological and physiological role of reactive oxygen species–the good, the bad and the ugly. Acta Physiol 214:329–348

Chapter 6
Antioxidant and Age-Related Diseases

6.1 Antioxidant

Antioxidant refers to any components that at low concentration compared to an oxidizable substrate (Kurutas 2016). Antioxidants have an ability to prevent or delay the oxidation of the substrate (Santos-Sánchez et al. 2019). The previous study has demonstrated that antioxidants reduce the malignant transformation, decrease DNA mutations, and reduce oxidative stress, as well as prevent cell damage (Poljsak et al. 2013). In support of this, several epidemiological studies have shown that antioxidant was inversely associated with numerous age-related diseases including diabetes (Mancini et al. 2018), cardiovascular disease (CVD), and cancer (Aune et al. 2018). Nonetheless, at the sustained free radical circumstances, the defense systems capacity towards ROS could be overwhelmed and thereby resulting in disease occurrence (Godic et al. 2014).

The first types of antioxidant defense systems developed against oxidative damage are those that capture and block radicals that are generated and those that prevent ROS occurrence (Cheeseman and Slater 1993). In general, the repair intervention system is comprised of repairing oxidized lipids by acyltransferases, peroxidases, or phospholipases (Hitchon and El-Gabalawy 2004), removing oxidized proteins by proteolytic systems, and repairing oxidatively damaged nucleic acids by specific enzymes (Poljsak et al. 2013). It has been suggested that damage of repair systems could result in more age-related diseases and aging than moderate changes of the antioxidant defense's potential against ROS (Gems and Doonan 2009; Perez et al. 2009; Jang and Remmen 2009).

The study reported by Dröge (2002) has revealed that a shift in the balance between antioxidant and prooxidant is favorably of prooxidants under physiological circumstances, and therefore producing oxidative stress, that is requiring the intervention of endogenous antioxidant systems of the organism. Under this phenomenon, oxidative stress is increasing with age when the repair systems and endogenous

B. L. Tan, M. E. Norhaizan, *The Role of Antioxidants in Longevity and Age-Related Diseases*, https://doi.org/10.1007/978-3-030-82859-2_6

antioxidants cannot counteract effectively. Subsequently, several interventions inhibiting or limiting these aggressive factors are directed to reduce these diseases. Nonetheless, the utilization of synthetic antioxidants in age-related diseases including cancer is still subject to controversy (Cheeseman and Slater 1993; Halliwell and Gutteridge 1999; Godic et al. 2014).

The redox homeostasis of the cells is maintained by a complex endogenous antioxidant defense system including endogenous antioxidant enzymes, for instance, low molecular weight scavengers (lipoic acid, coenzyme Q, and uric acid), proteins (albumin, ceruloplasmin, transferrin, and ferritin), non-enzymatic compounds (glutathione), and complex endogenous antioxidant enzymes (glutathione peroxidase (GPx), catalase, and SOD) (Poljsak et al. 2013). The previous study revealed that exogenous antioxidants present in vegetables and fruits could counterpart the endogenous antioxidative defense activity (Pisoschi and Pop 2015). Antioxidants such as anthocyanidins, flavonoids, phenolics (hydroxycinnamic and cinnamic acid derivatives, hydroxybenzoic and benzoic acids), carotenoids, and vitamins E and C, are regarded as the predominant exogenous antioxidants (Xu et al. 2017; Lourenço et al. 2019; Neha et al. 2019). Many studies reported by Tan et al. (2018a) have demonstrated that age-related diseases and oxidative stress can be modulated by diet-rich antioxidants. In support of this, the recent study showed that diets rich in legumes, nuts, vegetables, fruits, and whole grains are inversely related to CVD risk (Hemler and Hu 2019). Several epidemiological studies, for instance, Nurses' Health Study (NHS) (Stampfer et al. 1993), European paradox study (Bellizzi et al. 1994), Harvard Health Professionals Follow-Up Study (HPFS) (Rimm et al. 1993), and World Health Organization (WHO)/Multinational Monitoring of Trends and Determinants in Cardiovascular Disease (MONICA) study (Gey and Puska 1989) have demonstrated that antioxidant was inversely related to age-related diseases. This finding could be attributed to the unique complex of bioactive compounds, which can protect against oxidative stress that can lead to inflammation (Tan et al. 2015a, b; Tan and Norhaizan 2017). In this regard, the antioxidant capacity in natural products has drawn great deals of interest among scientists in industry and academia to prevent age-related diseases.

6.2 Natural Antioxidants

Natural antioxidants, for instance, carotenoids, phenolic compounds, and vitamins are predominantly present in vegetables and fruits (Septembre-Malaterre et al. 2018). Natural antioxidants present in low levels within cells, in which they can effectively decrease free radicals to protect against age-related diseases (Tan et al. 2018a; Tan and Norhaizan 2019a). Because food tissues are (or were) living, and thus they are under constant oxidative stress from prooxidants, ROS, and free radicals produced both endogenously (transition metals and H_2O_2) and exogenously (light and heat). In this regard, most of the tissues have developed antioxidant systems to control secondary breakdown products, oxidation intermediates, lipid oxidation catalysts, and free radicals (Nimse and Pal 2015). These antioxidant components including tocopherols, carotenoids, phenolic acids, and flavonoids can act as reductants, scavenge free radicals, and suppress Fe^{3+}/ascorbic acid-induced oxidation (Lü et al. 2010).

Natural antioxidants possess the ability to suppress oxidation of breakdown products, lipid peroxidation, and oxidative stress (Yadav et al. 2016). It can function either synergistically or individually to remove free radicals produced during oxidative metabolism to mediate the balance between antioxidants and oxidants (AlBasher et al. 2020). Nevertheless, when there is an excessive production of reactive nitrogen and oxygen species (RNS and ROS, collectively known as RONS), it may breakdown in the delicate physiological balance and thereby results in oxidative stress (Sies and Jones 2020). Subsequently, this oxidative stress may contribute to several chronic diseases such as diabetes, cancer, and CVD (Liguori et al. 2018; Tan et al. 2018b; Hayes et al. 2020).

Plant polyphenols are the components holding one or more phenolic rings derived from the secondary metabolism of plants (Lin et al. 2016). Polyphenols are present in many drinks and foods of plant origin including chocolate, wine, beer, tea, coffee, fruits, and vegetables (Fukushima et al. 2014). Phenolic compounds are not only being present in many foods naturally (Table 6.1), they can also be extracted from their respective source and thereby added into the foods to provide antioxidant and coloring effects (Soto-Vaca et al. 2012; Caporaso et al. 2018). Previous studies have illustrated the effectiveness of phenolic compounds in the prevention of lipid oxidation (Socrier et al. 2019). Plant extracts high in polyphenolic components were shown to be a safe additive because they can easily reserve from natural sources and hinder lipid oxidation effectively (Difonzo et al. 2018). In recent decades, plant polyphenols extracts have been increasingly utilized in the food industry to retard lipid oxidation and prolong their shelf-life (Zhang et al. 2021). Nonetheless, the plant extracts derived from different plants may exhibit considerable variations in their antioxidant activity (Hung 2016). This could be due to several factors, for instance, soil type, light, temperature, climate, growing season, growing environment, genotype, or variety, which may alter both phenolic or antioxidant components and antioxidant activity in the plants (Shamloo et al. 2017; Kabtni et al. 2020). Several studies have reported that the quantity of bioactive compounds and their activities could be affected by species variation, environmental, and genetic (Alfaro et al. 2013; Pohl et al. 2019; Ruiz-López et al. 2019; Zalazar et al. 2019; Przybylska-Balcerek et al. 2020). A study reported by Alfaro et al. (2013) has revealed that genotype and growing season had a significant effect on dry matter, antioxidant activity, and polyphenol components of murtilla fruits from three genotypes. Herbs and spices are usually added to food for their medicinal mixtures and improved their flavor (El-Sayed and Youssef 2019). Herbs and spices possess high amounts of phenolic compounds that have strong H-donating activity (Ulewicz-Magulska and Wesolowski 2019). Several of these herbs and spices contain high values of oxygen radical absorbance capacity (ORAC) (Table 6.2). The previous study demonstrated that plant-derived components (rosmaridiphenol, rosmariquinone, rosmanol, and carnosol) show a better antioxidant activity compared

Table 6.1 Phenolic antioxidants from selected natural sources

Category	Natural sources	Phenolic compounds	References
Fruits	Apple peel	Chlorogenic acid, epicatechin, phloridzin, catechin, hyperoside, and quercitrin	Mihailović et al. (2018)
	Apple pulp	Chlorogenic acid, epicatechin, phloridzin	Mihailović et al. (2018)
	Strawberry	Anthocyanidins, flavanols, flavonols, phenolic acids, hydroxybenzoic acids, hydroxycinnamic acids	Miller et al. (2019)
	Citrus	Luteolin, kaempferol, quercetin, apigenin, diosmetin, chrysoeriol, isorhamnetin, vicenin 2, lucenin 2, vitexin, isovitexin, isoquercitrin, diosmin, rutin, scoparin, neodiosmin, eriodyctiol, hesperetin, naringenin, hesperidin, naringin	Brito et al. (2014)
	Mango	Chlorogenic acid, gallic acid, vanillic acid, protocatechuic acid	Palafox-Carlos et al. (2012)
	Black currants	Anthocyanins, hydroxycinnamic acid, flavonols, delphinidin-3-*O*-glucoside, delphinidin-3-*O*-rutinoside, myricetin-3-*O*-glucoside	Zheng et al. (2012)
	Sweet cherry	Anthocyanins, flavonols, chlorogenic acids, hydroxycinnamic acid, flavan-3-ols	Martini et al. (2017)
	Persimmon	Gallic acid, (+)-catechin, quercetin, luteolin, kaempferol	Fu et al. (2016)
	Guava	Gallic acid, (+)-catechin, (−)-epicatechin, quercetin, luteolin, kaempferol	Fu et al. (2016)
	Sweetsop	Gallic acid, (+)-catechin, chlorogenic acid, ferulic acid, quercetin, luteolin	Fu et al. (2016)
	Pineapple	Epicatechin and felinic acid	Jianwen and Chenghai (2006)
	Plum	Cyanidin 3-rutinoside, flavonols (rutin and quercetin 3-glucoside)	Slimestad et al. (2009)
	Blueberry	Gallic acid, p-coumaric acid, ferulic acid, ellagic acid, catechin, myricetin, quercetin, kaempferol	Sellappan et al. (2002)
	Apricot	Gallic acid, procatechin (acid), 4-aminobenzoic acid, chlorogenic acid, catechin, caffeic acid, epicatechin, vanillin, p-coumaric acid, rutin, ferulic acid, salicylic acid, resveratrol, quercetin	Sochor et al. (2010)
	Kiwi	Protocatechuic acid, chlorogenic acid, caffeic acid, rutin, p-hydroxybenzoic acid, quercetin	Wang et al. (2018d)

(continued)

Table 6.1 (continued)

Category	Natural sources	Phenolic compounds	References
Vegetables	Broccoli	Flavonoids and hydroxycinnamoyl acid derivatives	Vallejo et al. (2004)
	Leafy vegetables	Salicylic acid, vanillic acid, gallic acid, caffeic acid, chlorogenic acid, p-coumaric acid, ferulic acid, m-coumaric acid, isoquercetin, rutin	Khanam et al. (2012)
	Carrot	3′-caffeoylquinic acid, 5′-caffeoylquinic acid, caffeic acid, cis-5′-caffeoylquinic acid, 4′p-coumaroylquinic acid, hydroxycinnamic derivative, 3′4′-dicafferoylquinic acid, 3′5′-dicafferoylquinic acid	Zhang and Hamauzu (2004)
	Chinese cabbage	Flavonols (quercetin and kaempferol), phenolic acids (sinapic, ferulic, p-coumaric, and caffeic acids), cyanidin	Lee et al. (2018a)
	Curly kale	Quercetin-3-sinapoyl-diglucoside-7-diglucoside, kaempferol-3-sinapoyl-diglucoside-7-diglucoside, quercetin, kaempferol, p-coumaric, ferulic, sinapic, and caffeic acid, hydroxycinnamic acids, flavonoid, quercetin-3-disinapoyl-triglucoside-7-diglucoside	Olsen et al. (2009)
	Iceberg lettuce	Caffeic acid, chlorogenic acid, phaseolic acid, chicoric acid, isochlorogenic acid, luteolin-7-O-glucuronide, quercetin-3-O-glucuronide, quercetin-3-O-galactoside, quercetin-3-O-glucoside, quercetin-3-O-(6″-malonyl)-glucoside, syringin	Mai and Glomb (2013)
	Potato	Chlorogenic, caffeic, coumaric, protocatechuic, vanillic, ferulic, syringic, *p*-coumaric, sinapic, gallic acids	Akyol et al. (2016)
	Artichoke	Apigenin 7-O-glucoside, chlorogenic acid	Alarcón-Flores et al. (2014)
	Spinach	Ferulic acid	Alarcón-Flores et al. (2014)
	Tomato	4-caffeoylquinic acid, 5-caffeoylquinic acid, dicaffeoylquinic acids I and II, tricaffeoylquinic acid, quercetin trisaccharide, quercetin-3-*O*-rutinoside, kaempferol-3-*O*-rutinoside, naringenin, naringenin chalcone	Ribas-Agustí et al. (2012)

(continued)

Table 6.1 (continued)

Category	Natural sources	Phenolic compounds	References
Cereals and legumes	Soybean	Gallic acid, pyrogallol, homogentisic acid, chlorogenic acid, catechin, vanillic acid, caffeic acid, syringic acid, ρ-coumaric acid, rutin, ferulic acid, naringin, hesperidin, *o*-coumaric acid, myricetin, quercetin, *trans*-cinnamic acid, naringenin, kaempferol, hesperetin, biochanin A	Kim et al. (2006)
	Oat	Avenanthramides A, avenanthramides C, caffeic acid, quercetin-3,4′-O-di-beta-glucopyranoside derivative, 6-hydroxykaempferol 3,5,7,4′-tetramethyl ether 6-rhamnoside	Rao et al. (2019)
	Wheat grain	p-hydroxybenzoic acid, vanillic acid, syringic acid, p-coumaric acid, ferulic acid, four ferulic acid derivatives, apigenin	Hernández et al. (2011)
Herbs and spices	Basil	Rosmarinic, chicoric, caffeic, caftaric acid	Kwee and Niemeyer (2011)
	Garlic	Caffeic acid, quercetin	Alarcón-Flores et al. (2014)
	Ginger	Pyrogallol, p-hydroxybenzoic acid, ferulic acid, vanillin, p-coumaric acid, gallic acid, caffeic acid	Tohma et al. (2017)
	Mint	Rosmarinic acid, neoponcirin, narirutin, chlorogenic acid, biochanin A, caffeic acid, apigenin, hesperetin, naringenin	Tang et al. (2016)
	Onion	Isorhamnetin, kaempferol, luteolin, quercetin	Leighton et al. (1992); Miean and Mohamed (2001); Lanzotti (2006)
	White pepper	Decaffeoylacteoside, forsythoside D, vanillin, norbergenin, N-trans coumaroyltaramine, protocatechuic aldehyde, tribulusamide A, moupinamide, 1,5-bis(4-hydroxy-3 methoxyphenyl)-1,4 pentadien-3-one, moracin H, cishinokiresinol, moracin O, kukoamine A, moupinamide, 6-gingerol, moracin C, 2-octylphenol, octahydrocurcumin, 2,6-di-tert-butyl-4-hydroxy toluene, 6-gingerol	Olalere et al. (2019)
	Black pepper	Forsythoside D, decaffeoylacteoside, vanillin, norbergenin, N-trans Coumaroyltaramine, moupinamide, tribulusamide A, cishinokiresinol, moupinamide, 1,5-bis(4-hydroxy-3 methoxyphenyl)-1,4 pentadiene-3-one, N-dihydro- caffeoyltyramine, moracin C, 2-octylphenol, octahydrocurcumin, kukoamine A, 2,6-di-tert-butyl-4-hydroxytoluene, 6-gingerol	Olalere et al. (2019)

to butylated hydroxyanisole (BHA) (Richheimer et al. 1996; Carvalho et al. 2005). In a study by Cheng et al. (2019) focusing on various flaxseed phenolic extracts on the oxidative and physical stability of flaxseed oil-in-water nanoemulsions, flaxseed lignan extract and secoisolariciresinol were shown to inhibit lipid oxidation in nanoemulsion, suggesting that both flaxseed lignan extract and secoisolariciresinol are good plant-based antioxidants for improving the stability of flaxseed oil nanoemulsions. A further study reported by de Florio Almeida et al. (2017) showed that the antioxidant effect of lyophilized bee pollen extract was effective in retarding lipid oxidation in pork sausage.

In addition, the plant extracts have also been demonstrated their anticancer ability. For example, sapodilla leaf extracts have been demonstrated their abilities to inhibit different cancer cells including human hepatocellular carcinoma (HepG2) (Tan et al. 2018c), human cervical cancer (HeLa) (Tan et al. 2018d), and human colorectal cancer (HT-29) cells (Tan and Norhaizan 2019b). This favorable effect is more likely due to the presence of phenolic compounds including gallic acid, p-coumaric acid, ferulic acid, caffeic acid, and vanillic acid (Tan et al. 2018c, d; Tan and Norhaizan 2019b). The anticancer ability of sapodilla leaf extracts is modulated via a few mechanisms including apoptosis, cell cycle arrest, and caspase-dependent pathway (Tan et al. 2018c, d; Tan and Norhaizan 2019b).

6.3 Mechanism of Action of Antioxidants

Antioxidants can act at successive stages of the oxidative radical process, namely initiation, propagation, and chain termination (Kurutas 2016). The detailed explanations of the steps, which are part of the radical sequences is particularly centered on the propagation and initiation, implied that lipid oxidation in cell membranes can be activated by exogenous chemical and physical factors, including endogenous enzyme systems (cytochrome P450, uncoupled nitric oxide synthase, XO, and NADPH oxidase), ionization radiation, UV-light, smoking, air pollution, as well as the ETC in mitochondria (Yin et al. 2011).

The previous study has revealed that propagation step peroxidation initiates when oxygen reacts with carbon-centered radicals, occurring near or at the diffusion-controlled rate. The propagation, occurring at most oxidations following a radical mechanism, and usually at a slow rate, is represented by the transfer of a hydrogen atom to the chain carrying peroxyl radical (Pisoschi and Pop 2015). Peroxyl free radicals can react to carbon-carbon double bonds. In particular, the conjugated dienes may subject to peroxyl addition. Furthermore, the intramolecular radical substitution on peroxide and radical cyclization reactions may occur, resulting in cyclic peroxides, while the polyunsaturated lipids involve in the peroxidation process (Yin et al. 2011).

In addition, antioxidants can react by quenching singlet oxygen ($^1O_2$) or breaking the chain of a radical sequence, scavenging chain initiating radicals such as peroxyl $ROO^•$, alkoxyl $RO^•$, or hydroxyl $OH^•$, trapping aggressive ROS including H_2O_2 or

Table 6.2 ORAC amounts of selected herbs and spices

Herbs and spices	Total ORAC (μm TE/100 g)
Basil (fresh)	4805
Oregano (fresh)	13,970
Sage (fresh)	32,004
Thyme (fresh)	27,426
Basil (dried)	61,063
Thyme (dried)	157,380
Cinnamon (ground)	131,420
Clove (ground)	290,283
Ginger (ground)	39,041
Nutmeg (ground)	69,640
Turmeric (ground)	127,068
Sage (ground)	119,929
Oregano (dried)	175,295
Rosemary (dried)	165,280
Black pepper	34,053

Source: USDA (2010)
ORAC oxygen radical absorbance capacity, *TE* trolox equivalent

superoxide anion radical, removing prooxidative metal ions, and depleting molecular oxygen (Nimse and Pal 2015).

The primary antioxidants are the chain-breaking antioxidants that can scavenge radical species; while secondary antioxidants are UV radiation absorbers or oxidative enzyme inhibitors, metal chelators, peroxide decomposers, and singlet oxygen quenchers (Simitzis 2018). In general, secondary antioxidants may show synergistic effects in combination with primary antioxidants via different mechanisms. The mechanisms include (1) quenching molecular oxygen; (2) chelating pro-oxidative transition metal cations; (3) regenerating primary antioxidants through hydrogen donation; and (4) stabilizing primary antioxidants by creating an acidic environment (Tan et al. 2018a).

Antioxidant controls the autoxidation by suppressing the generation of free radicals or by disrupting the propagation of free radicals through several mechanisms (Gholamian-Dehkordi et al. 2017). This compound facilitates in breaking the autoxidative chain reaction, preventing the formation of peroxides, scavenging the species that initiate the peroxidation, and quenching $\bullet O_2^-$ (Gaschler and Stockwell 2017). Among the mechanism of action, interruption of the free radical chain reaction is the most effective antioxidant (Nimse and Pal 2015). It contains aromatic or phenolic rings that allow antioxidants to donate H• to the free radicals formed during oxidation. Subsequently, the radical intermediate is stabilized by the resonance delocalization of the electron within the aromatic ring (Wojtunik-Kulesza et al. 2016). Notably, both natural botanicals (phenolics (flavonoids)) and synthetic antioxidants (propyl gallate, butylated hydroxytoluene (BHT), and BHA) function in this manner (Brewer 2011). The botanicals extracts exert antioxidant activity and usually quench free radical oxygen with phenolic compounds (Xu et al. 2017).

Antioxidants play a prominent role in the termination of oxidative chain reactions by removing the free radical intermediates (Gholamian-Dehkordi et al. 2017). Chain-breaking antioxidant varies in their antioxidative effectiveness based on the physical location within the food (aqueous phase, emulsion interfaces, and proximity to membrane phospholipids) and chemical characteristics (Jacobsen et al. 2019). The chemical potential of the solubility in oil and an antioxidant affect its accessibility to peroxy radicals particularly in the amphiphilic character, emulsion, micellar system, and in the membrane, which is required for the effectiveness in the system (Watanabe et al. 2010).

The effectiveness of antioxidants is linked to the antioxidant solubility, the ease with which the antioxidant is destroyed or lost (heat susceptibility and volatility), oxidation-reduction potential, rate constant, and activation energy (Brewer 2011). Furthermore, chain and inhibitor propagation reactions are both exothermic (Lazarovici et al. 2004). When the R:H and A:H bond dissociation energy is increased, the antioxidant efficiency decreases, and the activation increases (Brewer 2011). However, when the bond energy decrease, the antioxidant efficiency increases (Brewer 2011).

The presence of bivalent transition metal ions, particularly Fe^{2+}, can catalyze oxidative processes and thereby resulted in the formation of hydroxyl radicals, as well as can decompose hydroperoxides through Fenton reactions. Through these reactions, chelating metals can effectively decrease oxidation (Goddard et al. 2012). For example, food such as red meat containing significant levels of these transition metals and thus is susceptible to metal-catalyzed reactions (Domínguez et al. 2019).

A study reported by Fang et al. (2018) revealed that cellular redox status is critical for mitochondrial function and ROS-mediated signaling. The reduction of intracellular GSH levels markedly induces mitochondrial membrane depolarization and triggers mitochondrial ROS production (Lohan et al. 2018). Activation of the Nrf2/antioxidant response element (ARE) pathway is crucial for the mediation of the intracellular GSH in response to stress and antioxidant defense enzyme (Liu et al. 2018). *N*-acetylcysteine restores ARE-related gene expression and reverses GSH depletion to basal levels (Limón-Pacheco et al. 2007). Indeed, appropriate intracellular amounts of ROS are crucial for the physiological redox signaling by regulating and activating endogenous defenses to protect cells from oxidative, electrophilic, and nitrosative stress (Moldogazieva et al. 2018). Administration of exogenous antioxidants reduces exercise-triggered improvements in insulin sensitivity and antioxidant transcriptional activity (Ji et al. 2006), implied the crucial of ROS-triggered endogenous antioxidant enzymes in restoring physiological redox balance. It has been suggested that Nrf2 can protect oxidative stress in aging (de Oliveira et al. 2018). The development of age-related diseases has been associated with the reduction of Nrf2 activity (Cuadrado et al. 2018).

6.4 Reversal of Age-Related Changes by Antioxidants

Coronary heart disease has become one of the leading causes of disability and death worldwide. The predominant underlying cause of the disease is atherosclerosis, which occurs in the large or medium-size arteries, and thereby resulted in necrosis or cell death and loss of vascular functions (Geovanini and Libby 2018). Inflammation and oxidative stress are prominent factors responsible for injury and endothelial dysfunction (Incalza et al. 2018). The vascular cells are vital sites for oxidative stress and ROS production (Sena et al. 2018). Many naturally occurring antioxidants are being evaluated for their potential to enhance vascular integrity and regenerate vascular cells by restoring the cellular functions of diseased vessels (Shafi et al. 2019). Indeed, natural antioxidants regulate oxidative stress in different types of vascular cells including progenitor, stem, and endothelial cells (Kumar et al. 2009; Shaban et al. 2017; Pereira et al. 2017; Abdel-Daim et al. 2018; Su et al. 2018). The data from *in vitro* and animal studies revealed that antioxidants mediate oxidative stress via a few mechanisms including modulation of proinflammatory vascular adhesion molecules (E-selectin, ICAM-1, and VCAM-1), NF-κB, and MAPK signaling pathways (Li et al. 2014a). Nonetheless, the studies examined natural antioxidants, for example, polyphenols and vitamins (E, D, and C), and the regeneration potential of vascular cells are remained obscure (Tan et al. 2018a). Natural antioxidants are being investigated for their potential to improve the reendothelialization process by enhancing rapid endothelial cell growth and stimulating adhesion molecules (Shafi et al. 2019) (Fig. 6.1). The previous study has revealed the new approach involving several vitamins on direct reprogramming of progenitor or stem cells into endothelial cells (Cimmino et al. 2018). A study by Rodrigo et al. (2007) and Wong et al. (2014) further supported that vitamins E, D, and C improved the reendothelialization process and thereby leading to vascular repair following vascular injury. This process is more likely due to the cell proliferation and growth (Ulrich-Merzenich et al. 2007), which enhanced the vascular endothelial cells functions (Rodrigo et al. 2007; Mazidi et al. 2017; Jamali et al. 2018). Table 6.3 summarizes the effect of natural antioxidants on the vasculature. Although antioxidant pharmacological properties were demonstrated in various studies, there is still a lack of clinical evidence on the specific molecular targets and benefits of dietary antioxidants on health (Dalle-Donne et al. 2006). It has been suggested that diets containing natural antioxidants may provide vascular regenerating properties (Huang 2018). Collectively, dietary antioxidants may play an essential role in preventing vascular disorders.

In addition to the effect observed in vascular disease, dietary intake of plant-based foods, for example, whole grains, vegetables, and fruits are related to the positive health effects (Kim et al. 2019). Numerous epidemiological and case-control studies have revealed that consumption of phytochemical-rich foods was negatively associated with the incidence of the prostate (Hoang et al. 2018), pancreas (Morrison et al. 2021), lung (Cohen et al. 2019), colon (Lee et al. 2018b), and breast (Sharif et al. 2021) cancers. The phytochemical-rich diets prevent cancers via a few mechanisms including modulation of carcinogen detoxification/activation, cell differentiation, DNA repair, cell cycle, and apoptosis by tumor-suppressor genes and xenobiotic-metabolizing enzymes (Tan et al. 2018b) (Fig. 6.1). In this regard, alleviating the oxidative stress-related tumorigenesis is one of a mode of action to explain this anticancer potential (Chikara et al. 2018). Indeed, oxidative stress contributes to all stages of tumorigenesis either through an indirect mechanism by mediating the cell signaling transduction or directly caused in the DNA damage (Tan et al. 2018b). Hence, decreasing oxidative stress may play a pivotal role in chemoprevention. Diet rich in phytochemicals has emerged as a promising chemopreventive potential towards different types of cancers (Tan et al. 2018b) due to its less or without significant undesirable effects on healthy tissues (Tan et al. 2018a). Some phytochemicals are not only exerting anticancer potential, but they are also demonstrated their anti-invasive, anti-migratory, and anti-proliferative properties on cancer cells, suggesting that a phytochemical-rich diet may act as adjuvant therapy in cancer (Tan and Norhaizan 2019c).

Dietary antioxidant not only reduces CVD and cancer but it also exerts a beneficial impact on glucose homeostasis and insulin sensitivity (Straub et al. 2019). The Rotterdam Study, a population-based cohort, has demonstrated an inverse relationship between total dietary antioxidant capacity and type 2 diabetes (van der Schaft et al. 2019). The study further revealed that high total dietary antioxidant capacity is negatively linked to the risk of prediabetes in men (van der Schaft et al. 2019). Notably, a study by van der Schaft et al. (2019) found a significant association between dietary antioxidant capacity and homeostasis model assessment of insulin resistance (HOMA-IR) or type 2 diabetes in both prediabetes and normoglycemia individuals. Chronic hyperglycemia and prolonged hyperinsulinemia, along with increased RNS and ROS levels, may result in the impairment of pancreatic β-cell

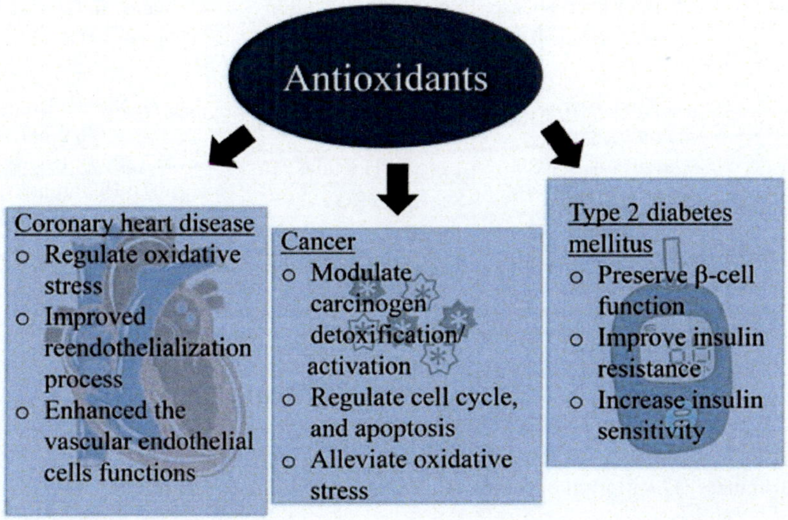

Fig. 6.1 Effect of antioxidants on coronary heart disease, cancer, and type 2 diabetes mellitus

Table 6.3 Effect of natural antioxidants on the vasculature

Antioxidants	Food rich in antioxidants	Findings	References
Flavonoids	Plants, berries, honey	Downregulate iNOS expression	Di Tomo et al. (2012); Zeng et al. (2017); Abdel-Daim et al. (2018); Mozos et al. (2018)
Trace elements	Dietary source from animal	Scavenge ROS, neovascularization	Shen et al. (2016); Zeng et al. (2017)
Polyphenols	Peaches, red wine, green tea, cocoa	↓ Blood cholesterol	Basu and Lucas (2007); Zhen et al. (2012); Cong et al. (2018)
		Reendothelialization	
		↓ Blood pressure	
		Suppresses ox-LDL	
		Regulates TLR4 pathway	
		↑ Antioxidant enzymes	
Lycopene, β-carotene, carotenoids	Yellow, green, red fruits and vegetables	↓ ox-LDL	Oak et al. (2003); Hibino et al. (2012); Boydens et al. (2015); Wang et al. (2018a)
		↓ Blood cholesterol	
		Preserve endothelial function	
		↓ iNOS	
		↑ NO	
		↓ TNF-α	
		↑ Antioxidant enzymes	
		Scavenge ROS	
Vitamin E (α-tocopherol)	Broccoli, spinach, nuts, vegetable oils	Preserve endothelial function	Martin-Nizard et al. (1998); Azzi (2007); Rodrigo et al. (2007); Lee et al. (2018a)
		↓ Blood cholesterol	
		Suppress ox-LDL	
		↑ Antioxidant enzymes	
Vitamin D	Beef liver, cheese, egg yolk, fish	Reendothelialization	Wong et al. (2014); Mazidi et al. (2017)
		Suppress NADPH oxidase	
		↑ Antioxidant enzymes	
Vitamin C	Kiwi, berries, pineapple, papaya, mango, citrus fruits	Preserve endothelial function	Ulrich-Merzenich et al. (2002, 2007, 2009); Heller et al. (2004); Rodrigo et al. (2007); Honarbakhsh and Schachter (2009); Cimmino et al. (2018)
		↓ NADPH oxidase	
		Scavenge ROS	
		↑ SOD	
		Reendothelialization	
		↓ Blood pressure	
		Suppress ox-LDL	
		Preserve endothelial function	

iNOS inducible nitric oxide synthase, *NADPH* nicotinamide adenine dinucleotide phosphate, *NO* nitric oxide, *ox-LDL* oxidized low-density lipoprotein, *ROS* reactive oxygen species, *SOD* superoxide dismutase, *TLR4* toll-like receptor 4, *TNF-α* tumor necrosis factor-α and subsequently contribute to the pathogenesis of type 2 diabetes (Prentki and Nolan 2006). Dietary antioxidants have been suggested to decrease the risk of type 2 diabetes by preserving the β-cell function (Prentki and Nolan 2006) (Fig. 6.1). Such finding indicates that a diet high in antioxidant capacity possesses protective effects against type 2 diabetes (Tan et al. 2018a). Despite the mechanism underlying dietary antioxidants' protective effect and type 2 diabetes requires further elucidation, most of the experimental studies hypothesized that the protective role is linked to both β-cell dysfunction and insulin resistance (van der Schaft et al. 2019). Individuals with type 2 diabetes have relatively high oxidative stress markers and reduced total antioxidant status compared to healthy individuals (Rani and Mythili 2014). Similarly, several findings have corroborated these data and found that high intakes of tea, vegetables, and fruits, which are known to be rich in antioxidant components, are inversely linked to type 2 diabetes (Maritim et al. 2003; Li et al. 2014b; Yang et al. 2014; Mancini et al. 2018). The preventive role of fruits and vegetables and tea toward type 2 diabetes is more likely due to the synergistic effect of the antioxidant components (Mancini et al. 2018). Compared to those who never consume alcohol, individuals who consume high alcoholic beverages are positively linked to diabetes (Carlsson et al. 2005). Whereas consumption of moderate alcohol is linked to a reduced risk of diabetes compared to none consumption individuals (Carlsson et al. 2005). Intriguingly, wine was found to be linked to the decreased risk of diabetes; while none of the associations was reported on spirits or beer and diabetes (Zenebe et al. 2001; Arranz et al. 2012). The favorable effect of wine could be attributed to the polyphenols and phenolic acids, which exert antioxidant activity that prevail over the deleterious effect of the alcohol itself (Howard et al. 2004; Hodge et al. 2006).

6.5 Natural Products as a Protective Agent against Age-Related Diseases

Substantial evidence has shown that oxidative stress and obesity-associated age-related diseases can be modulated by antioxidant-rich in natural products (Tan et al. 2018b). A unique complex of bioactive compounds can protect against oxidative stress (Tan et al. 2015a, b; Tan and Norhaizan 2019a). In this regard, the development and progression of these diseases could be prevented by changing dietary habits (Tan et al. 2018b).

6.5.1 Fruits and Vegetables

Fruits and vegetables contain an abundance of dietary fibers, vitamins, and minerals. High consumption of fruits and vegetables is inversely related to the incidence and mortality of obesity-associated diseases, for instance, cancer, type 2 diabetes, and CVD (Saad et al. 2017). This favorable effect has been accredited to antioxidant vitamins such as vitamin C, vitamin E, and β-carotene (Leenders et al. 2014). In general, hydrophilic antioxidants contributed to nearly 85% of total antioxidants in fruits and vegetables (Jędrejek et al. 2017). Vitamins C and E and β-carotene are crucial for the proper regulation of physiological function (Hecht et al. 2016). Vitamin E has been demonstrated to modulate oxidative-antioxidant balance, yet vitamin C can improve antioxidant protection (Williamson 2013). Bright-colored fruits and vegetables are rich in β-carotene (Rodriguez-Casado 2016). A previous study has found that β-carotene exerts ability in reducing low-density lipoprotein (LDL)-cholesterol oxidation and maintaining the immune system via the regulation of antioxidant enzymes (Rodriguez-Casado 2016). Furthermore, other dietary compounds, for instance, flavonoids have preventive effects against oxidative stress. Flavonoids possess several biological activities including antimicrobial action, antioxidant activity, anti-inflammatory activity, and antitumor effects (Priviero et al. 2017).

Anthocyanins are one of the flavonoids abundantly found in fruits and vegetables (Azevedo da Silva et al. 2014; Faria et al. 2014; de Oliveira et al. 2015; Zanotti et al. 2015). They are one of the most widespread natural pigments in the plant kingdom (Li et al. 2017a). It has been demonstrated that anthocyanins in blackcurrant, blackberry, and blueberry (0–20 μg/mL for 12 h) are potent inflammatory mediators in the alleviation of lipopolysaccharides (LPS)-induced (100 ng/mL for 3 h) NF-κB translocation to the nucleus and thus mediates inflammatory responses in RAW 264.7 macrophages (Lee et al. 2014). In this regard, this finding showed that blackcurrant, blackberry, and blueberry exert their anti-inflammatory effects in macrophages, possibly by inhibiting nuclear translocation of NF-κB independent of the Nrf2. Data from *in vivo* study using 24 adult male Wistar rats further revealed that anthocyanins-rich extract from Kamchatka honeysuckle berry (2 g/kg for 4 weeks) can alleviate the disturbances in glucose and lipid metabolism (Fig. 6.2), which are the fundamental risk factors for CVD and diabetes (Jurgoński et al. 2013) (Table 6.4). Additionally, many studies demonstrated that an anthocyanin-rich diet is related to weight loss. Anthocyanins (cyanidin 3-glucoside) (2 g/kg for 12 weeks) promote the secretion of adipocytokine (leptin and adiponectin) and uncoupling protein 2 (UCP2), stimulate adipocyte fatty acid-binding protein (aP2), upregulate lipoprotein lipase (LPL), and promote peroxisome proliferator-activated receptor gamma (PPAR-γ) expression in isolated rat adipocytes (Tsuda et al. 2004) (Table 6.4). In a human study, intakes of fruits and vegetables (≥500 g/day for 16 weeks) significantly decreased the sagittal abdominal diameter, body weight, and waist circumference (Fig. 6.2) in obese and overweight individuals aged 35–65 years (BMI >27 kg/m^2) (Järvi et al. 2016). Table 6.5 shows the effects of fruits and vegetables on human health.

Low intake of fruits and vegetables is a major risk factor for CVD (Seferidi et al. 2019). A study reported by Aune et al. (2017) revealed that nearly 1.3 million CVD deaths could have been prevented in 2013 worldwide when the intakes of fruits and vegetables were at 800 g/day (10 servings/day). The European Prospective Investigation into Cancer and Nutrition-Netherlands study involving 34,560 participants (26% men and 74% women) aged 20–69 years suggested that consumption of pure fruit juice up to 7 glasses/week but not ≥8 glasses were significantly reduced the risk of coronary heart disease (CHD) and CVD. The data further revealed the highest three quintiles of fruit consumption were significantly related to lower CVD incidence (Scheffers et al. 2019). Wood et al. (2017) further showed that encapsulated fruit and vegetable juice concentrate (6 capsules for 8 weeks/day) could reduce the systolic blood pressure, plasma TNF-α, LDL-cholesterol, and TC in overweight and obese adults aged ≥40 years (BMI ≥28 kg/m^2). The favorable effect could be attributed to the alteration of gene expression through several signaling pathways, for instance, NF-κB related genes and adenosine monophosphate-activated protein kinase (AMPK) (Wood et al. 2017). Collectively, fruits and vegetables protect against CVD by reducing oxidative stress, ameliorating ischemia/reperfusion injury, alleviating inflammation, suppressing thrombosis, regulating blood pressure, protecting vascular endothelial function, and modulating lipid metabolism (Buil-Cosiales et al. 2016; Zhao et al. 2017).

Emerging evidence has suggested that fruits and vegetables are related to a lower incidence of type 2 diabetes. Data from systematic review and meta-analysis of

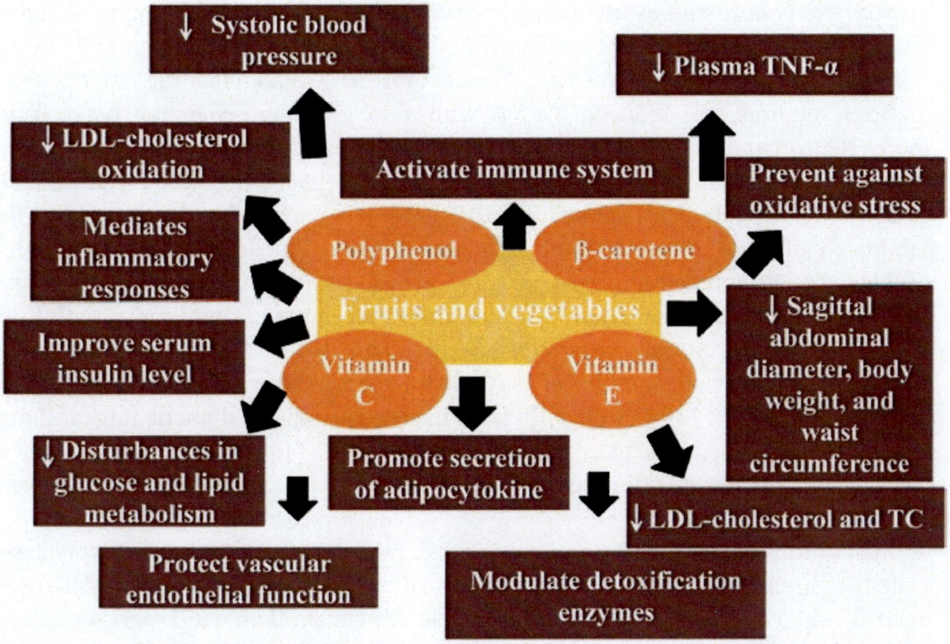

Fig. 6.2 Biological mechanism of fruits and vegetables in relation to several diseases. *LDL* low-density lipoprotein, *TC* total cholesterol, *TNF-α* tumor necrosis factor-α

Table 6.4 The effects of fruits and vegetables on age-related diseases in *in vitro* and *in vivo* studies

Age-related diseases	*In vitro* and *in vivo* models	Treatment	Findings	References
CVD and diabetes	Adult male Wistar rats	Anthocyanins-rich extract from Kamchatka honeysuckle berry (2 g/kg for 4 weeks)	↓ Glucose and lipid metabolism disturbances	Jurgoński et al. (2013)
Type 2 diabetes mellitus	Human intestinal Caco-2 cells	Berry extract (0.125% (w/v) for 16 h)	↓ SGLT1 and GLUT2 mRNA expression	Alzaid et al. (2013)
			↓ Facilitated glucose uptake	
Type 2 diabetes mellitus	Alloxan-induced diabetic rats (150 mg/kg body weight)	Water extract of garden strawberry (50 mg/kg body weight) for 45 days	Improve serum insulin level	Abdulazeez and Ponnusamy (2016)
Obesity	Isolated rat adipocytes	Cyanidin 3-glucoside (2 g/kg for 12 weeks)	↑ Secretion of adipocytokine (leptin and adiponectin) and UCP2, aP2, LPL, and PPAR-γ expression	Tsuda et al. (2004)

aP2 adipocyte fatty acid-binding protein, *CVD* cardiovascular disease, *GLUT2* glucose transporter 2, *LPL* lipoprotein lipase, *UCP2* uncoupling protein 2, *PPAR-γ* peroxisome proliferator-activated receptor gamma, *SGLT1* sodium-dependent glucose transporter 1

prospective cohort studies involving 194,019 participants and 13,013 type 2 diabetes mellitus cases showed that intakes of berries were associated with an 18% reduction of type 2 diabetes mellitus risk. The data further revealed that the risk of type 2 diabetes mellitus was reduced by 5% with a 17 g/day increment of berry fruits intake (Guo et al. 2016). Data from *in vitro* study revealed that berry extract (0.125% (w/v) for 16 h) significantly decreased the sodium-dependent glucose transporter 1 (SGLT1) and glucose transporter 2 (GLUT2) mRNA expression and reduced the facilitated glucose uptake in human intestinal Caco-2 cells (Alzaid et al. 2013) (Table 6.4). The concentrations of berry extract used are consistent with the amounts of polyphenols that are found in the intestinal lumen after consumption of polyphenol-rich fruits or beverages (Williamson 2013). Furthermore, the oral administration of water extract of garden strawberry (50 mg/kg body weight) for 45 days could improve serum insulin level in alloxan-induced diabetic rats (150 mg/kg body weight) (Abdulazeez and Ponnusamy 2016) (Table 6.4). This biological effect highlights the role of unique complexes of bioactive constituents in fruits and vegetables.

In addition to the effects mentioned above, fruit consumption is negatively associated with cancer risk. A human study involving 90,476 premenopausal women aged 27–44 years old from the Nurses' Health Study II in 1991 and 44,223 of women in 1998 revealed that high intakes of fruits (2.9 servings/day), particularly grapes, banana, and apple during adolescence and kale and oranges during early

Table 6.5 The effects of fruits and vegetables on human health

Metabolic ailments	Subjects	Treatment	Findings	References
Obesity	Obese and overweight individuals	≥500 g/day for 16 weeks	↓ Sagittal abdominal diameter, body weight, and waist circumference	Järvi et al. (2016)
CVD	Patients with coronary heart disease and stroke combined	800 g/day (10 servings/day)	↓ Risk of CVD	Aune et al. (2017)
CVD	Overweight and obese adults	Encapsulated fruit and vegetable juice concentrate (6 capsules for 8 weeks/day)	↓ Systolic blood pressure, plasma TNF-α, LDL-cholesterol, and TC	Wood et al. (2017)
CVD	Two cohort studies (34,560 participants, 26% men and 74% women aged 20–69 years)	Up to 7 glasses/week but not ≥8 glasses of pure fruit juice	↓ Risk of CHD and CVD	Scheffers et al. (2019)
Type 2 diabetes mellitus	5 prospective cohort studies (194,019 participants and 13,013 type 2 diabetes mellitus cases)	Berries	↓ by 5% with a 17 g/day increment	Guo et al. (2016)
Breast cancer	90,476 premenopausal women (aged 27–44 years old from the Nurses' Health Study II in 1991) and 44,223 of women in 1998	Fruits (2.9 servings/day), particularly grapes, banana, and apple during adolescence; kale and oranges during early adulthood	↓ 25% risk of breast cancer	Farvid et al. (2016)

CHD coronary heart disease, *CVD* cardiovascular disease, *LDL-C* low-density lipoprotein cholesterol, *TC* total cholesterol, *TNF-α* tumor necrosis factor-α

adulthood was associated with a 25% lower risk of breast cancer (Farvid et al. 2016). Research evidence has suggested that phytochemicals in fruits and vegetables can have overlapping and complementary mechanisms of action, such as hormone metabolism, scavenging of oxidative agents, modulation of detoxification enzymes, activation of the immune system, and modulation of gene expression in apoptosis and cell proliferation (Waladkhani and Clemens 1998). The short-chain fatty acids are produced when the dietary fiber in fruits and vegetables undergo fermentation by gut microbiota. The short-chain fatty acids such as propionate (Yang et al. 2017), butyrate (Han et al. 2017), and acetate (Schug et al. 2016) may have protective effects against cancers. Collectively, bioactive compounds in fruits and vegetables might be a useful nutritional intervention for the attenuation of age-related diseases (Alissa and Ferns 2017). However, the bioactive components responsible for the effects that we stated above require further elucidation.

6.5.2 *Whole Grains*

Given the role of diet, long-term intake of staple food with functional properties as a daily dietary habit has gained more acceptability rather than drug treatment in a disease setting. Whole grains are rich in dietary fiber with some vitamins, minerals, and phytochemicals. Grains contain various bioactive compounds beneficial to human health besides their nutritional contents. Among these components, arabinoxylans are a significant source of fiber. The anti-inflammatory effect and antioxidant activities in polyphenols could be attributed to their chemical structures (Zhang and Tsao 2016). Similar to other secondary metabolites, polyphenols are the frontline defense of plants. They act as a predominant non-nutrient bioactive group in the human diet, primarily found in plants (Marilo et al. 2018). A highly conjugated system with several hydroxyl groups and aromatic characteristic make these polyphenols a good hydrogen atom and electron donor by neutralizing free radicals (Tsao 2010). These bioactive compounds may serve as a preventative strategy in combating oxidative stress by suppressing the proinflammatory signaling transductions (Tan et al. 2018b).

Whole grains have beneficial antioxidants and biochemical effects on oxidative stress-induced diseases (Table 6.6). Data from randomized, double-blind, controlled cross-over trial involving 14 middle-aged obese and diabetic adults (age = 38 ± 2 years; BMI = 34.0 ± 1.1 kg/m^2) revealed that intakes of whole grains (50 g per 1000 kcal) for 8 weeks reduced risk of diabetes by modulating peripheral insulin resistance and postprandial blood glucose (Malin et al. 2018). A human study conducted in healthy men (age = 21 ± 2 years; BMI = 21.4 ± 1.0 kg/m^2) revealed that plasma glucose response during oral glucose tolerance test (OGTT) after the intake of cooked barley kernels (86 g) was 29% smaller compared to those who consume 105 g white wheat bread (Priebe et al. 2010). The improvement in plasma glucose response is more likely due to the fermentation of soluble fiber in whole grains (Higgins 2012). The dietary fibers and phenolic are metabolized by the gut microbiota and converted into short-chain fatty acids (Fardet 2010) to facilitate gut permeability (Suzuki et al. 2008), immune homeostasis (Furusawa et al. 2013), lipid and glucose metabolism (Priebe et al. 2010), and gut hormone secretion (Wichmann et al. 2013). Figure 6.3 showed the mechanisms of whole grains involved in the modulation of age-related diseases.

Notably, data from a randomized cross-over trial involving 60 Danish adults at risk of developing metabolic syndrome revealed that intake of whole grains (179 ± 50 g/day) for two 8-week dietary intervention periods did not significantly alter glucose homeostasis. The study further demonstrated that consumption of whole grain decreased body weight, CRP, and IL-6 levels (Roager et al. 2019). This finding implies that whole grain can reduce systemic low-grade inflammation compared to a refined grain diet. In another study, Hajihashemi and Haghighatdoost (2019) evaluated the whole grain in relation to inflammatory markers and proinflammatory cytokine production. The data involving 13 randomized clinical trials with 466 participants demonstrated that whole grain consumption significantly reduced the serum concentration of IL-6 and CRP but did not result in a significant decrease in serum concentration of TNF-α. Similarly, a systematic review and dose-response meta-analysis included 19 cohort studies involving 1,041,692 participants and 96,710 deaths in total suggested that intake of 28 g/day whole grain could reduce the total mortality by 9%, CVD mortality by 14%, and cancer mortality (3%) (Zhang et al. 2018). Hollænder et al. (2015) further demonstrated that intakes of whole grains (28–213 g/day for 6–8 weeks) reduced TC and LDL-cholesterol compared to those who never consume whole grains, suggesting the cardioprotective potential of whole grains intake. A similar dietary intake was also found to improve glucose homeostasis and postprandial levels (Hollænder et al. 2015; Marventano et al. 2017). By contrast, refined grain intakes (3 servings/day), particularly from white rice were found to increase the risk of type 2 diabetes mellitus (Aune et al. 2013a). Such findings imply that coarsely milled grains or whole kernels tend to have a lower glycemic index (GI) and contain a higher amount of phytochemicals and fibers with antioxidant and anti-inflammatory properties compared to refined grains.

Table 6.6 The effects of whole grains on human health

Metabolic ailments	Study conditions	Treatment	Findings	References
Type 2 diabetes mellitus	Cross-over design (Healthy men)	Cooked barley kernels (86 g)	↓ 29% of OGTT plasma glucose response compared to those who consume 105 g white wheat bread	Priebe et al. (2010)
Type 2 diabetes mellitus	16 cohort studies	Refined grain intakes (3 servings/day)	↑ Risk of type 2 diabetes mellitus	Aune et al. (2013a)
Type 2 diabetes mellitus	Randomized, double-blind, controlled cross-over trial (obese and diabetic adults)	(50 g per 1000 kcal) for 8 weeks	↓ Risk of diabetes	Malin et al. (2018)
Type 2 diabetes mellitus	Randomized cross-over trial (60 Danish adults at risk of developing metabolic syndrome)	179 ± 50 g/day for two 8-week dietary intervention periods	No effect on glucose homeostasis ↓ Body weight, CRP, and IL-6 levels	Roager et al. (2019)
CVD and cancer	19 cohort studies	28 g/day	↓ Total mortality by 9%, CVD mortality by 14%, and cancer mortality (3%)	Zhang et al. (2018)
CVD	24 randomized controlled studies	28–213 g/day for 6–8 weeks	↓ Total cholesterol, and LDL cholesterol	Hollænder et al. (2015)

CVD cardiovascular disease, *CRP* C-reactive protein, *IL-6* interleukin-6, *LDL* low-density lipoprotein, *OGTT* oral glucose tolerance test

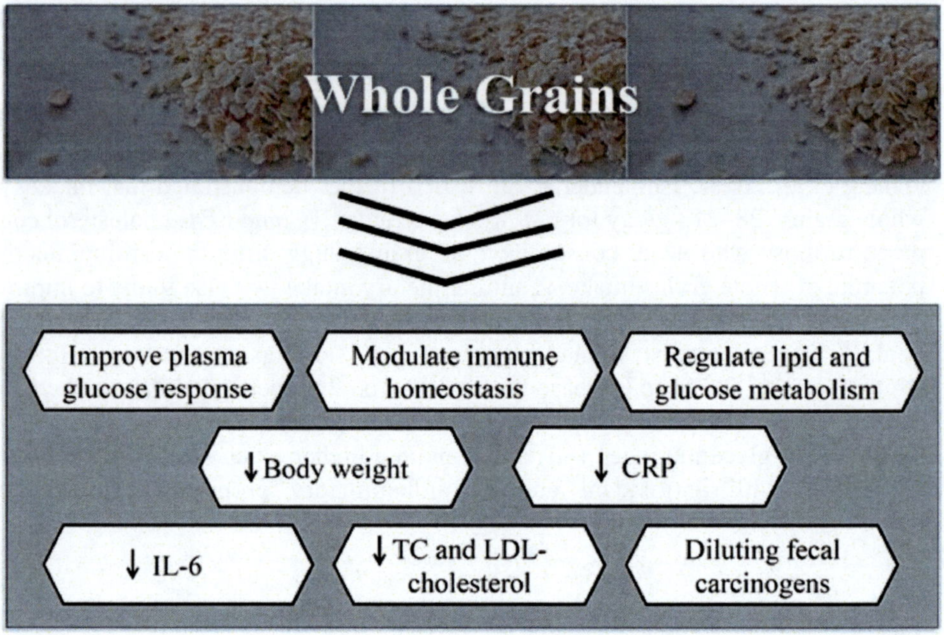

Fig. 6.3 The mechanisms of whole grains involved in the modulation of age-related diseases. *CRP* C-reactive protein, *IL-6* interleukin-6, *LDL* low-density lipoprotein, *TC* total cholesterol

The implications of dietary fibers in relation to age-related diseases have also been widely studied in a human study (Slavin 2013). Data from the meta-analysis and systematic review involving 4635 adults of 58 randomized controlled trials included a study from inception to February 28, 2018, and 185 prospective studies from inception to April 30, 2017 suggest a 15–30% decrease in colorectal cancer, type 2 diabetes, stroke incidence and mortality, and CHD when daily intake of dietary fiber was 25–29 g (Reynolds et al. 2019). Indeed, food containing fibers need to be chewed before passing via the stomach and small bowel, which may affect lipid absorption, insulin and glucose responses, and satiety (Grundy et al. 2016). The previous study suggested that the whole foods that require chewing can retain much of their physical structures in the gut, and are more likely to enhance satiety via different mechanisms, and ultimately modulate the lipid and carbohydrate metabolism and weight loss (Mattes and Dreher 2010; Tan et al. 2018b). In the large bowel, the fibers are almost completely broken down by the resident microflora in an anaerobic reaction known as fermentation (Hur and Lee 2015). In addition, the gut microbiota also plays a crucial role in human health, such as xenobiotics metabolism, vitamin synthesis, and gut immune system development. The microbiome digested carbohydrates and fibers and breakdown in the small bowel. The availability of fiber in the diet may enhance the metabolism of the gut microbiome and thus protect against colorectal cancer (Shanahan et al. 2017). Substantial evidence highlights that whole grain foods could reduce the risk of colorectal cancer by diluting fecal carcinogens, increasing stool bulk, reducing transit time, and hence decreasing the interaction between carcinogens and the lining of the colorectum (Slavin 2013). What is apparent from studies on whole grains is that it could reduce oxidative stress and therefore prevent or at least delay oxidative stress-associated complications. However, further studies are required to explore the types of whole grains and the bioavailability of their bioactive compounds and metabolites in subjects with one or more risk factors for chronic diseases through long-term interventional studies.

6.5.3 Milk and Dairy Products

Milk has become an important dietary component in the diet for nearly 6 billion people (FAO 2012). Many dairy products have been produced and consumed worldwide including butter, kefir, yogurt, cheese, and cream. Therefore, the impact of dairy products and milk on human health is vitally important (Table 6.7 and Fig. 6.4). Randomized clinical trials involving 903 healthy adolescents showed that intake of milk and dairy products for at least 2 serving/day (1 serving = 125 g of yogurt, 200 mL of milk, or 28 g of cheese) can significantly reduce the body fat and promote weight loss (Abreu et al. 2012a, b). Of all the milk components, dairy protein is the most common nutrient found in milk and dairy products. Dairy protein has been suggested as a reducer for body weight and adipose mass (Vergnaud et al. 2008; Bendtsen et al. 2013) in both diabetic patients (Shahar et al. 2007) and overweight and obese individuals (Faghih et al. 2011; Josse et al. 2011; Sanders 2012; Abargouei et al. 2012). Whey protein appears to be substantially effective in weight control (Pal et al. 2010; Sousa et al. 2012) by increasing satiety and decreasing appetite (Sousa et al. 2012). In addition, whey protein can also suppress gastric (Pal et al. 2010) and promote glucagon-like peptide 1 (GLP-1) (Brubaker and Anini 2003; Hall et al. 2003) and glucose-dependent insulinotropic polypeptide (GIP) secretion (Samra et al. 2007), accompanied by the inhibition of ghrelin secretion (Bowen et al. 2006), and subsequently contribute synergistically to the weight control.

The bioactive components in milk theoretically exert chemopreventive activity. Data from cohort and case-control studies have demonstrated that consumption of milk fat-containing dairy products was inversely associated with breast (>3 servings/day of total dairy food) and colon (400 g/day of total dairy products and 200 g/day of milk intake) cancers (Dong et al. 2011; Pala et al. 2011; Zhang et al. 2011; Aune et al. 2012; Chagas et al. 2012). This protective effect can be ascribed to vitamin D, calcium, and conjugated linoleic acid (Norat and Riboli 2003; Średnicka-Tober et al. 2016). Calcium has been suggested to suppress cell growth, activate apoptosis and differentiation in the mammalian gland and gastrointestinal tract, interact with fatty acids and biliary salts in the intestine, and reduce noxious effects in the mucosa (Visioli and Strata 2014). However, Bermejo et al. (2019) showed that high consumption of whole milk increased the risk of bladder cancer. The calcium may bind to vitamin D and insulin-like growth factor-1 (IGF-1) and promote the risk of prostate cancer (Giovannucci 1998; Rodriguez et al. 2003). In this regard,

Table 6.7 The effects of milk and dairy products on human health

Metabolic ailments	Study conditions	Treatment	Findings	References
Obesity	Randomized clinical trials	At least 2 serving/day (1 serving = 125 g of yogurt, 200 mL of milk, or 28 g of cheese)	↓ Body fat and promote weight loss	Abreu et al. (2012a, b)
Cancer	Cohort and case-control studies	>3 servings/day of total dairy food	↓ Breast cancer	Dong et al. (2011); Pala et al. (2011); Zhang et al. (2011); Aune et al. (2012); Chagas et al. (2012)
		400 g/day of total dairy products and 200 g/day of milk intake	↓ Colon cancer	
Cancer	Meta-analysis of 32 prospective studies	Total dairy products (400 g/day), total milk (200 g/day), low-fat milk (200 g/day), cheese (50 g/day), and dietary calcium (400 mg/day)	↑ Risk of total prostate cancer	Mandair et al. (2014); Aune et al. (2015)
Cancer	Meta-analysis of 26 epidemiologic studies (8 cohort and 18 case-control)	High consumption of whole milk (~220 mL/day)	↑ Risk of bladder cancer	Bermejo et al. (2019)
Cancer	Meta-analysis of 26 epidemiologic studies (8 cohort and 18 case-control)	Medium consumption of total dairy products (~345 g/day), medium and high consumption of milk (medium = ~227 mL/day; high = ~336 mL/day) and fermented dairy products (medium = ~67 g/day; high = ~160 g/day)	↓ Risk of bladder cancer	Bermejo et al. (2019)
Type 2 diabetes mellitus	Randomized clinical trials (Type 2 diabetic subjects)	Whey protein (27.6 g) in the diet (breakfast = 102 g white wheat bread or lunch = 52.2 g instant potato powder and 50 g meatballs)	↓ 21% of blood glucose response	Frid et al. (2005)
Type 2 diabetes mellitus	Meta-analysis of 17 cohort studies	200 g low-fat dairy products/day, 400 g total dairy products/day, or 50 g cheese/day	↓ Risk of type 2 diabetes	Aune et al. (2013b)
Type 2 diabetes mellitus	Randomized double-blind placebo-controlled clinical trial (60 diabetic patients)	Probiotic fermented milk (kefir) (600 mL/day for 8 weeks)	↓ HbA1c level compared to control group	Ostadrahimi et al. (2015)

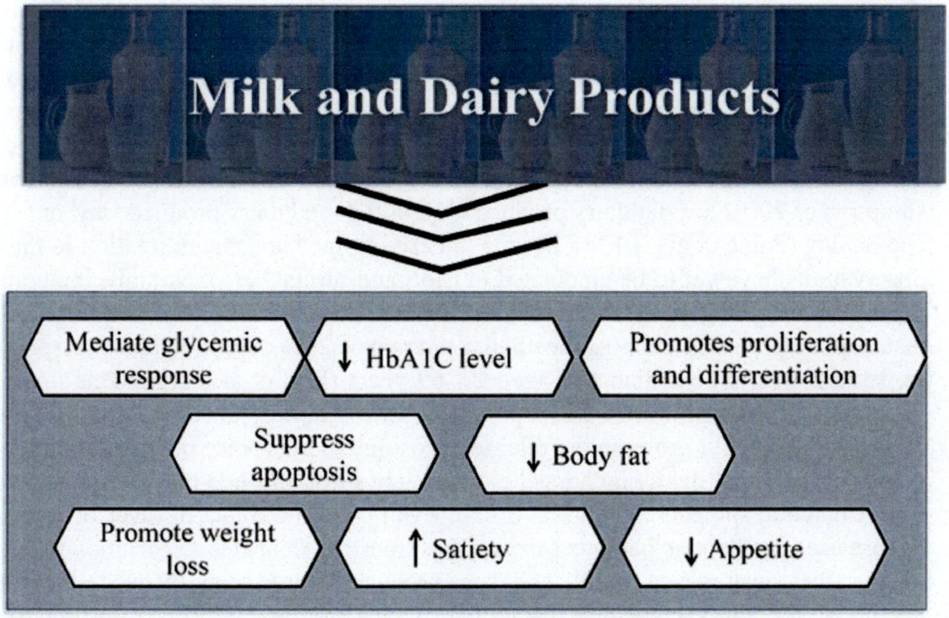

Fig. 6.4 The mechanisms of milk and dairy products involved in the modulation of age-related diseases

intakes of total dairy products (400 g/day), total milk (200 g/day), low-fat milk (200 g/day), cheese (50 g/day), and dietary calcium (400 mg/day) were associated with increased risk of total prostate cancer (Mandair et al. 2014; Aune et al. 2015). Although research evidence demonstrated that dairy intake was positively related to prostate cancer, not all data showed such a link. Huncharek et al. (2008) did not find an association between prostate cancer and dairy or milk intake.

Besides the bioactive components, relatively little attention has been paid to bioactive hormones. The extent and the nature of the role of IGF-1 in the development of numerous diseases are highlighted here. IGF-1 is increased proportionally with the presence of casein in milk. In particular, milk protein promotes postprandial hyperinsulinemia and shifts the growth hormone (GH)/IGF-1 axis, and subsequently leads to an irreversible increase of IGF-1 serum levels (Ludwig 2002). In addition, it has been demonstrated that more than 90% of this hormone is bound to IGF-binding protein 3 (IGFBP-3), the rest is bound to IGFBP-1, -2, -4, and -6. IGF-1 receptor (IGF-1R) involves in the regulation of many signal transductions. Because of their tyrosine kinase activity, IGF-1R can form a heterodimer with the insulin receptor. The IGF-1R-mediated signal transduction pathway mainly via the activation of Ras/Raf/MAP kinase signaling cascade concomitantly with the stimulation of PI3K signaling pathway, suggesting that these pathways may stimulate lipogenesis and cellular proliferation and suppress apoptosis. Since IGF-1 promotes proliferation and differentiation and suppresses apoptosis, implies that this hormone may influence the growth of tumors (Duan et al. 2010). In support of this, several studies demonstrated a positive correlation between high IGF-1 serum levels and the risk of colorectal, prostate, esophageal, and breast cancers (Shiratsuchi et al. 2011; Walsh and Damjanovski 2011; Doyle et al. 2012; Cao et al. 2015). In short, consumption of milk and dairy products does not consistently increase the risk of cancers, the evidence in favor or against such effect remains too limited to draw a firm conclusion.

In addition to the effects mentioned above, data from a meta-analysis of 17 cohort studies have shown an inverse relationship between risk of type 2 diabetes and consumption of 200 g low-fat dairy products/day, 400 g total dairy products/day, or 50 g cheese/day (Aune et al. 2013b), but the underlying mechanisms in relation to these observations have yet to be elucidated (Visioli and Strata 2014). Notably, including whey protein (27.6 g) in the diet (breakfast = 102 g white wheat bread or lunch = 52.2 g instant potato powder and 50 g meatballs) can reduce 21% of blood glucose response in patients with type 2 diabetes, aged 27–69 years (Frid et al. 2005). This finding suggests that the addition of whey protein to meals rapidly absorbed and digested carbohydrates could promote the release of insulin and decrease postprandial blood glucose excursion after a lunch meal consisting of meatballs and mashed potatoes in type 2 diabetic subjects (Frid et al. 2005). Even though the fact in favor or against lactose used by diabetic patients is scant, the American Diabetes Association recommended the consumption of milk and dairy products. This is partially due to the milk has a relatively low GI and the dairy proteins which are mainly from casein, and thus mediate glycemic response through stimulation of certain plasma amino acids and incretins (Gunnerud et al. 2012). A randomized double-blind placebo-controlled clinical trial involving 60 diabetic patients aged 35–65 years revealed that intake of probiotic fermented milk (kefir) (600 mL/day for 8 weeks) significantly decreased the HbA1c level compared to the control group, suggesting that probiotic fermented milk can be useful as adjuvant therapy in the treatment of diabetes (Ostadrahimi et al. 2015). Although limited available evidence to draw a firm conclusion, previous studies suggest that milk and dairy products may be potential to some population segments. Collectively, further studies are warranted to evaluate the implication of specific types of dairy products in different populations on metabolic ailments and any gender-specific recommendations.

6.5.4 Olive and Fish Oils

Research evidence revealed the beneficial effects of dietary monounsaturated fatty acids (MUFA) in regulating body weight and cardiometabolic risk factors (Viitasalo et al. 2016). The main active components in olive are oleic acid (18:1 n-9), squalene, and phenolic compounds such as hydroxytyrosol, oleuropein, and tyrosol (Roselló-Soto et al. 2015). Both oleuropein and hydroxytyrosol have established bioactivity, primarily associated with their antioxidant properties. These compounds appear to play a role in several diseases in preclinical studies, especially in cardiovascular and metabolic disorders (Bulotta et al. 2014).

Adherence to the Mediterranean dietary pattern is related to a lower risk of obesity and metabolic ailments (Álvarez-Pérez et al. 2016). Olive phenolics act as α-glucosidase and α-amylase inhibitors to hinder carbohydrate digestion and thus suppressing carbohydrate uptake and absorption (Collado-González et al. 2017) (Fig. 6.5). Both *in vitro* and *in vivo* studies have demonstrated that olive polyphenols can suppress pre-adipocyte differentiation, inhibit lipogenesis, promote lipolysis, and induce adiponectin secretion (Fig. 6.5), possibly mediated via inhibition of adipogenic gene expressions such as leptin, FAS, CCAAT-enhancer-binding proteins (C/EBPα), and PPAR-γ, and upregulation the genes related to the inhibition of adipogenesis like GATA2, GATA3, WNT3A, SFRP5, HES1, and SIRT (Higa et al. 2014; Shen et al. 2014; Stefanon and Colitti 2016).

Furthermore, intakes of olive oil can also reduce insulin resistance. Dysfunction of pancreatic β-cell is a major determinant for the development of type 2 diabetes, while endoplasmic reticulum (ER) stress plays a crucial role in the β-cell failure. The bioactive components in olive oil can attenuate ER stress-mediated apoptosis and unfolded protein response via inhibition of JNK phosphorylation (Lee et al. 2016).

In addition to the effects observed in olive oil, fish oil may also emerge as a key element for the modulation of age-related diseases. The main elements in fish oil are n-3 polyunsaturated fatty acids (PUFA) including docosahexaenoic acid (DHA) and eicosapentaenoic acid (EPA). Fish oil supports various body functions and reduces several diseases such as diabetes, cancer, CVD, and other conditions, for instance, inflammation by downregulating CRP, TNF-α, and IL-6 expression (Ellulu et al. 2015).

Observational studies involving young male adults (18–25 years old) have demonstrated that daily fish oil supplements (up to 3 g omega-3 fatty acid for 12 weeks) can reduce serum TG effectively (Roke et al. 2015). Omega-3 fatty acids in fish oil are ligands for PPAR, which may induce changes in cardiometabolic markers (Binia et al. 2017). Patients with coronary artery disease supplemented with fish oil (≥1000 mg/day for at least 1 month) improved intermediate-density lipoprotein cholesterol, TG, and very-low-density lipoprotein (VLDL)-cholesterol (Franzese et al. 2015). Table 6.8 summarizes the effects of fish oil on age-related diseases in *in vivo* and human studies.

Omega-3 PUFAs are not only improved lipid profiles, they also having beneficial effects on type 2 diabetes. Several studies as reported by Kröger and Schulze (2014) have shown that long-chain n-3 PUFA from dietary fish intake can reduce the risk of type 2 diabetes. In this regard, a key mode of action to explain this relationship is via the reduction of 7α-hydroxylase, 3-hydroxy-3-methylglutaryl-CoA, and LDL-cholesterol (Scicchitano et al. 2014). Consistent with the study reported by Kröger and Schulze (2014), Gao et al. (2017) also found that diet enriched with fish oil (1–4 g/day for 4–24 weeks) enhanced insulin sensitivity among people with at least one symptom of metabolic disorders. The insulin-sensitizing effects of n-3 PUFA may be partially due to the anti-inflammatory properties which are mediated by G-protein coupled receptor GPR120 (Oh et al. 2014).

In addition to the effects mentioned above, fish oil has the potential to protect against cancers. Dietary administration of fish oil for 21 weeks at 15 mg/kg body weight of azoxymethane/2% dextran sodium sulfate (AOM/DSS)-induced rats can

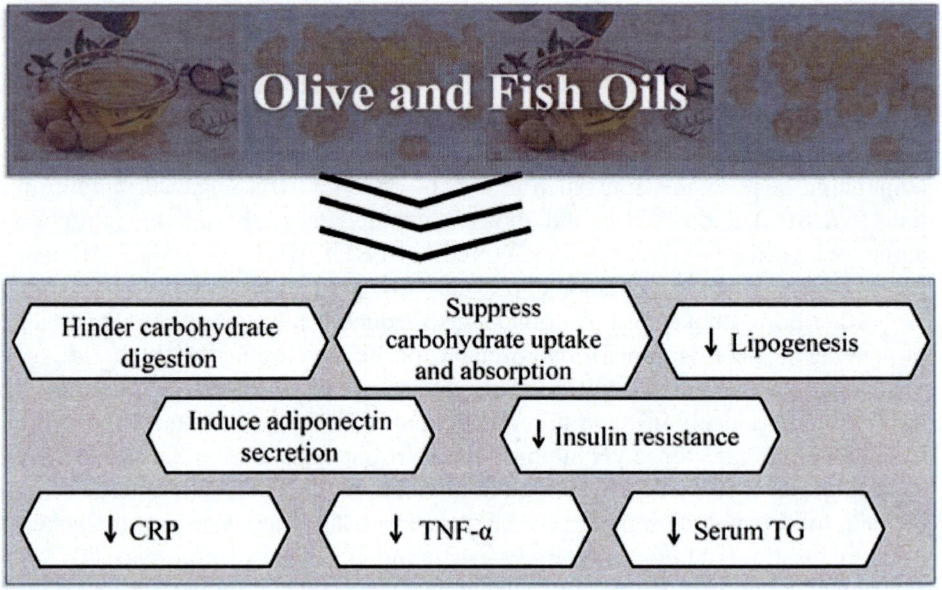

Fig. 6.5 The mechanisms of olive and fish oils involved in the modulation of age-related diseases. *CRP* C-reactive protein, *TG* triglycerides, *TNF-α* tumor necrosis factor-α

significantly reduce the mean colon tumor volume and tumor multiplicity (Djuric et al. 2017) (Table 6.8). Fish oil modulates cancer via several mechanisms including promotion of BAD-dependent apoptosis via PI3K/AKT survival pathway, suppression of raft-associated signal transduction, activation of Nrf2, and resolution of inflammation by D-resolvins (RvD2 and RvD1), protectin (PD1), and E-resolvins (RvE1 and RvE2) through lipoxygenase (LOX) and COX pathways (Aucoin et al. 2017). Collectively, the available research evidence suggests that MUFA and other main active components appear to be favorable for its beneficial effects, suggesting that including olive oil or fish oil in a diet may extend cardioprotective effects and reduce other metabolic disorders beyond those defined for a contemporary healthy dietary pattern. Further studies are warranted to elucidate whether the antioxidant or its PUFA content is responsible for its beneficial effects.

6.6 Dietary Antioxidants and Age-Related Diseases

6.6.1 Curcumin

Curcumin is a bioactive compound extracted from turmeric rhizome *Curcuma longa* L., is a perennial herbaceous plant of the ginger family (Zingiberaceae) (Priyadarsini 2014). It is widely utilized in India, China, and Southeast Asia for medicinal purposes and in food colorants (Kocaadam and Şanlier 2017). Curcumin usage dates

Table 6.8 The effects of fish oil on age-related diseases in *in vivo* and human studies

Metabolic ailments	*In vivo* models and human studies	Treatment	Findings	References
CVD	Young male adults	Fish oil supplements (up to 3 g omega-3 fatty acid for 12 weeks)	↓ Serum TG	Roke et al. (2015)
CVD	Patients with coronary artery disease	Fish oil (≥1000 mg/day for at least 1 month)	Improved intermediate-density lipoprotein cholesterol, TG, and VLDL-cholesterol	Franzese et al. (2015)
Type 2 diabetes mellitus	People with at least one symptom of metabolic disorders	Diet enriched with fish oil (1–4 g/day for 4–24 weeks)	Enhanced insulin sensitivity	Gao et al. (2017)
Colon cancer	AOM/DSS-induced rats	Fish oil for 21 weeks on 15 mg/kg body weight	↓ Mean colon tumor volume and tumor multiplicity	Djuric et al. (2017)

AOM azoxymethane, *CVD* cardiovascular disease, *DSS* dextran sodium sulfate, *TG* triglycerides, *VLDL* very-low-density lipoprotein

back more than 2500 years in Asia, particularly in traditional Indian Medicine (Ayurveda) (Gupta et al. 2013). It has been used for the treatment of numerous diseases including stress, depression, rheumatism, eye infections, skin diseases, burns, and wounds (Singh 2007; Hatcher et al. 2008; Qin et al. 2009). Several studies reported by Farooqui and Farooqui (2019) and Sharifi-Rad et al. (2020) evaluated the pharmacological and biological activity of curcumin, which having a polyphenol structure, in relation to several receptors, growth factors, transcription factors, enzymes, kinases, and cytokines. The data showed that curcumin has some hypoglycemic (Rivera-Mancía et al. 2018), hepatoprotective (Khan et al. 2019), renoprotective (Huang et al. 2020), immunomodulatory (Yang et al. 2020), antioxidant (Boroumand et al. 2018), anti-inflammatory (Zhang et al. 2015), and antimicrobial effects (Zhu et al. 2015).

The data from *in vitro* studies have demonstrated the anticancer potential of curcumin in reducing oxidative stress by mediating several signaling pathways (Wang et al. 2018b; Lin et al. 2019). Curcumin was shown its cytotoxicity against androgen-independent prostate cancer cell lines (PC-3 and DU-145) and androgen-dependent prostate cancer cell line (LNCaP) (Jordan et al. 2016). The animal model studies further revealed the chemopreventive potential of curcumin in chemical-induced animal models of the liver (Damiano et al. 2021), colon (Byun et al. 2015), and skin (Tsai et al. 2012) cancers. This favorable effect could be attributed to its antioxidant effect. Yet, the chemopreventive ability of curcumin is mediated via several mechanisms such as triggers heme oxygenase (HO) phase II enzymes and Nrf2/Keap signaling pathway (Fig. 6.6), the transcriptional regulator for detoxification of ROS

(Muhammad et al. 2018; Ashrafizadeh et al. 2020). However, a study reported by Huang et al. (1997) has shown that feeding mice with curcumin showed a little or no effect on 7,12-dimethylbenz[a]anthracene (DMBA)-induced breast carcinogenesis and 4-(methylnitosamino)-1-(3-pyridyl)-1-butanone (NNK)-induced lung cancer. The data further showed that poor circulating bioavailability of curcumin may contribute to the lack of the inhibition of breast and lung cancers (Huang et al. 1997).

The anticancer activity of curcumin was not only observed in both *in vivo* and *in vitro*, the efficacy of curcumin has also been demonstrated in clinical trials. Data from a phase II clinical study involving 44 colon cancer lesions patients showed that curcumin (4 g/day for 30 days) reduced 40% number of lesions (Carroll et al. 2011). In another clinical study involving 199 localized prostate cancer patients, with an average age of 74 years, who are given an oral capsule containing a mixture of turmeric (100 mg), broccoli (100 mg), pomegranate (100 mg), and green tea (100 mg) for 6 months showed a median increase in prostate-specific antigen (PSA) was 63.8% lower than the placebo group on primary active surveillance and those experiencing a PSA relapse after radiotherapy (Thomas et al. 2014). Indeed, several initial stage studies have shown a favorable response in the reduction of symptoms and adverse outcomes related to radiotherapy (Amir et al. 2018). A study that analyzed 50 patients with neck and head cancer undergoing radiotherapy found that cream containing turmeric (2 g) and sandal wood oil (2 mL) (5 times/day for 2 weeks) is effective in preventing radiodermatitis (Palatty et al. 2014). Indeed, phase I and II clinical trials have shown that curcumin is safe and exhibit some therapeutic effects in certain cancer (Cheng et al. 2001; Tan and Norhaizan 2019c). The data showed that consumption of curcumin (500–8000 mg/day for 3 months) may improve the histology of precancerous lesions in resected uterine cervical intraepithelial neoplasms, intestinal metaplasia, oral leucoplakia, and bladder cancer patients (Cheng et al. 2001). Likewise, patients with advanced pancreatic cancer who received 8 g curcumin reduced COX-2 and NF-κB expression (Dhillon et al. 2008). The implication of curcumin combination chemotherapy has also been evaluated in several clinical studies (Table 6.9). The phase I escalated clinical trial measured the tolerability and feasibility of combination curcumin and docetaxel in relation to metastatic or advanced breast cancer among 14 patients. The patients were given 500 mg/day of curcumin for 7 days and increased until dose-limiting toxicity. The study showed that combination therapy of curcumin (500 mg/day) with docetaxel (100 mg/m^2) show a promising biological response by decreasing carcinoembryonic antigen (CEA) tumor marker (Bayet-Robert et al. 2010). The data further revealed that the maximum concentration of curcumin that could be tolerated is 6000 mg/day for 7 consecutive days in every 3 weeks when combined with the standard dosage of docetaxel (Bayet-Robert et al. 2010). Another phase I clinical study involving 21 gemcitabine-resistant pancreatic cancer patients revealed that oral administration of 8 g/day curcumin and gemcitabine [(1000 mg/m^2 on day 1 and 8) and (60 mg/m^2 of S-1 orally for 14 consecutive days every 3 weeks)] was demonstrated no dose-limiting toxicities (Kanai et al. 2011). Such findings implied that administration of curcumin (8 g/day) orally is feasible and safe in pancreatic cancer patients (Kanai et al. 2011). The efficacy and safety of curcumin were further

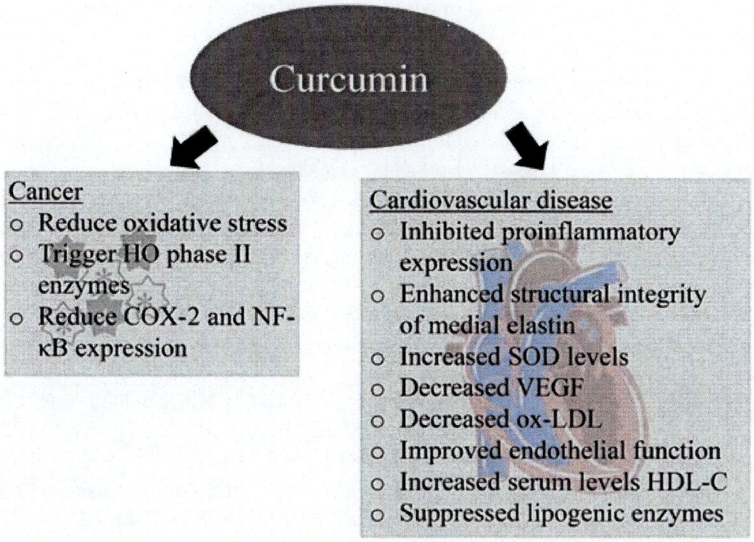

Fig. 6.6 Mechanisms of curcumin in the modulation of cancer and cardiovascular disease. *COX-2* cyclooxygenase-2, *HDL-C* high-density lipoprotein cholesterol, *HO* heme oxygenase, *NF-κB* nuclear factor-kappa B, *ox-LDL* oxidized low-density lipoprotein, *SOD* superoxide dismutase, *VEGF* vascular endothelial growth factors

evaluated in a phase II trial in metastatic and advanced pancreatic cancer patients. The study found that a combination of gemcitabine (10 mg/m^2) and curcumin (2000 mg/die continuously (4 capsules, each of 500 mg/day)) showed the overall survival and median progression-free survival were 10.2 and 8.4 months, respectively. This finding suggests that combination therapy of curcumin and gemcitabine is efficient and safe to translate a good response rate in first-line therapy of advanced pancreatic cancer (Pastorelli et al. 2018). While for colorectal liver metastases, James et al. (2015) found that administration of 5 μM curcumin could improve FOLFOX-based chemotherapy (5 μM 5-FU and 2 μM oxaliplatin) in colorectal liver metastasis patients. Collectively, the synergistic effects played by curcumin combination chemotherapeutic drugs on cancer worth further investigation in large clinical trials.

Curcumin is not only exerted anticancer effects (Adiwidjaja et al. 2017), existing research evidence has been demonstrated the benefit of curcumin in treating osteoporosis (Jiang et al. 2021), CVD (Pourbagher-Shahri et al. 2021), arthritis, liver problem (Nabavi et al. 2014), neurological disorders (Bagheri et al. 2020), diabetes (Shome et al. 2016), and chronic inflammatory disease (He et al. 2015), suggesting the numerous functional potential of curcumin. Specifically, substantial studies have shown that curcumin exerts preventive effects toward the diabetic and stroke cardiovascular complications, aortic aneurysm, atherosclerosis, myocardial infarction, drug-induced cardiotoxicity, heart failure, and cardiac hypertrophy (Aggarwal and Harikumar 2009; Campbell and Fleenor 2018; Farkhondeh and Samarghandian 2016; Karuppagounder et al. 2017). Aortic aneurysm, one of the cardiovascular

Table 6.9 Clinical studies conducted in curcumin combination chemotherapy

Subjects	Sample size	Treatment	Outcomes	References
Metastatic or advanced breast cancer patients	14	Curcumin (500 mg/day) and escalated until a dose-limiting toxicity + docetaxel (100 mg/ m^2) for 7 days every 3 weeks	↓ CEA tumor marker	Bayet-Robert et al. (2010)
Gemcitabine-resistant pancreatic cancer patients	21	8 g/day curcumin and gemcitabine (1000 mg/m^2 on day 1 and 8) and (60 mg/m^2 of S-1 orally for 14 consecutive days every 3 weeks)	Median survival time was 161 days and 1-year survival rate was 19%	Kanai et al. (2011)
Chronic myeloid leukemia	50	Turmeric powder (5 g 3 times/ day) + imatinib (400 mg twice/day) for 6 weeks	↓ Nitric oxide levels	Ghalaut et al. (2012)
Colorectal liver metastases	12	5 μM curcumin +5 μM 5-FU + 2 μM oxaliplatin	Curcumin enhance FOLFOX-based chemotherapy	James et al. (2015)
Metastatic and advanced pancreatic cancer patients	44	Gemcitabine (10 mg/m^2) and curcumin (2000 mg/die continuously (4 capsules, 500 mg/day each))	Overall survival and median progression-free survivals were 10.2 and 8.4 months, respectively	Pastorelli et al. (2018)
Metastatic colorectal cancer patients	28	FOLFOX +2 g oral curcumin/ day	The addition of daily oral curcumin to FOLFOX chemotherapy was tolerable and safe	Howells et al. (2019)

CEA carcinoembryonic antigen, *FOLFOX* folinic acid/5-fluorouracil/oxaliplatin, *5-FU* 5-fluorouracil

phenomenon, can cause death in case of rupture (Chen et al. 2011). In the last few decades, several key elements have been suggested in the development of aortic aneurysms such as loss of smooth muscle cells, destructive connective tissue remodeling, and chronic inflammation in the aortic wall (Guo et al. 2006). A study reported by Parodi et al. (2006) showed that feeding curcumin (100 mg/kg/day for 14 days) inhibited proinflammatory expression, enhanced the structural integrity of medial elastin, and decreased the aortic diameter enlargement via suppression of AP-1 and NF-κB. The data from animal study further demonstrated that pretreatment of curcumin (100 mg/kg/day for 4 weeks) alleviated the stimulation of ERK signaling pathway, mediated the production of pro-inflammatory mediators and macrophage infiltration, decreased the development and progression of the aortic aneurysm, and elevated SOD levels in apoE-deficient ($ApoE^{-/-}$) mice induced with a subcutaneous infusion of angiotensin II (Ang II) (Hao et al. 2014). In line with this, Li et al. (2017b) also found that curcumin (100 mg/kg/day for 4 weeks) decreased the production of VEGF and neovascularization, and reduced the size of thoracic aortic aneurysm in rats. Nonetheless, there was limited clinical trial evidence evaluated on the beneficial effects of curcumin on aortic aneurysms. In the context of atherosclerosis, emerging evidence has suggested that curcumin exerts hypolipidemic activity along with the anti-inflammatory and antioxidant effects, which may lead to the reduction of atherosclerosis (Panahi et al. 2018). The remarkable antioxidant activity of curcumin decreases the production of ox-LDL and lipid peroxidation, which in turn leading to the reduction of inflammatory response. Ultimately, these lead to the progression of atherosclerosis (Panahi et al. 2018). Endothelial dysfunction is demonstrated in patients with CVD such as atherosclerosis (Karimian et al. 2017). In this regard, endothelial cells are potential targets for curcumin to protect against atherosclerosis and enhance endothelial function (Santos-Parker et al. 2017). Pretreatment of curcumin (1, 0.5, and 0.1 μM) significantly decreased the adhesion of monocytes to activated human umbilical vein endothelial cells (HUVECs) via suppression of gene expression implicated in adhesion, for instance, *E*-selectin, ICAM-1, and VCAM-1, which were mediated by TNF-α stimulation (Coban et al. 2012). In another study using a TNF-α stimulation of the HUVECs model showed that pre-exposure of curcumin (0.5–1 μM) decreased monocyte adhesion and endothelial permeability in both flow and static circumstances (Monfoulet et al. 2017). Data from a randomized controlled double-blind prospective study involving 59 healthy adults revealed that curcumin oral supplementation (200 mg/day for 8 weeks) showed a concentration-mediated improvement in endothelial function, suggesting that it may reduce the risk of CVD in healthy adults (Oliver et al. 2016). Similarly, Akazawa et al. (2012) also found that postmenopausal women who received 150 mg/day theracurcumin for 8 weeks improved flow-mediated dilation (FMD). By contrast, it has been shown that postmenopausal and older women (mean 60 years old) have low baseline FMD (˂3%), implied that increased risk of CVD (Akazawa et al. 2012). A 12-week randomized double-blind placebo-controlled study on 118 type 2 diabetes patients showed that curcuminoids (1000 mg/day + piperine 10 mg/day) increased serum levels of high-density lipoprotein cholesterol (HDL-C) and decreased lipoprotein(a) (Lp(a)), non-HDL-C, and TC (Panahi et al. 2017b). Importantly, Lp(a) is a plasma lipoprotein comprising of apolipoprotein(a) covalently interact to apolipoprotein B-100, which is linked to the increased risk of CVD, given the scarcity of pharmacological approach to decrease its activity (van der Valk et al. 2016; Panahi et al. 2017b). In this regard, curcuminoids may represent a plausible strategy to alleviate the complications related to the increased expression of Lp(a) in diabetes mellitus. Such finding indicates that curcuminoids supplementation may reduce the serum levels of atherogenic lipid indices in dyslipidemic patients with type 2 diabetes (Panahi et al. 2017b). While, a randomized clinical trial analyzed of 60 overweight and obese female adolescent revealed that curcumin supplementation (500 mg/day, containing 95% turmeric extract, for 10 weeks) can improve the lipid profiles such as HDL-C levels and TG/HDL-C ratio (Saraf-Bank et al. 2019). Curcumin is believed to have hypolipidemic activity by suppressing lipogenic enzymes and expression such as peroxisome proliferator-activated receptor-α, acetyl-CoA carboxylase, and sterol-regulatory element-binding protein-1, downregulating fatty acid synthase, and increasing lipoprotein lipase and fatty acid β-oxidation (Sahebkar 2014; Panahi et al. 2016). Through modulating ATP-binding cassette subfamily G member 1 (ABCG1) expression, curcumin may increase serum HDL-C levels by stimulating HDL-dependent lipid efflux (Peschel et al. 2007).

Several studies have reported that curcumin supplementation exerts its benefit in decreasing insulin resistance and fasting glucose levels (Chuengsamarn et al. 2012; Na et al. 2013). A study reported by Saraf-Bank et al. (2019) revealed that consumption of curcumin (500 mg/day, containing 95% turmeric extract, for 10 weeks) can significantly increase insulin levels. It has been suggested that curcumin may improve pancreatic function and promote insulin secretion from pancreatic cells (Ghorbani et al. 2014) (Fig. 6.7). A similar dietary intake was also found to increase insulin secretion, possibly by stimulating the glucagon-like peptide-1 secretion (Kato et al. 2017). Nevertheless, some surveys failed to identify any significant decreasing effect on glycemic indices in non-diabetic patients (Tang et al. 2008; Panahi et al. 2016). One of the largest randomized double-blinded placebo-controlled trials with curcumin involving 240 pre-diabetic participants for 9-month period was that of Chuengsamarn et al. (2012). In this phenomenon, curcumin was used for preventive purposes in the pre-diabetic population. Notably, Chuengsamarn et al. (2012) did not identify an association between the curcumin-treated group (1500 mg of curcuminoids/day) and diabetes mellitus after 9 months of treatment. Curcumin has also been suggested to be beneficial in several parameters including improved insulin resistance, increased adiponectin levels, decreased C-peptide, insulin, and glucose concentration, and lowered waist circumference and body weight (Chuengsamarn et al. 2012). The previous study suggests that a protective role of curcumin on diabetes by lowering the circulating IL-6 levels (Usharani et al. 2008; Na et al. 2014). Increased inflammatory cytokine IL-6 levels have been identified as an independent predictor of type 2 diabetes in the European population (Spranger et al. 2003).

Diabetes nephropathy represents one of the severe microvascular complications (Sagoo and Gnudi 2020). Nearly 40% of diabetic patients develop diabetic nephropathy (Tervaert et al. 2010; Schernthaner 2011). The nephroprotective potential of curcumin has been evaluated in human studies. The study reported by Yang et al. (2015) showed that curcumin promotes renal function by modulating the urinary microalbumin and blood urea nitrogen (BUN). The nephroprotective effect is more likely due to the stimulation of the Nrf2 antioxidative system in the lymphocytes of type 2 diabetes mellitus patients treated with curcumin. This mechanism of action has also been demonstrated in animal models of diabetes (Kim et al. 2016). However, the findings in clinical trials are inconsistent to a certain degree. Jiménez-Osorio et al. (2016) did not identify any positive effect on Nrf2 and proteinuria in patients with nondiabetic or diabetic proteinuric chronic kidney disease who received curcumin (320 mg/day for 8 weeks). Similarly, Panahi et al. (2017a) also did not identify an association of curcuminoids and glycosylated hemoglobin or glucose levels in type 2 diabetes mellitus patients who received curcuminoids (1000 mg/day) mixed with piperine for 12 weeks. From the study reviewed, administration of curcumin increased SOD1 activity and total antioxidant capacity and decreased

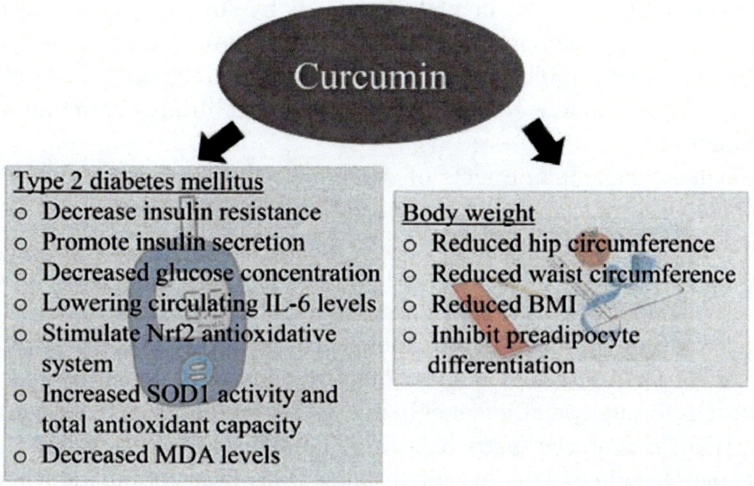

Fig. 6.7 Mechanisms of curcumin in the modulation of type 2 diabetes mellitus and body weight. *BMI* body mass index, *IL-6* interleukin-6, *MDA* malondialdehyde, *Nrf2* nuclear factor erythroid 2-related factor 2, *SOD1* superoxide dismutase 1

malondialdehyde (MDA) levels (Panahi et al. 2017a). Data from *in vitro* study has revealed that curcuminoids can suppress the activity of the cytochromes, for instance, CYP3A4, CYP2E1, CYP2D6, CYP2C9, CYP2C19, CYP2B6, CYP1B1, and CYP1A2, and upregulated the activity of CYP1A1 (Bahramsoltani et al. 2017). Furthermore, clinical studies have demonstrated that curcumin triggers CYP2A6 and suppresses CYP1A2 (Bahramsoltani et al. 2017). This finding suggests a potential pharmacological activity between conventional drugs and curcuminoids that are metabolized by them. For example, gliquidone, glipizide, gliclazide, glimepiride, glibenclamide, and tolbutamide are anti-diabetic drugs metabolized by CYP2C9; while, glibenclamide is metabolized by CYP3A4 (Holstein and Beil 2009). A study reported by Sakunthala Devi et al. (2015) found that curcumin improved the apparent volume of distribution, the mean residence time, and the half-life at the steady manner of glibenclamide, suggesting that such effects are likely due to the reduced metabolism of drugs modulated by the suppression of hepatic and intestinal CYP3A4 (Sakunthala Devi et al. 2015). The beneficial effect of a combination of curcumin and the antidiabetic drug has also been described by Rivera-Mancía et al. (2018). Notably, co-administration of curcumin may increase the bioavailability of glibenclamide (Neerati et al. 2014). The bioavailability of glibenclamide can be influenced by permeability glycoprotein (P-gp)-modulated efflux mechanism (Neerati et al. 2014). In another study, administration of curcumin and gliclazide showed a suppressive reduction in glucose levels for both diabetic and normal rats but no pharmacokinetic interaction was found in rabbits (Vatsavai and Kilari 2016).

In addition, clinical trials have also found that curcumin (500 mg/day, containing 95% turmeric extract, for 10 weeks) reduced hip circumference, waist circumference, and body mass index (BMI) (Saraf-Bank et al. 2019) (Fig. 6.7). A crucial underlying mode of action that has been proposed to explain an inverse association between curcumin and BMI or body weight is by stimulating the metabolic rate (Ejaz et al. 2009) via suppression of adipocytic transcriptional activity including PPAR-γ, and thereby inhibition of preadipocyte differentiation (Ejaz et al. 2009; Zhao et al. 2011). Table 6.10 summarizes the clinical trials of curcumin on metabolic ailments.

Although the beneficial effects of curcumin were demonstrated *in vivo* and *in vitro*, several clinical trials and experimental studies have revealed that the systemic bioavailability of orally administered curcumin is low. In particular, curcumin shows a rapid elimination in the urine, bile, and feces and the serum concentration often does not reach >0.1% of the intake (Anand et al. 2007), which limits the therapeutic potential of curcumin/curcuminoids. Therefore, researchers have attempted to explore the pharmacological and biological activity of curcumin and overcome its drawbacks through an efficient delivery system including co-administration with piperine (Panahi et al. 2017a, b), nanoformulation, or nanoencapsulation (de Souza Ferreira and Bruschi 2019), and phytosomes (a lecithin formulation) (Appendino et al. 2011). A study reported by Steigerwalt et al. (2012) evaluated the lecithinized curcumin delivery system in relation to diabetic retinopathy and microangiopathy. The data showed that lecithinized curcumin (200 mg/day for 4 weeks) with original medication had positive microcirculatory effects on patients suffering from diabetic retinopathy and microangiopathy (Steigerwalt et al. 2012). Emerging curcumin nanoformulations have been developed in recent decades (Wong et al. 2019) (Table 6.11). The studies are mainly focused on shielding curcumin from hydrolysis inactivation as well as enhancing curcumin's solubility and bioavailability (Karthikeyan et al. 2020). A study reported by Sun et al. (2013) found that curcumin solid lipid nanoparticles showed the inhibition of cancer cell proliferation and extended cellular uptake with enhanced chemical stability and dispersibility of the drug. Curcumin solid lipid nanoparticles have demonstrated a high solubility and improved drug release compared to curcumin alone. Curcumin solid lipid nanoparticles significantly triggered the apoptosis in breast adenocarcinoma cells (MDA-MB-231), suggesting that this formulation is useful for cancer treatment (Bhatt et al. 2018). A recent study further demonstrated that a combination of curcumin solid lipid nanoparticles and doxorubicin is used to overcome Pgp-modulated chemoresistance in triple-negative breast cancer cells (Fathy Abd-Ellatef et al. 2020). P-gp is a protein that expels xenobiotics from the intracellular space (Choi 2005), in which the expression and activity are suppressed by curcuminoids in human cell cultures (Romiti et al. 1998; Anuchapreeda et al. 2002; Chearwae et al. 2004). This nanoformulation appears to be safe and effective due to the less toxicity and high biocompatibility (Fathy Abd-Ellatef et al. 2020). In the context of diabetes mellitus, the hypoglycemic effects of curcumin are contradictory along with all the studies, despite most of the studies suggest that curcumin exert a potential effect in many diseases (Nelson et al. 2017). This favorable effect could be attributed to the suppressive effect on glycogen synthase kinase-3β (GSK-3β) that regulates the upstream and downstream factors and pathophysiology of a wide variety of diseases, for instance, Alzheimer's disease, malaria, cancer, and diabetes mellitus (Bustanji et al. 2009; Saraswati et al. 2018). Nonetheless, Nelson et al. (2017) stated

Table 6.10 Clinical studies of curcumin on age-related diseases

Subjects	Sample size	Duration/Intervention	Outcomes	References
Resected uterine cervical intraepithelial neoplasms, intestinal metaplasia, oral leucoplakia, and bladder cancer patients	25	500–8000 mg/day for 3 months/curcumin	Improved histology of precancerous lesions	Cheng et al. (2001)
Advanced pancreatic cancer patients	25	8 g/day until disease progression, with restaging every 2 months/curcumin	↓ COX-2, NF-κB expression	Dhillon et al. (2008)
Colon cancer lesions patients	44	4 g/day for 30 days/ curcumin	↓ 40% number of lesions	Carroll et al. (2011)
Postmenopausal women	32	8 weeks/150 mg/day theracurcumin	Improved FMD	Akazawa et al. (2012)
Patients with neck and head cancer	50	5 times/day for 2 weeks/ sandal wood oil (2 mL) and turmeric (2 g) based cream	Prevent radiodermatitis	Palatty et al. (2014)
Localized prostate cancer patients	199	6 months/mixture of turmeric (100 mg), broccoli (100 mg), pomegranate (100 mg), and green tea (100 mg)	The median increase in PSA was 63.8% lower than the placebo group	Thomas et al. (2014)
Healthy adults	59	8 weeks/curcumin (200 mg/day for 8 weeks)	Improved endothelial function	Oliver et al. (2016)
Dyslipidemia in diabetic patients	118	12 weeks/curcuminoids (1000 mg/day + piperine 10 mg/day)	↑ HDL-C	Panahi et al. (2017b)
			↓ Lp(a), non-HDL-C, and TC	
Non-alcoholic fatty liver disease patients	228	8 weeks/≥1000 mg/day curcumin	↓ ALT and AST levels	Mansour-Ghanaei et al. (2019)
Overweight and obese female adolescent	60	10 weeks/500 mg/day, containing 95% turmeric extract	Improved lipid profiles (HDL-C levels and TG/ HDL-C ratio)	Saraf-Bank et al. (2019)
Overweight and obese female adolescent	60	10 weeks/500 mg/day, containing 95% turmeric extract	↑ insulin levels	Saraf-Bank et al. (2019)
Overweight and obese female adolescent	60	10 weeks/500 mg/day, containing 95% turmeric extract	↓ Hip circumference, waist circumference, and BMI	Saraf-Bank et al. (2019)
Mild-to-moderate UC	70	8 weeks/1500 mg/day curcumin	↓ Serum hs-CRP concentration and ESR levels	Sadeghi et al. (2020)
			No significant changes in the TNF-α levels	

ALT Alanine aminotransferase, *AST* aspartate aminotransferase, *BMI* body mass index, *COX-2* cyclooxygenase-2, *ESR* erythrocyte sedimentation rate, *FMD* flow-mediated dilation, *HDL-C* high-density lipoprotein cholesterol, *hs-CRP* high-sensitivity C-reactive protein, *Lp(a)* lipoprotein(a), *NF-κB* nuclear factor-kappa B, *PSA* prostate-specific antigen, *TC* total cholesterol, *TG* triglycerides, *TNF-α* tumor necrosis factor alpha, *UC* ulcerative colitis

Table 6.11 Summary of nanoformulations curcumin on age-related diseases *in vitro* and *in vivo* studies

Age-related diseases	Co-delivery system	Cell lines and animal models	Outcomes	References
Breast cancer	Solid lipid nanoparticles	MDA-MB-231 cells	↓ Cancer cell proliferation ↑ Apoptosis	Sun et al. (2013); Bhatt et al. (2018)
Liver cancer	Dendrosomal curcumin	HepG2 and Huh7 cells (18 μM for 24, 48, and 72 h)	↑ Apoptosis in a time-dependent manner	Montazeri et al. (2016)
Skin cancer	2-FA/CA/CUR@ CMC-CA NGs (13.1 μg/mL for 48 h)	MEL-39 cell lines	↑ Percentage of apoptotic cells	Priya et al. (2020)
CVD	CUR encapsulated by carboxymethyl chitosan nanoparticle-peptide (5 mg/kg body weight)	Rat (*Rattus norvegicus*) model	Produced regression of cardiac hypertrophy	Ray et al. (2016)
CVD	CurNisNp (10 and 21 mg/kg)	Guinea pigs	Prevented the increment in hypertrophy index	Nabofa et al. (2018)
CVD	CCNP (100, 150, and 200 mg/kg) for 15 days	Rats	Prevent CK-MB leakage from cardiomyocytes	Boarescu et al. (2019)
Neurodegenerative disease	Dendrosomal nanocurcumin	Male C57BL/6 mice (12.5 mg/kg/day) for 6 weeks	↓ Accumulation and activation of astrocytes and microglia in corpus callosum of cuprizone treated mice	Motavaf et al. (2020)
Neurodegenerative disease	CUR-HA-palmitate (CUR-loaded) nanoparticles (3.8 μM for 30, 60, and 120 min)	STHdh$^{111/111}$ Huntington's disease striatal-derived cells	↓ Cell susceptibility to apoptosis	Pepe et al. (2020)
Type 2 diabetes mellitus	Curcumin nanoparticles (30 mg/mL or 60 mg/mL for 30 days)	Male albino rats	↓ Glucose level, MDA levels ↑ TAC	Abu-Taweel et al. (2020)
Type 2 diabetes mellitus	Nanocurcumin (100 and 200 mg/kg)	Male Wistar rats	↓ Insulin resistance, serum levels of FBS, VLDL, LDL, TG, and cholesterol	Shamsi-Goushki et al. (2020)

(continued)

Table 6.11 (continued)

Age-related diseases	Co-delivery system	Cell lines and animal models	Outcomes	References
Rheumatoid arthritis	Curcumin-loaded oil–water nanoemulsions or curcumin suspension (orally administered 50 mg/kg body weight once a day for 14 days)	Male Sprague Dawley rats	TNF-α and IL-1β levels in synovial fluid of curcumin-loaded oil–water nanoemulsions-treated group were more than two-fold lower than the suspension group	Zheng et al. (2015)

CA casein, *CCNP* CUR nanoparticles, *CK-MB* creatine kinase myocardial band, *CMC* carboxymethyl cellulose, *CUR* curcumin, *CurNisNp* CUR and nisin (antimicrobial peptide) based poly lactic acid nanoparticles, *FA* folic acid, *FBS* fasting blood sugar, *HepG2* hepatocarcinoma cells, *Huh7* hepatocarcinoma cells, *IL-1β* interleukin-1beta, *LDL* low-density lipoprotein, *MDA-MB-231* breast adenocarcinoma cells, *MDA* malondialdehyde, *MEL-39* melanoma cancer cell line, *NGs* nanogels, *TAC* total antioxidant capacity, *TG* triglycerides, *TNF-α* tumor necrosis factor-α, *VLDL* very-low-density lipoprotein

that the stability of curcumin should be revised carefully before considering GSK-3β activity as a therapeutic target of curcumin. So far, many curcumin nanoformulations have been developed to enhance the effectiveness of curcumin, tissue specificity, and cellular uptake (Gera et al. 2017). However, most of the nanoformulation studies are only performed in the preclinical models. The efficacy and toxicology of curcumin nanoformulation in humans have to be researched and performed in comparative randomized clinical trials before introducing to the pharmaceutical market.

6.6.2 *Epigallocatechin Gallate*

Green tea has become the second largest drink after water (La et al. 2019). It is one of the popular beverages consumed worldwide (Namita et al. 2012). Green tea is produced by exposing the fresh leaves of *Camellia sinensis* to hot steam or heat after plucking to minimal polyphenol oxidation (Gianfredi et al. 2018). The predominant bioactive compounds of green tea are polyphenolic components and caffeine. (−)-epigallocatechin-3-gallate (EGCG) is the most abundant catechins found in green tea (Du et al. 2012), representing nearly a third of the solid content (Potenza et al. 2007). In particular, catechins comprised 30–42% of the total dry weight of green tea (Dufresne and Farnworth 2001). In general, a cup of green tea containing nearly 150–200 mg of total flavonoids, in which 90–100 mg is catechins (Hodgson and Croft 2010).

EGCG is the physiologically active polyphenols, known for its broad spectrum of health benefits, including anti-atherogenic, anti-inflammatory, and anticancer activity (Chu et al. 2017; Yamagata 2020). Substantial studies have revealed that green tea consumption possesses a protective effect against several types of cancers (Abe and Inoue 2020; Musial et al. 2020). Preclinical studies from *in vivo* and *in vitro* studies have demonstrated that green tea polyphenols exert anticarcinogenic activity via suppression of cell proliferation and induction of apoptosis (Man et al. 2020; Yoshimura et al. 2019; Chen et al. 2020). An *in vitro* study showed that co-treated with EGCG (50 μM) for 24 h improved the sensitivity of human colon carcinoma cell line (DLD1 and HCT-116) to 5-fluorouracil (5-FU) (La et al. 2019). This study further showed that EGCG improved the activity of 5-FU by modulating the glucose-regulated proteins 78 (GRP78)/NF-κB/multidrug resistance mutation 1 (MDR1) signaling pathway in colorectal cancer cells (La et al. 2019). The data from xenografts model have also shown a similar effect, in which co-treatment with EGCG (25 mg/kg) and 5-FU (20 mg/kg) for 14 days markedly suppressed the tumor growth (La et al. 2019), implied that EGCG may synergistically enhance the activity of 5-FU by reducing the tumor burden via downregulation of drug resistance-associated protein-MDR1 (La et al. 2019). Data from a meta-analysis involving 8 cohort studies and 5 case-control studies suggested that drinking 5 cups of green tea/day can reduce breast cancer risk (Gianfredi et al. 2018) (Table 6.12). Such anticancer effect has been accredited to its antioxidant activity. Evidence from experimental studies shows that long-term exposure to polyphenols may decrease oxidative stress and chronic inflammation, and thereby suppressed the diffusion, reproduction, and proliferation of cancer cells (Mao et al. 2017; Ma et al. 2018). Importantly, accumulating evidence from *in vivo* studies highlights the inhibitory effects of EGCG on triple-negative breast cancer (TNBC) cell growth (Bimonte et al. 2020). TNBC is accounting for 15–20% of breast cancer and does not express human epidermal growth factor receptor 2 (HER2) negative (neu) markers, progesterone receptor (PR), and estrogen receptor (ER) (Bianchini et al. 2016). A high concentration of EGCG (100 μM) inhibits cell proliferation (Fig. 6.8) by inducing apoptosis via stimulation of ROS production in TNBC (Hs578T) cell lines (Braicu et al. 2011). The animal model study further showed that feeding TNBC mouse model with green tea polyphenols (1%) and EGCG (1 mg/animal in 100 μL of distilled water) for 10 weeks induced apoptosis and reduced tumor growth (Thangapazham et al. 2007).

With regard to the potential of EGCG in the treatment of Alzheimer's disease, compelling evidence from *in vivo* and *in vitro* studies has suggested that EGCG may play a neuroprotective role in Alzheimer's disease (Cascella et al. 2017; Wei et al. 2019). Wei et al. (2019) found that feeding rats with EGCG (100 mg/kg body weight for 4 weeks) significantly reduced β-amyloid ($A\beta_{1-42}$) plaque formation in brains and alleviated cognitive impairments in the aging rats with cognitive impairments (Fig. 6.8). In the brain, EGCG act as an anti-inflammatory, the signaling pathway modulatory agent, and a potent antioxidant. In a further study focused on inflammation outcome, Biasibetti et al. (2013) showed that EGCG (10 mg/kg/day for 4 weeks) improved the cognitive deficit by mediating NO metabolites in rats. Data from

Table 6.12 The effects of EGCG on age-related diseases in clinical studies

Age-related diseases	Study conditions	Treatment	Outcomes	References
Breast cancer	Observational study (8 cohort studies and 5 case-control studies)	5 cups of green tea/ day	↓ Breast cancer risk	Gianfredi et al. (2018)
Cognitive dysfunction	Non-blinded, non-placebo controlled design, pilot study	2 g/day for 3 months	Improved cognitive performance	Ide et al. (2014)
Alzheimer' disease	Double-blind, placebo-controlled, cross-over study	Single dose (300 mg)	↑ Theta, beta, and alpha activity, particularly in central and midline frontal regions	Scholey et al. (2012)
Type 2 diabetes mellitus	Meta-analysis of 27 randomized controlled trials	Green tea catechins ranged from 80 to 1344 mg/day for 3 weeks to 12 months	↓ Fasting blood glucose level	Xu et al. (2020a)
			No effect on HOMA-IR, HbA1c, and FBI	
Type 2 diabetes mellitus	Meta-analysis of 8 randomized controlled trials	Green tea (400–10,000 mg/day for 8 weeks-9 months)	↓ CRP levels	Asbaghi et al. (2019)
			No effect on plasma levels of MDA and TAC	
Type 2 diabetes mellitus	Meta-analysis of 11 randomized controlled trials	Green tea catechins (varied from $<$400 mg/day to ≥400 mg/day for $<$8 weeks to ≥8 weeks)	No effect on plasma CRP levels	Serban et al. (2015)
CVD	Meta-analysis of 10 cohort studies and seven case-control studies	3 cups per day (1 cup = 237 mL)	↓ Incidence rate of myocardial infarction by 11%	Peters et al. (2001)
CVD	Prospective cohort study	>2 cups/day, about 7 oz/day for 10 years	↓ Mortality risk of CVD by 22–33%	Kuriyama et al. (2006)
CVD	Meta-analysis of 9 randomized controlled studies	Drinking ≥3 cups of tea/day (green or black)	↓ 21% risk of ischemic stroke compared to those consuming <1 cup/day	Arab et al. (2009)

(continued)

Table 6.12 (continued)

Age-related diseases	Study conditions	Treatment	Outcomes	References
CVD	Randomized, double-blind, placebo-controlled clinical trial	1500 mg of decaffeinated green tea extract/day for 4 weeks	↓ Total cholesterol levels	Lu and Hsu (2016)
CVD	Randomized, double-blind, cross-over and placebo-controlled clinical trial	Green tea extract (one capsule 30 min after meal, 3 times/day for 6 weeks), EGCG amounted to 856.8 mg/day	↓ LDL-C levels ↑ Leptin No effect on TC, TG, and HDL-C levels	Huang et al. (2018)
CVD	Systematic review and meta-analysis analyzed of 31 randomized controlled trials	Green tea (varied from 80 to 2488.7 mg/day for 3 weeks to 12 months)	↓ LDL-C, TC, and TG levels No effect on HDL-C	Xu et al. (2020b)
Obesity	Randomized controlled trials	EGCG (800 mg/day for 8 weeks)	No changes in body fat percentage and BMI	Brown et al. (2009)
Obesity	Meta-analysis of 8 randomized controlled trials	EGCG (varied from 300 to 800 mg/day for 8–12 weeks and 300–600 mg/day for 2–3 days)	↑ Energy expenditure ↓ Respiratory quotient	Kapoor et al. (2017)

BMI body mass index, *CRP* C-reactive protein, *CVD* cardiovascular disease, *EGCG* (−)-epigallocatechin-3-gallate, *FBI* fasting blood insulin, *HDL-C* high-density lipoprotein cholesterol, *HOMA-IR* homeostasis model assessment of insulin resistance, *LDL-C* low-density lipoprotein cholesterol, *MDA* malondialdehyde, *TAC* total antioxidant capacity, *TC* total cholesterol, *TG* triglycerides

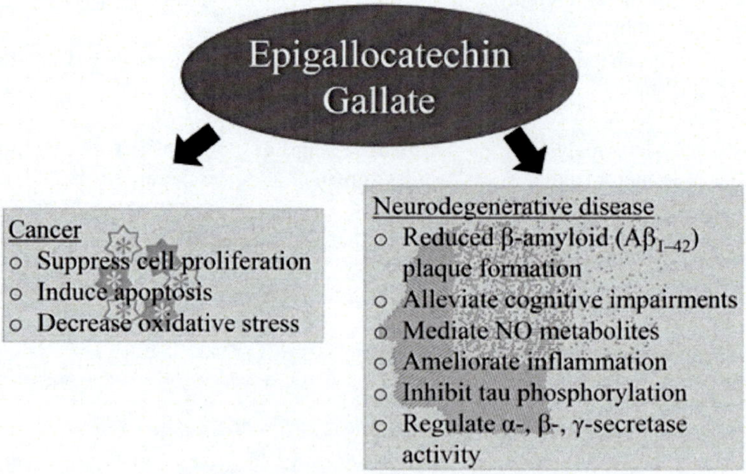

Fig. 6.8 Mechanisms of epigallocatechin gallate in the modulation of cancer and neurodegenerative disease. *NO* nitric oxide

systematic review and meta-analysis of the animal model included randomized controlled trials from inception to November 2019 have demonstrated that EGCG is inversely associated with Alzheimer's disease (Zhang et al. 2020). EGCG is thought to protect against Alzheimer's disease by suppressing acetylcholinesterase (AChE) activity, ameliorating inflammation, inhibiting tau phosphorylation, and regulating α-, β-, γ-secretase activity (Zhang et al. 2020). These findings are in line with the earlier study reported *in vitro* for MC65 cell lines, in which EGCG (20 μM) downregulates Aβ levels by reducing the nuclear translocation of c-Abl and improving the endogenous amyloid precursor protein (APP) proteolysis (Lin et al. 2009). Indeed, tau accumulation promotes load on clearance in the machinery of neurons, which may collapse, and ultimately leading to neuronal death (Keller et al. 2000; Keck et al. 2003). Notably, EGCG (64.2 μM) was shown to dissolve tau fibrils and oligomers and inhibit the heparin-triggered full-length tau aggregation (Sonawane et al. 2020). The vital role played by EGCG such as quenching the oligomers and disassembling tau filaments, implied that EGCG may be a promising tool for the amelioration of Alzheimer's disease. Consistent with the findings from preclinical studies, the evidence from a clinical study has revealed that consumption of green tea (2 g/day for 3 months) significantly improved cognitive performance in subjects with cognitive dysfunction (Ide et al. 2014) (Table 6.12). Another double-blind, placebo-controlled, cross-over study consistently reported a neurocognitive potential of administration single dose of EGCG (300 mg) (Scholey et al. 2012). From the study reviewed, Scholey et al. (2012) found that EGCG stimulates theta, beta, and alpha activity, particularly in central and midline frontal regions. In general, alpha activity is characterized by large rhythmic waves that are linked to relaxation and less in the active cognitive process. Beta activity is alleviated by movement and stimulates behavioral arousal and focused attention. Theta activity is mainly linked to quiet wakefulness (Cantero et al. 2003). Further, EGCG treatment can also decrease self-rated stress and increased self-rated calmness (Scholey et al. 2012), suggesting that EGCG may provide more attentive and relaxed conditions after consuming green tea. An extensive body of systematic review of the cohort, cross-sectional, cross-over studies, and randomized controlled trials has demonstrated a beneficial role of green tea supplementation (Mancini et al. 2017). From the study reviewed, it showed that green tea can influence brain function, cognition, and psychopathological symptoms (Mancini et al. 2017). Such finding highlights the neuroprotective effect of green tea may be modulated partly through synergistic/additive effects of the bioactive compounds in green tea (Mancini et al. 2017).

In addition to the effects demonstrated in cancer and Alzheimer's disease, the beneficial impact of EGCG in humans has also been observed on type 2 diabetes (Fig. 6.9). The animal model study showed that treatment of EGCG (5 mg/kg/day) for 4 days intraperitoneally decreased β-cells response to the elevation of glucose in diabetic rats (Yun et al. 2006). From the study reviewed, EGCG increases insulin-immunoreactivity in β-cells and stimulates the loss of islet cell mass (Yun et al. 2006), suggesting that EGCG had prooxidant properties in the β-cells. A recent meta-analysis with 27 randomized controlled trials involving 2194 subjects has revealed a significant role of green tea consumption on glycemic control (Xu et al.

2020a) (Table 6.12). It showed that adults who consume green tea for 3 weeks to 12 months with green tea catechins ranged from 80 to 1344 mg/day can significantly reduce fasting blood glucose levels (Xu et al. 2020a). The improvements in these indices could be attributed to the tea catechins, which are one of the prominent antioxidants to alleviate oxidative stress (Mustata et al. 2005). Tea catechins are thought to decrease carbohydrate absorption from the intestine through the suppression of α-glucosidase, α-amylase, and intestinal sucrose (Collins et al. 2007). Glucose metabolism and insulin sensitivity were improved in rodents after supplemented with tea catechins and thus provide a useful approach for decreasing the type 2 diabetes risk (Wolfram et al. 2006). Further, tea catechins may also suppress hepatic gluconeogenesis by modulating protein-tyrosine phosphorylation and altering gluconeogenic gene expression in the mouse liver (Waltner-Law et al. 2002). In another meta-analysis involving 614 type 2 diabetes patients have shown that green tea (400–10,000 mg/day for 8 weeks-9 months) was effective in reducing the CRP levels (Asbaghi et al. 2019), which is known as an important inflammatory mediator to predict the incidence of CVD (Ridker 2003; Eslampour et al. 2019). This finding indicates that green tea may decrease the risk of CVD in type 2 diabetes mellitus patients by reducing CRP levels. However, a meta-analysis of 11 randomized controlled trials revealed that green tea catechins (varied from $<$400 mg/day to ≥400 mg/day for $<$8 weeks to ≥8 weeks) had no significant effect on plasma CRP levels (Serban et al. 2015). Intriguingly, both of the studies of meta-analyses showed an inconsistent finding of the effects of green tea on CRP levels. This could be partly due to the different participants included in these meta-analyses. Indeed, most of the meta-analyses had no specific on the types of subjects, for instance, healthy individuals or patients with certain types of metabolic ailments. Further, the human interindividual polymorphism may affect the metabolic rate and bioavailability of the tea flavonoids (Serban et al. 2015). In addition, Xu et al. (2020a) did not identify a significant effect of green tea (80–1344 mg/day for 3 weeks–12 months) on homeostasis model assessment of insulin resistance (HOMA-IR), HbA1c, and fasting blood insulin (FBI). A study by Asbaghi et al. (2019) also found that green tea consumption (400–10,000 mg/day for 8 weeks to 9 months) had no significant effect on plasma levels of MDA and total antioxidant capacity in type 2 diabetes mellitus patients aged 52 years old and above.

EGCG supplementation not only decreases cancer, neurodegenerative disorder, and diabetes but also induces protective cardiometabolic effects (Fig. 6.9). Meta-analysis of population studies concerning tea consumption suggests that drinking 3 cups of tea/day decreases the risk of ischemic stroke and myocardial infarction by 21% and 11%, respectively (Peters et al. 2001; Arab et al. 2009) (Table 6.12). A recent systematic review and meta-analysis analyzed of 31 randomized controlled trials from inception to September 2019 including 3321 individuals showed that green tea consumption (varied from 80 to 2488.7 mg/day for 3 weeks to 12 months) significantly reduced low-density lipoprotein cholesterol (LDL-C), total cholesterol (TC), and TG levels (Xu et al. 2020b). Nonetheless, Xu et al. (2020b) did not identify an association between green tea consumption and HDL-C. Consistent with the study reported by Xu et al. (2020b), Lu and Hsu (2016) and Huang et al. (2018) also

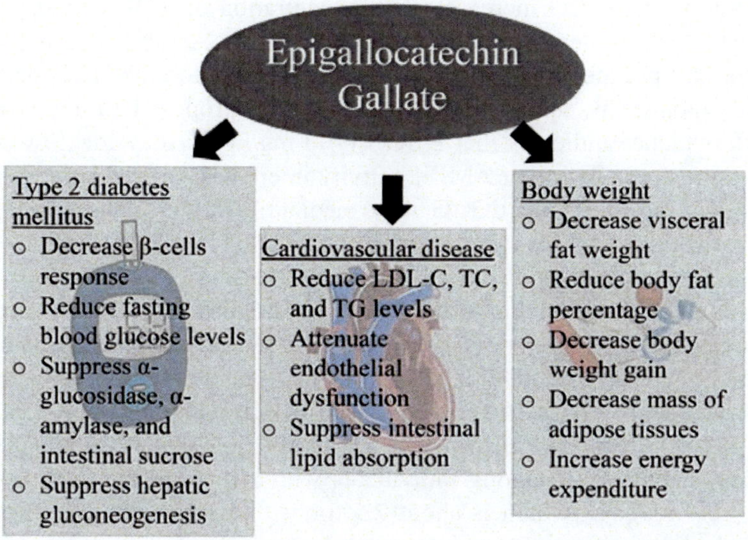

Fig. 6.9 Mechanisms of epigallocatechin gallate in the modulation of type 2 diabetes mellitus, cardiovascular disease, and body weight. *LDL-C* low-density lipoprotein cholesterol, *TC* total cholesterol, *TG* triglycerides

found that no association between green tea intake and HDL-C levels. Further, some observational study has emerged to suggest that green tea is negatively linked to the risk of CVD (Xu et al. 2020b). The previous study stated that compared to the non-tea drinkers, the mortality risk of CVD was reduced by 22–33% in middle-aged individuals from Japan who consuming green tea (>2 cups/day, about 7 oz/day for 10 years) (Kuriyama et al. 2006). Green tea is believed to have hypolipidemic activity by attenuating the endothelial dysfunction triggered by ox-LDL through the Jagged-1/Notch signaling pathway in human umbilical vein endothelial cells and thereby suppressing the atherosclerotic plaque formation (Wang et al. 2018c). Further, green tea may also suppress intestinal lipid absorption by interrupting micelle formation (Koo and Noh 2007).

Several studies have revealed that genetically obese or high-fat diet (HFD) rodents who consume EGCG or green tea extract are negatively associated with adiposity or body weight gain compared to those who never receive EGCG (Li et al. 2018). Compelling evidence has corroborated these data and found that dietary supplementation of EGCG (3.2 g/kg diet for 16 weeks) significantly decreased visceral fat weight, body fat percentage, and body weight gain (Fig. 6.9) in mice compared to those who never receive EGCG (Bose et al. 2008). Further, Lee et al. (2009) compared different groups of C57BL/6J mice fed with HFD treated with 0.2 or 0.5% EGCG (w/w) and a high-fat control diet for 8 weeks. The data showed that EGCG significantly decreased the mass of adipose tissues and body weight (Lee et al. 2009). Consistent with the studies reported by Lee et al. (2009), Li et al. (2018) also found that feeding HFD mice with EGCG (50 mg/kg and 100 mg/kg per day for 20 weeks) reduced epididymal fat and obesity. From the study reviewed, administration of EGCG increased the concentration of HDL-C and reduced the levels of LDL-C, cholesterol, and TG (Li et al. 2018). An emerging role of EGCG in relation to lipid metabolism has been demonstrated and the effect of AMPK on lipolysis, lipogenesis, and adipogenesis has been proposed (Yang et al. 2016). Research evidence indicates that EGCG (100 mg/kg per day for 20 weeks) is of benefit in the upregulation of AMPK activity in epididymal and subcutaneous adipose tissues, indicates that EGCG may mediate the lipid metabolism partly via AMPK activation in adipose tissues (Li et al. 2018). Moreover, data from systematic review and meta-analysis of randomized controlled trials have demonstrated that EGCG supplementation (varied from 300 to 800 mg/day for 8–12 weeks and 300 to 600 mg/day for 2–3 days) moderately increase energy expenditure and decrease respiratory quotient (Kapoor et al. 2017) (Table 6.12). Despite some evidence have demonstrated an inverse association between EGCG and body weight gain, not all studies showed such a link. Brown et al. (2009) did not identify the changes in body fat percentage and BMI among individuals who consumed high concentration of EGCG (800 mg/day), which is about 2.5 times that applied in other randomized controlled trials for 8 weeks of the period. In this regard, the intakes of as low as 300 mg of EGCG/day may have the potential to stimulate fat metabolism (Kapoor et al. 2017). Indeed, the concentration of 300 mg EGCG/day is listed in the established list of plants authorized in food supplements in French regulatory (Guillaume 2014). It has also been reported as an acceptable dosage by the Japanese Consumer Affairs Agency under the health claim laws of the labeling system (White Paper on Consumer Affairs 2015). Overall, the aforementioned findings suggest that EGCG may become a useful approach to improve age-related chronic diseases.

6.6.3 Resveratrol

Resveratrol is a stilbenoid known as *trans*-3,4′,5-trihydroxystilbene, comprised of two aromatic rings which are bound with a methylene bridge (Salehi et al. 2018). Resveratrol is found naturally in over 70 different plant species including legumes, pines, and grapevines (Soleas et al. 1997). In brief, resveratrol is a naturally occurring polyphenol, which is widely found in pomegranates, soybeans, and peanuts (Sinha et al. 2016) (Fig. 6.10). Botrytis cinerea infection in grapes can cause the exclusive synthesis of resveratrol in the grape skins and leaf epidermis (Hasan and Baek 2013). Indeed, grape skins are not undergone fermentation during white wine production, and thus only red wines contain remarkable levels of resveratrol (Fuhrman et al. 2001). Resveratrol exists as *cis*/*trans* isoforms, in which the *trans* isomer is the predominant form (Gambini et al. 2015).

Substantial evidence suggests that resveratrol exhibits a broad range of health-promoting effects including neuroprotective, vasorelaxant, cardioprotective, anti-carcinogenic, immune-modulator, anti-hyperlipidemic, antioxidant, and anti-inflammatory activity (Bo et al. 2013; Andrade et al. 2018; Colica et al. 2018; Pannu and Bhatnagar 2019; Xiao et al. 2019). Resveratrol acts as a potent antioxidant by reducing oxidative stress, one of the predominant contributors to chronic diseases, via several redox-related signaling pathways (Petrella et al. 2020; Wang et al. 2020). For instance, resveratrol promotes the phosphatase and tensin homolog (PTEN) and subsequently leading to the downregulation of Akt phosphorylation, and ultimately increased the antioxidant enzymes such as SOD and catalase (Ingles et al. 2014). Further, resveratrol is also enhanced the antioxidant defense system by mediating antioxidant enzymes via decreased ERK stimulated by ROS (Singh and Vinayak 2017). A systematic review and meta-analysis of 6 randomized clinical trials involving 299 participants revealed that supplementation of resveratrol (varied from ≤500 mg/day and ≥500 mg/day of resveratrol for 1.5 to 5 months) increased the circulating GPx levels; whereas no association was reported between resveratrol supplementation and MDA, total antioxidant capacity, or SOD levels (Omidian et al. 2020). In a further study focused on oxidative stress markers, Seyyedebrahimi et al. (2018) showed that patients with type 2 diabetes mellitus who received resveratrol (800 mg/day for 2 months) had no effect on MDA levels. Despite resveratrol is absorbed rapidly, yet the systemic bioavailability is ˂1% (Goldberg et al. 2003; Walle et al. 2004). In general, the antioxidant activities of resveratrol include (1) inducing autophagy through transcription factor EB (TFEB)-dependent or mammalian target of rapamycin (mTOR)-dependent pathway (Zhou et al. 2019); (2) upregulating the gene expression involved in mitochondrial energy biogenesis, primarily via PTEN/Akt, ERK/p38 MAPK, and AMPK/Sirtuin 1 (SIRT1)/Nrf2 signaling pathways (Ungvari et al. 2011); (3) enhancing endogenous antioxidant enzymes (Tung et al. 2014); (4) scavenging free radicals (Cicero et al. 2019); (5) decreasing RNS/ROS production (Lorenz et al. 2003; Zhu et al. 2020).

The antidiabetic potential of resveratrol in diabetes mellitus has been described by Oyenihi et al. (2016) and Öztürk et al. (2017). Emerging research evidence on animal models indicates that resveratrol exerts glucose-lowering effects by improving clinical and biochemical parameters of diabetes-induced liver injury (Hamadi et al. 2012; Hussein and El-Maksoud 2013), diabetes-induced CVD (Turan et al. 2012; Guo et al. 2014; Mohammadshahi et al. 2014), diabetes-induced hypertension (Allah and El-Debakey 2010; Vella et al. 2015), diabetic retinopathy (Hua et al. 2011; Kubota et al. 2012; Kim et al. 2012; Soufi et al. 2012), diabetic neuropathy (Kumar et al. 2007, 2013; Bastianetto et al. 2015), and diabetic nephropathy (Xu et al. 2014; Wu et al. 2012; Elbe et al. 2015). These findings implied that resveratrol may mediate pancreatic β cells and improve insulin action (Fig. 6.11), and thus prevent the complications of diabetes mellitus (Szkudelski and Szkudelska 2015).

A randomized, double-blind, placebo-controlled clinical trial evaluated resveratrol administration in relation to diabetic nephropathy. The study showed that resveratrol supplementation (500 mg/day for 90 days) significantly decreased the mean of urine albumin/creatinine ratio and increased serum antioxidant enzymes in patients with type 2 diabetes and albuminuria (Sattarinezhad et al. 2019). From the study reviewed, resveratrol administration failed to show any benefit in improving the estimated glomerular filtration rate (eGFR) and serum creatinine level (Sattarinezhad et al. 2019). The finding in this study suggests that resveratrol may be an effective adjunct to angiotensin receptor blockers (ARBs) for decreasing

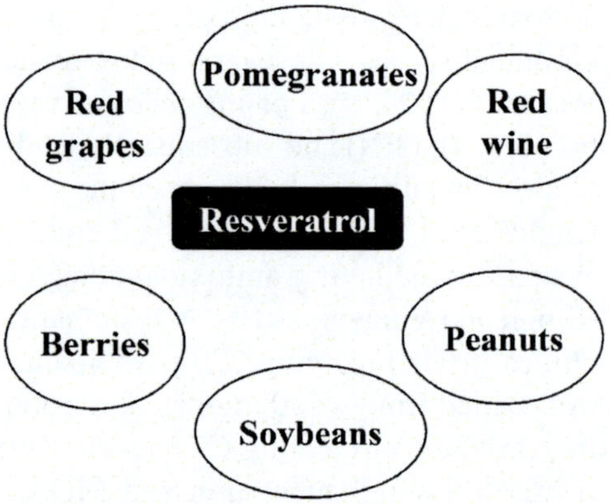

Fig. 6.10 Sources of dietary resveratrol

urinary albumin excretion in patients with diabetic nephropathy. Resveratrol has shown a hypoglycemic effect by activating the SIRT1 signaling pathway (Yun et al. 2012). The previous study has stated that administration of resveratrol significantly reduced SIRT1 expression and activity *in vitro* and *in vivo* models of diabetes mellitus (Lagouge et al. 2006; Milne et al. 2007; De Kreutzenberg et al. 2010; Yar et al. 2011). Resveratrol regulates glucose homeostasis by activating the AMPK. Indeed, AMPK mediates several prominent intracellular processes including cellular homeostasis, mitochondrial function, and energy metabolism (Herzig and Shaw 2018). Under hyperglycemic circumstances, the dysregulation of AMPK activity is linked to hyperglycemic-related tissue damage and insulin resistance, suggesting an interrelated stimulation of the AMPK and type 2 diabetes mellitus (Oyenihi et al. 2016). Table 6.13 summarizes the clinical studies of resveratrol on several age-related diseases.

Besides its effects on diabetes mellitus, a beneficial effect of resveratrol supplementation has also been demonstrated on the CVD risk (Fig. 6.11). In randomized, double-blind, cross-over trial involving 50 healthy adult smokers revealed that resveratrol (500 mg/day for 30 days) significantly increased total antioxidant status and decreased TG and CRP levels (Bo et al. 2013). A 1-year follow-up study including 75 patients further supported that intake of 8 mg of resveratrol for the first 6 months and a double concentration for the next 6 months demonstrated that resveratrol-rich grape supplement improved the fibrinolytic status and inflammation in patients who were on statins and at high CVD risk (Tomé-Carneiro et al. 2012). Similarly, a double-blind, placebo-controlled trial including 40 post-infarction Caucasian patients showed that intakes of resveratrol (10 mg/day for 3 months) decreased LDL-C levels, improved endothelial function, enhanced left ventricle diastolic function, and protected against unfavorable hemorheological changes in patients with coronary artery disease (Magyar et al. 2012), suggesting the cardioprotective potential of resveratrol. Although most of the studies have reported a negative link between resveratrol and CVD risk, not all findings showed such a link. Compared to those who received 300 mg and placebo group, a high concentration of resveratrol (1000 mg/day for 90 days) increased the biomarkers of CVD risk

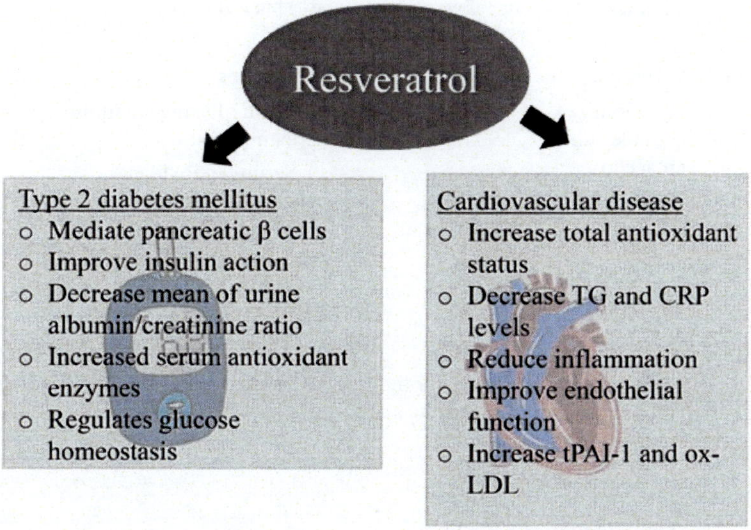

Fig. 6.11 Mechanisms of resveratrol in the modulation of type 2 diabetes mellitus and cardiovascular disease. *CRP* C-reactive protein, *ox-LDL* oxidized low-density lipoprotein, *TG* triglycerides, *tPAI-1* total plasminogen activator inhibitor-1

including total plasminogen activator inhibitor-1 (tPAI-1), soluble vascular cell adhesion molecule-1 (sVCAM-1), soluble intercellular adhesion molecule-1 (sICAM-1), soluble E-selectin 1 (sE-selectin 1), and ox-LDL in overweight individuals aged 65 years old and above (Mankowski et al. 2020). Consistent with the study reported by Mankowski et al. (2020), administration of 150 mg/day of resveratrol for 4 weeks in overweight individuals also failed to show any benefit in improving the metabolic risk markers, for instance, inflammation or endothelial function, which are linked to cardiovascular health risk (van der Made et al. 2015, 2017). Inconclusive findings could be due to the pharmacokinetics and bioavailability of resveratrol, health status of the subjects, varying of the gut microbiota, and the concentration of resveratrol. Other factors such as the form of administration (gel caps, powder, and tablet), lifestyle, gender, and age may also confound the findings. In this regard, future studies should be conducted in a similar study design and compared in the same study (Ramírez-Garza et al. 2018; Shaito et al. 2020).

Emerging evidence has suggested that resveratrol protects against neural degeneration and injuries in dementia and Alzheimer's disease. In the brain, resveratrol acts as a potent anti-oxidative and anti-inflammatory agent by modulating different signaling pathways including peroxisome proliferator-activated receptor gamma coactivator 1-alpha (PGC-1α), SIRT1, and AMPK (Kulkarni and Cantó 2015). These metabolic regulators have been demonstrated to be involved in the onset of neurodegenerative disorders (Pasinetti et al. 2015). Indeed, brain tissues are susceptible to oxidative stress due to their low concentration of antioxidants, less regenerative capability, high amounts of peroxidizable fatty acids, and high rate of oxygen consumption (Salim 2017). Free radicals have been considered as the underlying factor

Table 6.13 The clinical studies of resveratrols on age-related diseases

Age-related diseases	Study conditions	Treatment	Outcomes	References
Type 2 diabetes mellitus	Randomized, double-blind, placebo-controlled	500 mg/day for 90 days	↓ Mean of urine albumin/creatinine ratio ↑ Serum antioxidant enzymes No effect on eGFR and serum creatinine level	Sattarinezhad et al. (2019)
CVD	Randomized, double blind, cross-over study	500 mg/day for 30 days	↑ Total antioxidant status ↓ TG and CRP levels	Bo et al. (2013)
CVD	Triple-blinded, randomized, parallel, dose-response, placebo-controlled, 1-year follow-up trial	8 mg for the first 6 months and a double concentration for the next 6 months	Improved the fibrinolytic status and inflammation	Tomé-Carneiro et al. (2012)
CVD	Double-blind, placebo controlled trial	10 mg/day for 3 months	↓ LDL-C levels Improved endothelial function Enhanced left ventricle diastolic function Protected against unfavorable hemorheological changes	Magyar et al. (2012)
CVD	Randomized placebo-controlled trial	150 mg/day for 4 weeks	No effect on inflammation and endothelial function	van der Made et al. (2015, 2017)
CVD	Randomized, placebo-controlled trial	1000 mg/day for 90 days	↑ Biomarkers of CVD risk such as tPAI-1, sVCAM-1, sICAM-1, sE-selectin 1, and ox-LDL	Mankowski et al. (2020)
Alzheimer's disease	Randomized, placebo-controlled, double-blind, multi-site, phase 2 trial	500 mg/day with 500 mg increments every 13 weeks up to 52 weeks, ending with 1000 mg twice/day	↓ MMP-9 expression Attenuated the declines in cerebrospinal fluids Aβ42 and Aβ40 levels	Moussa et al. (2017)

(continued)

Table 6.13 (continued)

Age-related diseases	Study conditions	Treatment	Outcomes	References
Alzheimer's disease	Randomized, double-blind, placebo-controlled trial	500 mg/day with 500 mg escalation every 13 weeks up to 52 weeks, ending with 1000 mg twice/day	Attenuated the decline in plasma Aβ40 and cerebrospinal fluid Aβ40	Turner et al. (2015)
			↑ brain volume loss	
Colorectal Cancer	Intervention study	8 daily doses at 0.5 or 1.0 g prior to surgical resection	↓ Tumor cell proliferation by 5%	Patel et al. (2010)
Colorectal cancer	Phase I pilot clinical trial	Freeze-dried grape powder (80 g/day containing 0.07 mg of resveratrol for 2 weeks)	↓ Wnt target genes of the normal mucosa	Nguyen et al. (2009)
			No effect on the cancerous tissues	
Cancer	Prospective cohort study	24-h urinary resveratrol metabolites	No association between 24 h total urinary resveratrol metabolite and cancer or inflammatory markers such as IL-1β, IL-6, and CRP	Semba et al. (2014)
Prostate cancer	Randomized, placebo controlled clinical study	Two doses, 1000 mg/day for 4 months	↓ Serum levels of the androgen precursors DHEAS, DHEA, and androstenedione	Kjaer et al. (2015)
			No association between resveratrol and dihydrotestosterone, free testosterone, testosterone, circulating levels of PSA or prostate size	

CRP C-reactive protein, *CVD* cardiovascular disease, *eGFR* estimated glomerular filtration rate, *DHEA* dehydroepiandrosterone, *DHEAS* dehydroepiandrosterone-sulphate, *IL-1β* interleukin-1beta, *IL-6* interleukin-6, *LDL-C* low-density lipoprotein cholesterol, *MMP-9* matrix metalloproteinase-9, *ox-LDL* oxidized-low density lipoprotein, *PSA* prostate-specific antigen, *sE-selectin 1* soluble E-selectin 1, *sICAM-1* soluble intercellular adhesion molecule-1, *sVCAM-1* soluble vascular cell adhesion molecule-1, *TG* triglycerides, *tPAI-1* total plasminogen activator inhibitor-1 for the pathogenesis of brain aging and neurodegenerative diseases (Islam 2017). The previous study found that resveratrol possesses scavenging properties by reducing free radicals production in tissues (Fu et al. 2018). A study reported by Huang et al. (2011) showed that resveratrol (100 μM for 7 days) downregulated Aβ-induced iNOS expression (Fig. 6.12), which is involved in the HO-1 inhibition and Aβ-induced lipid peroxidation, and subsequently suppress Aβ-induced neurotoxicity in rats, suggesting that resveratrol may exert neuroprotectant against Aβ. Amyloid β peptide is one of the predominant components of amyloid plaques that interacts with several toll-like receptors (TLRs) including TLR4 and thereby stimulate microglial activation (Chen et al. 2017). The previous study has demonstrated that resveratrol can prevent the LPS-induced activation of microglial BV-2 cells and murine RAW 264.7 macrophages (Capiralla et al. 2012; Ge et al. 2019). Resveratrol was found to prevent the proinflammatory activity of Aβ on macrophages by suppressing stimulation of signal transducer and activator of transcription 3 (STAT3) and STAT1 and inhibiting the NF-κB activation by modulating the IκB and IKK phosphorylation (Capiralla et al. 2012). A study in the animal model has revealed that feeding mice with 350 mg/kg body weight of resveratrol for 15 days decreased cerebral amyloid deposition and reduced Aβ related to microglial activation (Capiralla et al. 2012). Despite the animal studies showed a promising result, the clinical trials evaluated the effect of resveratrol on Alzheimer's disease remain scarce. Clinical studies from Phase II trial have shown that resveratrol is beneficial to be consumed in patients with mild to moderate Alzheimer's disease (Turner et al. 2015; Moussa et al. 2017). The previous study showed that Alzheimer's disease patients who supplemented with resveratrol (500 mg/day with 500 mg increments every 13 weeks up to 52 weeks, ending with 1000 mg twice/day) downregulated matrix metalloproteinase-9 (MMP-9) expression (Moussa et al. 2017). Notably, a greater reduction of cerebrospinal fluids Aβ42 and Aβ40 expression for those who consumed placebo compared to the resveratrol group, implied that resveratrol attenuated the declines in cerebrospinal fluids Aβ42 and Aβ40 levels (Moussa et al. 2017). The decrease in MMP-9 levels suggested that resveratrol may protect the central nervous system by decreasing the permeability and thus limit the proinflammatory agents and infiltration of leukocytes into the brain (Moussa et al. 2017). Further, ameliorating of cerebrospinal fluid Aβ40 and Aβ42 levels in Alzheimer's disease patients who received resveratrol indicates a lower accumulation of Aβ in the brain (Thordardottir et al. 2017). Consistent with the study reported by Moussa et al. (2017), Turner et al. (2015) also found that resveratrol treatment (500 mg/day with 500 mg escalation every 13 weeks up to 52 weeks, ending with 1000 mg twice/day) attenuated the decline in plasma Aβ40 and cerebrospinal fluid Aβ40. From the study reviewed, resveratrol administration increased the brain volume loss compared to those who consumed placebo (Turner et al. 2015). Nonetheless, the underlying mechanisms of brain volume loss require further elucidation; the experimental study indicates that they are not related to functional or cognitive decline (Turner et al. 2015).

Numerous studies revealed that resveratrol administration is linked to the lower risk of several types of cancer. Resveratrol prevents cancers via a few mechanisms including suppress tumor proliferation, metastasis, and angiogenesis, induce

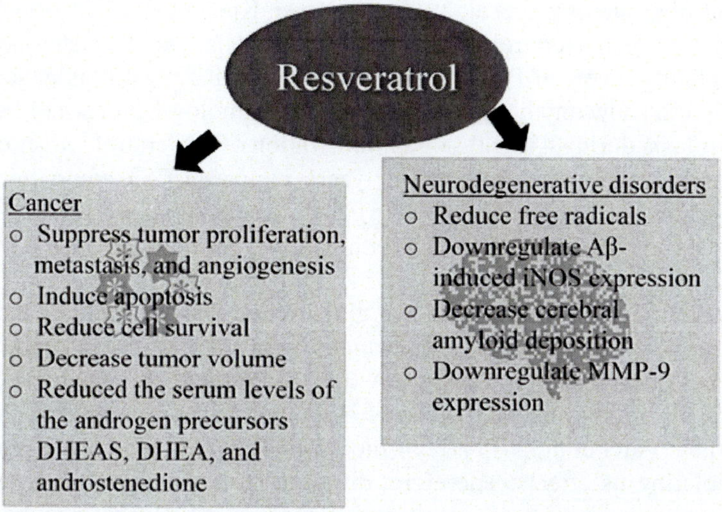

Fig. 6.12 Mechanisms of resveratrol in the modulation of cancer and neurodegenerative disorders. *DHEA* dehydroepiandrosterone, *DHEAS* dehydroepiandrosterone sulphate, *iNOS* inducible nitric oxide synthase, *MMP-9* matrix metalloproteinase-9

apoptosis, and reduce cell survival (Ferraz da Costa et al. 2018; Ko et al. 2017; Wu et al. 2018; Rodríguez-Enríquez et al. 2019) (Fig. 6.12). A study found that feeding 40 mg/kg/day of resveratrol orally for 30 days decreased tumor volume in cigarette smoke condensate transformed MCF-10A-Tr cells-mediated tumors in female xenograft BALB/c mice (Mohapatra et al. 2014). In a further study focused on ovarian cancer, Tan et al. (2016) showed that resveratrol (160 mg/kg intraperitoneally per day for 14 days) suppressed tumor regrowth after cisplatin in a xenograft mouse model of ovarian cancer compared to the vehicle control group, suggesting that resveratrol has the potential to prolong disease-free survival. In line with this, data from population-based studies further demonstrated that patients with colorectal cancer who consumed eight daily doses of resveratrol (0.5 or 1.0 g) before surgical resection reduced tumor cell proliferation by 5%, suggesting that administration of resveratrol at 0.5 or 1.0 g may sufficient elicit anticarcinogenic effect (Patel et al. 2010). The Wnt signaling pathway is known to be involved in the development of colon carcinogenesis (Kirsanov et al. 2020). Nguyen et al. (2009) found that freeze-dried grape powder (80 g/day containing 0.07 mg of resveratrol for 2 weeks) down-regulated the Wnt target genes of the normal mucosa, but had no effect on the cancerous tissues. This finding suggests that freeze-dried grape powder of resveratrol may prevent colon cancer rather than the treatment of colon cancer. In a further study focused on prostate cancer, Kjaer et al. (2015) evaluated the effects of resveratrol (two doses, 150 mg or 1000 mg/day for 4 months) on sex steroid hormones, PSA, and prostate size in middle-aged men with metabolic syndrome. The study showed that administration of 1000 mg resveratrol/day for 4 months significantly reduced the serum levels of the androgen precursors dehydroepiandrosterone sulphate (DHEAS), dehydroepiandrosterone (DHEA), and androstenedione compared to the control group (Kjaer et al. 2015). However, Kjaer et al. (2015) did not identify an association of resveratrol and dihydrotestosterone, free testosterone, testosterone, circulating levels of PSA, or prostate size. In this regard, evidence from this study failed to show any benefit in delaying the pathologic process of benign prostate hyperplasia during 4 months of administration of resveratrol (Kjaer et al. 2015). Intriguingly, a follow-up study for 9 years including 783 community-dwelling women and men aged 65 years and above had revealed that 24 h total urinary resveratrol metabolite was not associated with cancer or inflammatory markers such as IL-1β, IL-6, and CRP (Semba et al. 2014). Resveratrol is metabolized rapidly predominantly into sulfate conjugates and glucuronide that are excreted through urine (Ko et al. 2017). Due to the poor bioavailability of resveratrol, a high dosage (up to 5 g/day) has been used in the studies. A study found that participants who received resveratrol >1 g/day can lead to adverse outcomes such as abdominal pain, nausea, and diarrhea (Patel et al. 2011). Poor bioavailability of resveratrol is a major issue in extrapolating its effects. Therefore, different strategies have been developed to improve its bioavailability (Smoliga and Blanchard 2014), for instance, nanotechnology formulation (Cláudia Santos et al. 2019), micronized powders (Howells et al. 2011; Popat et al. 2013), using prodrug approach (Liang et al. 2013), and combination with phytochemical piperine (Johnson et al. 2011). Taken together, resveratrol might be a useful approach for ameliorating age-related diseases. The crucial role played by resveratrol on chronic diseases worth further elucidation in long-term randomized clinical trials.

References

Abargouei AS, Janghorbani M, Salehi-Marzijarani M et al (2012) Effect of dairy consumption on weight and body composition in adults: a systematic review and meta-analysis of randomized controlled clinical trials. Int J Obes 36:1485–1493

Abdel-Daim MM, Eltaysh R, Hassan A et al (2018) Lycopene attenuates tulathromycin and diclofenac sodium-induced cardiotoxicity in mice. Int J Mol Sci 19:E344

Abdulazeez SS, Ponnusamy P (2016) Antioxidant and hypoglycemic activity of strawberry fruit extracts against alloxan induced diabetes in rats. Pak J Pharm Sci 29:255–260

Abe SK, Inoue M (2020) Green tea and cancer and cardiometabolic diseases: a review of the current epidemiological evidence. Eur J Clin Nutr 75:865–876

Abreu S, Santos R, Moreira C et al (2012a) Association between dairy product intake and abdominal obesity in Azorean adolescents. Eur J Clin Nutr 66:830–835

Abreu S, Santos R, Moreira C et al (2012b) Milk intake is inversely related to body mass index and body fat in girls. Eur J Pediatr 171:1467–1474

Abu-Taweel GM, Attia MF, Hussein J et al (2020) Curcumin nanoparticles have potential antioxidant effect and restore tetrahydrobiopterin levels in experimental diabetes. Biomed Pharmacother 131:110688

Adiwidjaja J, Mclachlan AJ, Boddy AV (2017) Curcumin as a clinically promising anti-cancer agent: pharmacokinetics and drug interactions. Expert Opin Drug Metab Toxicol 13:953–972

Aggarwal BB, Harikumar KB (2009) Potential therapeutic effects of curcumin, the anti-inflammatory agent, against neurodegenerative, cardiovascular, pulmonary, metabolic, autoimmune and neoplastic diseases. Int J Biochem Cell Biol 41:40–59

Akazawa N, Choi Y, Miyaki A et al (2012) Curcumin ingestion and exercise training improve vascular endothelial function in postmenopausal women. Nutr Res 32:795–799

Akyol H, Riciputi Y, Capanoglu E et al (2016) Phenolic compounds in the potato and its byproducts: an overview. Int J Mol Sci 17:835

Alarcón-Flores MI, Romero-González R, Vidal JLM et al (2014) Determination of phenolic compounds in artichoke, garlic and spinach by ultra-high-performance liquid chromatography coupled to Tandem Mass Spectrometry. Food Anal Methods 7:2095–2106

AlBasher G, Abdel-Daim MM, Almeer R et al (2020) Synergistic antioxidant effects of resveratrol and curcumin against fipronil-triggered oxidative damage in male albino rats. Environ Sci Pollut Res 27:6505–6514

Alfaro S, Mutis A, Palma R et al (2013) Influence of genotype and harvest year on polyphenol content and antioxidant activity in murtilla (*Ugni molinae* Turcz) fruit. J Soil Sci Plant Nutr 13:67–78

Alissa EM, Ferns GA (2017) Dietary fruits and vegetables and cardiovascular diseases risk. Crit Rev Food Sci Nutr 57:1950–1962

Allah OMA, El-Debakey F (2010) Prophylactic resveratrol and telmisartan combination ameliorates experimentally-induced diabetic nephropathy in rats, focus on the pro-sclerotic cytokine, transforming growth factor-β1 (TGF-β1). Med J Cairo Univ 78:705–713

Álvarez-Pérez J, Sánchez-Villegas A, Díaz-Benítez EM et al (2016) Influence of a Mediterranean dietary pattern on body fat distribution: results of the PREDIMED–Canarias Intervention randomized trial. J Am Coll Nutr 35:568–580

Alzaid F, Cheung HM, Preedy VR et al (2013) Regulation of glucose transporter expression in human intestinal Caco-2 cells following exposure to an anthocyanin-rich berry extract. PLoS One 8:e78932

Amir M-BA, Faezeh G, Amirreza H et al (2018) Anti-cancer and radio-sensitizing effects of curcumin in nasopharyngeal carcinoma. Curr Pharm Des 24:2121–2128

Anand P, Kunnumakkara AB, Newman RA et al (2007) Bioavailability of curcumin: problems and promises. Mol Pharm 4:807–818

Andrade S, Ramalho MJ, Pereira MDC et al (2018) Resveratrol brain delivery for neurological disorders prevention and treatment. Front Pharm 9:1261

Anuchapreeda S, Leechanachai P, Smith MM et al (2002) Modulation of P-glycoprotein expression and function by curcumin in multidrug-resistant human KB cells. Biochem Pharmacol 64:573–582

Appendino G, Belcaro G, Cornelli U et al (2011) Potential role of curcumin phytosome (Meriva) in controlling the evolution of diabetic microangiopathy. A pilot study. Panminerva Med 53:43–49

Arab L, Liu W, Elashoff D (2009) Green and black tea consumption and risk of stroke. A meta-analysis. Stroke 40:1786–1792

Arranz S, Chiva-Blanch G, Valderas-Martínez P et al (2012) Wine, beer, alcohol and polyphenols on cardiovascular disease and cancer. Nutrients 4:759–781

Asbaghi O, Fouladvand F, Gonzalez MJ et al (2019) The effect of green tea on C-reactive protein and biomarkers of oxidative stress in patients with type 2 diabetes mellitus: a systematic review and meta-analysis. Complement Ther Med 46:210–216

Ashrafizadeh M, Ahmadi Z, Mohammadinejad R et al (2020) Curcumin activates the Nrf2 pathway and induces cellular protection against oxidative injury. Curr Mol Med 20:116–133

Aucoin M, Cooley K, Knee C et al (2017) Fish-derived omega-3 fatty acids and prostate cancer: a systematic review. Integr Cancer Ther 16:32–62

Aune D, Lau R, Chan DS et al (2012) Dairy products and colorectal cancer risk: a systematic review and meta-analysis of cohort studies. Ann Oncol 23:37–45

Aune D, Norat T, Romundstad P et al (2013a) Whole grain and refined grain consumption and the risk of type 2 diabetes: a systematic review and dose-response meta-analysis of cohort studies. Eur J Epidemiol 28:845–858

Aune D, Norat T, Romundstad P et al (2013b) Dairy products and the risk of type 2 diabetes: a systematic review and dose-response meta-analysis of cohort studies. Am J Clin Nutr 98:1066–1083

Aune D, Rosenblatt DAN, Chan DSM et al (2015) Dairy products, calcium, and prostate cancer risk: a systematic review and meta-analysis of cohort studies. Am J Clin Nutr 101:87–117
Aune D, Giovannucci E, Boffetta P et al (2017) Fruit and vegetable intake and the risk of cardiovascular disease, total cancer and all-cause mortality-a systematic review and dose-response meta-analysis of prospective studies. Int J Epidemiol 46:1029–1056
Aune D, Keum N, Giovannucci E et al (2018) Dietary intake and blood concentrations of antioxidants and the risk of cardiovascular disease, total cancer, and all-cause mortality: a systematic review and dose-response meta-analysis of prospective studies. Am J Clin Nutr 108:1069–1091
Azevedo da Silva N, Rodrigues E, Mercadante AZ et al (2014) Phenolic compounds and carotenoids from four fruits native from the Brazilian Atlantic forest. J Agric Food Chem 62:5072–5084
Azzi A (2007) Molecular mechanism of alpha-tocopherol action. Free Radic Biol Med 43:16–21
Bagheri H, Ghasemi F, Barreto GE et al (2020) Effects of curcumin on mitochondria in neurodegenerative diseases. Biofactors 46:5–20
Bahramsoltani R, Rahimi R, Farzaei MH (2017) Pharmacokinetic interactions of curcuminoids with conventional drugs: a review. J Ethnopharmacol 209:1–12
Bastianetto S, Ḿenard C, Quirion R (2015) Neuroprotective action of resveratrol. Biochim Biophys Acta Mol basis Dis 1852:1195–1201
Basu A, Lucas EA (2007) Mechanisms and effects of green tea on cardiovascular health. Nutr Rev 65:361–375
Bayet-Robert M, Kwiatowski F, Leheurteur M et al (2010) Phase I dose escalation trial of docetaxel plus curcumin in patients with advanced and metastatic breast cancer. Cancer Biol Ther 9:8–14
Bellizzi M, Franklin M, Duthie G et al (1994) Vitamin E and coronary heart disease: the European paradox. Eur J Clin Nutr 48:822–831
Bendtsen LQ, Lorenzen JK, Bendsen NT et al (2013) Effect of dairy proteins on appetite, energy expenditure, body weight, and composition: a review of the evidence from controlled clinical trials. Adv Nutr 4:418–438
Bermejo LM, López-Plaza B, Santurino C et al (2019) Milk and dairy product consumption and bladder cancer risk: a systematic review and meta-analysis of observational studies. Adv Nutr 10:S224–S238
Bhatt H, Rompicharla SV, Komanduri N et al (2018) Development of curcumin-loaded solid lipid nanoparticles utilizing glyceryl monostearate as single lipid using QbD approach: characterization and evaluation of anticancer activity against human breast cancer cell line. Curr Drug Deliv 15:1271–1283
Bianchini G, Balko JM, Mayer IA et al (2016) Triple-negative breast cancer: challenges and opportunities of a heterogeneous disease. Nat Rev Clin Oncol 13:674–690
Biasibetti R, Tramontina AC, Costa AP et al (2013) Green tea (−)epigallocatechin-3-gallate reverses oxidative stress and reduces acetylcholinesterase activity in a streptozotocin-induced model of dementia. Behav Brain Res 236:186–193
Bimonte S, Cascella M, Barbieri A et al (2020) Current shreds of evidence on the anticancer role of EGCG in triple negative breast cancer: an update of the current state of knowledge. Infect Agent Cancer 15:2
Binia A, Vargas-Martínez C, Ancira-Moreno M et al (2017) Improvement of cardiometabolic markers after fish oil intervention in young Mexican adults and the role of PPARα L162V and PPARγ2 P12A. J Nutr Biochem 43:98–106
Bo S, Ciccone G, Castiglione A et al (2013) Anti-inflammatory and antioxidant effects of resveratrol in healthy smokers a randomized, double-blind, placebo-controlled, cross-over trial. Curr Med Chem 20:1323–1331
Boarescu P-M, Chirilă I, Bulboacă AE et al (2019) Effects of curcumin nanoparticles in isoproterenol-induced myocardial infarction. Oxidative Med Cell Longev 2019, Article ID 7847142, 13 pages
Boroumand N, Samarghandian S, Hashemy SI (2018) Immunomodulatory, anti-inflammatory, and antioxidant effects of curcumin. J Herbmed Pharmacol 7:211–219

Bose M, Lambert JD, Ju J et al (2008) The major green tea polyphenol, (−)-epigallocatechin-3-gallate, inhibits obesity, metabolic syndrome, and fatty liver disease in high-fat-fed mice. J Nutr 138:1677–1683

Bowen J, Noakes M, Trenerry C et al (2006) Energy intake, ghrelin, and cholecystokinin after different carbohydrate and protein preloads in overweight men. J Clin Endocrinol Metab 91:1477–1483

Boydens C, Pauwels B, Decaluwe K et al (2015) Relaxant and antioxidant capacity of the red wine polyphenols, resveratrol and quercetin, on isolated mice corpora cavernosa. J Sex Med 12:303–312

Braicu C, Pilecki V, Balacescu O et al (2011) The relationships between biological activities and structure of flavan-3-ols. Int J Mol Sci 12:9342–9353

Brewer MS (2011) Natural antioxidants: sources, compounds, mechanisms of action, and potential applications. Compr Rev Food Sci Food Saf 10:221–247

Brito A, Ramirez JE, Areche C et al (2014) HPLC-UV-MS profiles of phenolic compounds and antioxidant activity of fruits from three citrus species consumed in Northern Chile. Molecules 19:17400–17421

Brown AL, Lane J, Coverly J et al (2009) Effects of dietary supplementation with the green tea polyphenol epigallocatechin-3-gallate on insulin resistance and associated metabolic risk factors: randomized controlled trial. Br J Nutr 101:886–894

Brubaker PL, Anini Y (2003) Direct and indirect mechanisms regulating secretion of glucagon-like peptide-1 and glucagon-like peptide-2. Can J Physiol Pharmacol 81:1005–1012

Buil-Cosiales P, Toledo E, Salas-Salvadó J et al (2016) Association between dietary fibre intake and fruit, vegetable or whole-grain consumption and the risk of CVD: results from the PREvención con DIeta MEDiterránea (PREDIMED) trial. Br J Nutr 116:534–546

Bulotta S, Celano M, Lepore SM et al (2014) Beneficial effects of the olive oil phenolic components oleuropein and hydroxytyrosol: focus on protection against cardiovascular and metabolic diseases. J Transl Med 12:219

Bustanji Y, Taha MO, Almasri IM et al (2009) Inhibition of glycogen synthase kinase by curcumin: investigation by simulated molecular docking and subsequent *in vitro*/*in vivo* evaluation. J Enzyme Inhib Med Chem 24:771–778

Byun S-Y, Kim D-B, Kim E (2015) Curcumin ameliorates the tumor-enhancing effects of a high-protein diet in an azoxymethane-induced mouse model of colon carcinogenesis. Nutr Res 35:726–735

Campbell MS, Fleenor BS (2018) The emerging role of curcumin for improving vascular dysfunction: a review. Crit Rev Food Sci Nutr 58:2790–2799

Cantero JL, Atienza M, Stickgold R et al (2003) Sleep-dependent oscillations in the human hippocampus and neocortex. J Neurosci 23:10897

Cao Y, Nimptsch K, Shui IM et al (2015) Prediagnostic plasma IGFBP-1, IGF-1 and risk of prostate cancer. Int J Cancer 136:2418–2426

Capiralla H, Vingtdeux V, Zhao H et al (2012) Resveratrol mitigates lipopolysaccharide- and Aβ-mediated microglial inflammation by inhibiting the TLR4/NF-κB/STAT signaling cascade. J Neurochem 120:461–472

Caporaso N, Formisano D, Genovese A (2018) Use of phenolic compounds from olive mill wastewater as valuable ingredients for functional foods. Crit Rev Food Sci Nutr 58:2829–2841

Carlsson S, Hammar N, Grill V (2005) Alcohol consumption and type 2 diabetes meta-analysis of epidemiological studies indicates a U-shaped relationship. Diabetologia 48:1051–1054

Carroll RE, Benya RV, Turgeon DK et al (2011) Phase IIa clinical trial of curcumin for the prevention of colorectal neoplasia. Cancer Prev Res (Phila) 4:354–364

Carvalho RN, Moura LS, Rosa PTV et al (2005) Supercritical fluid extraction from rosemary (*Rosmarinus officinalis*): kinetic data, extract's global yield, composition, and antioxidant activity. J Supercrit Fluids 35:197–204

Cascella M, Bimonte S, Muzio MR et al (2017) The efficacy of epigallocatechin-3-gallate (green tea) in the treatment of Alzheimer's disease: an overview of pre-clinical studies and translational perspectives in clinical practice. Infect Agent Cancer 12:36

Chagas CEA, Rogero MM, Martini LA (2012) Evaluating the links between intake of milk/dairy products and cancer. Nutr Rev 70:294–300

Chearwae W, Anuchapreeda S, Nandigama K et al (2004) Biochemical mechanism of modulation of human P-glycoprotein (ABCB1) by curcumin I, II, and III purified from Turmeric powder. Biochem Pharmacol 68:2043–2052

Cheeseman KH, Slater TF (1993) An introduction to free radical biochemistry. Br Med Bull 49:481–493

Chen J-H, Ishikawa T, Michiue T et al (2011) An autopsy case of unexpected sudden death due to rupture of a thoracic aortic aneurysm into the left lung. Legal Med 13:201–204

Chen G-f, Xu T-h, Yan Y et al (2017) Amyloid beta: structure, biology and structure-based therapeutic development. Acta Pharmacol Sin 38:1205–1235

Chen J, Zhang L, Li C et al (2020) Lipophilized epigallocatechin gallate derivative exerts anti-proliferation efficacy through induction of cell cycle arrest and apoptosis on DU145 human prostate cancer cells. Nutrients 12:92

Cheng AL, Hsu CH, Lin JK et al (2001) Phase I clinical trial of curcumin, a chemopreventive agent, in patients with high-risk or pre-malignant lesions. Anticancer Res 21:2895–2900

Cheng C, Yu X, McClements DJ et al (2019) Effect of flaxseed polyphenols on physical stability and oxidative stability of flaxseed oil-in-water nanoemulsions. Food Chem 301:125207

Chikara S, Nagaprashantha LD, Singhal J et al (2018) Oxidative stress and dietary phytochemicals: role in cancer chemoprevention and treatment. Cancer Lett 413:122–134

Choi C-H (2005) ABC transporters as multidrug resistance mechanisms and the development of chemosensitizers for their reversal. Cancer Cell Int 5:30

Chu C, Deng J, Man Y et al (2017) Green tea extracts epigallocatechin-3-gallate for different treatments. Biomed Res Int 2017, Article ID 5615647, 9 pages

Chuengsamarn S, Rattanamongkolgul S, Luechapudiporn R et al (2012) Curcumin extract for prevention of type 2 diabetes. Diabetes Care 35:2121–2127

Cicero AFG, Ruscica M, Banach M (2019) Resveratrol and cognitive decline: a clinician perspective. Arch Med Sci 15:936–943

Cimmino L, Neel BG, Aifantis I (2018) Vitamin C in stem cell reprogramming and cancer. Trends Cell Biol 28:698–708

Cláudia Santos A, Pereira I, Pereira-Silva M et al (2019) Nanotechnology-based formulations for resveratrol delivery: effects on resveratrol *in vivo* bioavailability and bioactivity. Colloids Surf B: Biointerfaces 180:127–140

Coban D, Milenkovic D, Chanet A et al (2012) Dietary curcumin inhibits atherosclerosis by affecting the expression of genes involved in leukocyte adhesion and transendothelial migration. Mol Nutr Food Res 56:1270–1281

Cohen A, Burgos-Aceves MA, Bar-Ziv N et al (2019) Cruciferous vegetables consumption and lung cancer prevention: epidemiological studies and molecular mechanisms. J Xiangya Med 4:21

Colica C, Milanović M, Milić N et al (2018) Systematic review on natural antioxidant properties of resveratrol. Nat Prod Commun 13:1195–1203

Collado-González J, Grosso C, Valentão P et al (2017) Inhibition of α-glucosidase and α-amylase by Spanish extra virgin olive oils: the involvement of bioactive compounds other than oleuropein and hydroxytyrosol. Food Chem 235:298–307

Collins QF, Liu HY, Pi J et al (2007) Epigallocatechin-3-gallate (EGCG), a green tea polyphenol, suppresses hepatic gluconeogenesis through 5′-AMP-activated protein kinase. J Biol Chem 282:30143–30149

Cong L, Zhang Y, Huang H et al (2018) DFMG reverses proliferation and migration of vascular smooth muscle cells induced by co-culture with injured vascular endothelial cells via suppression of the TLR4-mediated signaling pathway. Mol Med Rep 17:5692–5699

Cuadrado A, Manda G, Hassan A et al (2018) Transcription factor NRF2 as a therapeutic target for chronic diseases: a systems medicine approach. Pharmacol Rev 70:348–383

Dalle-Donne I, Rossi R, Colombo R et al (2006) Biomarkers of oxidative damage in human disease. Clin Chem 52:601–623

Damiano S, Longobardi C, Andretta E et al (2021) Antioxidative effects of curcumin on the hepatotoxicity induced by ochratoxin A in rats. Antioxidants 10:125

de Florio Almeida J, dos Reis AS, Heldt LFS et al (2017) Lyophilized bee pollen extract: a natural antioxidant source to prevent lipid oxidation in refrigerated sausages. LWT-Food Sci Technol 76:299–305

De Kreutzenberg SV, Ceolotto G, Papparella I et al (2010) Downregulation of the longevity-associated protein sirtuin 1 in insulin resistance and metabolic syndrome: potential biochemical mechanisms. Diabetes 59:1006–1015

de Oliveira PR, da Costa CA, de Bem GF et al (2015) Euterpe oleracea Mart.-derived polyphenols protect mice from diet-induced obesity and fatty liver by regulating hepatic lipogenesis and cholesterol excretion. PLoS One 10:e0143721

de Oliveira MR, da Costa FG, Basil FB et al (2018) Pinocembrin suppresses H2O2-induced mitochondrial dysfunction by a mechanism dependent on the Nrf2/HO-1 axis in SH-SY5Y cells. Mol Neurobiol 55:989–1003

de Souza Ferreira SB, Bruschi ML (2019) Improving the bioavailability of curcumin: is micro/nanoencapsulation the key? Ther Deliv 10:83–86

Dhillon N, Aggarwal BB, Newman RA et al (2008) Phase II trial of curcumin in patients with advanced pancreatic cancer. Clin Cancer Res 14:4491–4499

Di Tomo P, Canali R, Ciavardelli D et al (2012) Beta-carotene and lycopene affect endothelial response to TNF-alpha reducing nitro-oxidative stress and interaction with monocytes. Mol Nutr Food Res 56:217–227

Difonzo G, Pasqualone A, Silletti R et al (2018) Use of olive leaf extract to reduce lipid oxidation of baked snacks. Food Res Int 108:48–56

Djuric Z, Aslam MN, Simon BR et al (2017) Effects of fish oil supplementation on prostaglandins in normal and tumor colon tissue: modulation by the lipogenic phenotype of colon tumors. J Nutr Biochem 46:90–99

Domínguez R, Pateiro M, Gagaoua M et al (2019) A comprehensive review on lipid oxidation in meat and meat products. Antioxidants 8:429

Dong J-Y, Zhang L, He K et al (2011) Dairy consumption and risk of breast cancer: a meta-analysis of prospective cohort studies. Breast Cancer Res Treat 127:23–31

Doyle SL, Donohoe CL, Finn SP et al (2012) IGF-1 and its receptor in esophageal cancer: association with adenocarcinoma and visceral obesity. Am J Gastroenterol 107:196–204

Dröge W (2002) Free radicals in the physiological control of cell function. Physiol Rev 82:47–95

Du G-J, Zhang Z, Wen X-D et al (2012) Epigallocatechin gallate (EGCG) is the most effective cancer chemopreventive polyphenol in green tea. Nutrients 4:1679–1691

Duan C, Ren H, Gao S (2010) Insulin-like growth factors (IGFs), IGF receptors, and IGF-binding proteins: roles in skeletal muscle growth and differentiation. Gen Comp Endocrinol 167:344–351

Dufresne CJ, Farnworth ER (2001) A review of latest research findings on the health promotion properties of tea. J Nutr Biochem 12:404–421

Ejaz A, Wu D, Kwan P et al (2009) Curcumin inhibits adipogenesis in 3T3-L1 adipocytes and angiogenesis and obesity in C57/BL mice. J Nutr 139:919–925

Elbe H, Vardi N, Esrefoglu M et al (2015) Amelioration of streptozotocin-induced diabetic nephropathy by melatonin, quercetin, and resveratrol in rats. Hum Exp Toxicol 34:100–113

Ellulu MS, Khaza'ai H, Abed Y et al (2015) Role of fish oil in human health and possible mechanism to reduce the inflammation. Inflammopharmacology 23:79–89

El-Sayed SM, Youssef AM (2019) Potential application of herbs and spices and their effects in functional dairy products. Heliyon 5:e01989

Eslampour E, Ebrahimzadeh F, Abbasnezhad A et al (2019) Association between circulating irisin and C-reactive protein levels: a systematic review and meta-analysis. Endocrinol Metab (Seoul) 34:140–149

Faghih S, Abadi AR, Hedayati M et al (2011) Comparison of the effects of cows' milk, fortified soy milk, and calcium supplement on weight and fat loss in premenopausal overweight and obese women. Nutr Metab Cardiovasc Dis 21:499–503

Fang W-J, Wang C-J, He Y et al (2018) Resveratrol alleviates diabetic cardiomyopathy in rats by improving mitochondrial function through PGC-1α deacetylation. Acta Pharmacol Sin 39:59–73

FAO (2012) Food outlook—global market analysis. FAO, Rome

Fardet A (2010) New hypotheses for the health-protective mechanisms of whole-grain cereals: what is beyond fibre? Nutr Res Rev 23:65–134

Faria A, Fernandes I, Norberto S et al (2014) Interplay between anthocyanins and gut microbiota. J Agric Food Chem 62:6898–6902

Farkhondeh T, Samarghandian S (2016) Antidotal effects of curcumin against agents-induced cardiovascular toxicity. Cardiovasc Hematol Disord Drug Targets 16:30–37

Farooqui T, Farooqui A (2019) Chapter 2 - Curcumin: historical background, chemistry, pharmacological action, and potential therapeutic value. In: Farooqui T, Farooqui AA (eds) Curcumin for neurological and psychiatric disorders: neurochemical and pharmacological properties. Academic, London, pp 23–44

Farvid MS, Chen WY, Michels KB et al (2016) Fruit and vegetable consumption in adolescence and early adulthood and risk of breast cancer: population based cohort study. BMJ 353:i2343

Fathy Abd-Ellatef G-E, Gazzano E, Chirio D et al (2020) Curcumin-loaded solid lipid nanoparticles bypass P-glycoprotein mediated doxorubicin resistance in triple negative breast cancer cells. Pharmaceutics 12:96

Ferraz da Costa DC, Campos NPC, Santos RA et al (2018) Resveratrol prevents p53 aggregation *in vitro* and in breast cancer cells. Oncotarget 9:29112–29122

Franzese CJ, Bliden KP, Gesheff MG et al (2015) Relation of fish oil supplementation to markers of atherothrombotic risk in patients with cardiovascular disease not receiving lipid-lowering therapy. Am J Cardiol 115:1204–1211

Frid AH, Nilsson M, Holst JJ et al (2005) Effect of whey on blood glucose and insulin responses to composite breakfast and lunch meals in type 2 diabetic subjects. Am J Clin Nutr 82:69–75

Fu L, Lu WQ, Zhou XM (2016) Phenolic compounds and *in vitro* antibacterial and antioxidant activities of three tropic fruits: persimmon, guava, and sweetsop. Biomed Res Int 2016, Article ID 4287461, 9 pages

Fu S, Lv R, Wang L et al (2018) Resveratrol, an antioxidant, protects spinal cord injury in rats by suppressing MAPK pathway. Saudi J Biol Sci 25:259–266

Fuhrman B, Volkova N, Suraski A et al (2001) White wine with red wine-like properties: increased extraction of grape skin polyphenols improves the antioxidant capacity of the derived white wine. J Agric Food Chem 49:3164–3168

Fukushima Y, Tashiro T, Kumagai A et al (2014) Coffee and beverages are the major contributors to polyphenol consumption from food and beverages in Japanese middle-aged women. J Nutr Sci 3:e48

Furusawa Y, Obata Y, Fukuda S et al (2013) Commensal microbe-derived butyrate induces the differentiation of colonic regulatory t cells. Nature 504:446–450

Gambini J, Inglés M, Olaso G et al (2015) Properties of resveratrol: *in vitro* and *in vivo* studies about metabolism, bioavailability, and biological effects in animal models and humans. Oxidative Med Cell Longev 2015, Article ID 837042, 13 pages

Gao H, Geng T, Huang T et al (2017) Fish oil supplementation and insulin sensitivity: a systematic review and meta-analysis. Lipids Health Dis 16:131

Gaschler MM, Stockwell BR (2017) Lipid peroxidation in cell death. Biochem Biophys Res Commun 482:419–425

Ge Y-T, Zhong A-Q, Xu G-F et al (2019) Resveratrol protects BV2 mouse microglial cells against LPS-induced inflammatory injury by altering the miR-146a-5p/TRAF6/NF-κB axis. Immunopharmacol Immunotoxicol 41:549–557

Gems D, Doonan R (2009) Antioxidant defense and aging in *C.* elegans: is the oxidative damage theory of aging wrong? Cell Cycle 8:1681–1687

Geovanini GR, Libby P (2018) Atherosclerosis and inflammation: overview and updates. Clin Sci 132:1243–1252

Gera M, Sharma N, Ghosh M et al (2017) Nanoformulations of curcumin: an emerging paradigm for improved remedial application. Oncotarget 8:66680–66698

Gey KF, Puska P (1989) Plasma vitamins E and A inversely correlated to mortality from ischemic heart disease in cross-cultural epidemiology. Ann N Y Acad Sci 570:268–282

Ghalaut VS, Sangwan L, Dahiya K et al (2012) Effect of imatinib therapy with and without turmeric powder on nitric oxide levels in chronic myeloid leukemia. J Oncol Pharm Pract 18:186–190

Gholamian-Dehkordi N, Luther T, Asadi-Samani M et al (2017) An overview on natural antioxidants for oxidative stress reduction in cancers; a systematic review. Immunopathol Persa 3:e12

Ghorbani Z, Hekmatdoost A, Mirmiran P (2014) Anti-hyperglycemic and insulin sensitizer effects of turmeric and its principle constituent curcumin. Int J Endocrinol Metab 12:e18081

Gianfredi V, Nucci D, Abalsamo A et al (2018) Green tea consumption and risk of breast cancer and recurrence—a systematic review and meta-analysis of observational studies. Nutrients 10:1886

Giovannucci E (1998) Dietary influences of 1,25(OH)2 vitamin D in relation to prostate cancer: a hypothesis. Cancer Causes Contol 9:567–582

Goddard JM, McClements DJ, Decker EA (2012) Innovative technologies in the control of lipid oxidation. Lipid Techol 24:275–277

Godic A, Poljšak B, Adamic M et al (2014) The role of antioxidants in skin cancer prevention and treatment. Oxidative Med Cell Longev 2014, Article ID 860479, 6 pages

Goldberg DM, Yan J, Soleas GJ (2003) Absorption of three wine-related polyphenols in three different matrices by healthy subjects. Clin Biochem 36:79–87

Grundy MM-L, Edwards CH, Mackie AR et al (2016) Re-evaluation of the mechanisms of dietary fibre and implications for macronutrient bioaccessibility, digestion and postprandial metabolism. Br J Nutr 116:816–833

Guillaume MM (2014) Arrêté du 24 juin 2014 établissant la liste des plantes, autres que les champignons, autorisées dans les compléments alimentaires et les conditions de leur emploi; 2014. Nor: ERNC1406332A, JORF, 0163, 11922, T26. http://legifrance.gouv.fr/eli/arrete/2014/6/24/ERNC1406332A/jo/texte

Gunnerud U, Holst JJ, Ostman E et al (2012) The glycemic, insulinemic and plasma amino acid responses to equi-carbohydrate milk meals, a pilot-study of bovine and human milk. Nutr J 11:83

Guo DC, Papke CL, He R et al (2006) Pathogenesis of thoracic and abdominal aortic aneurysms. Ann N Y Acad Sci 1085:339–352

Guo R, Liu B, Wang K et al (2014) Resveratrol ameliorates diabetic vascular inflammation and macrophage infiltration in db/db mice by inhibiting the NF-κB pathway. Diab Vasc Dis Res 11:92–102

Guo X, Yang B, Tan J et al (2016) Associations of dietary intakes of anthocyanins and berry fruits with risk of type 2 diabetes mellitus: a systematic review and meta-analysis of prospective cohort studies. Eur J Clin Nutr 70:1360–1367

Gupta SC, Sung B, Kim JH et al (2013) Multitargeting by turmeric, the golden spice: from kitchen to clinic. Mol Nutr Food Res 57:1510–1528

Hajihashemi P, Haghighatdoost F (2019) Effects of whole-grain consumption on selected biomarkers of systematic inflammation: a systematic review and meta-analysis of randomized controlled trials. J Am Coll Nutr 38:275–285

Hall WL, Millward DJ, Long SJ et al (2003) Casein and whey exert different effects on plasma amino acid profiles, gastrointestinal hormone secretion and appetite. Br J Nutr 89:239–248

Halliwell B, Gutteridge JM (1999) Free radicals in biology and medicine, 3rd edn. Oxford University Press, Midsomer Norton

Hamadi N, Mansour A, Hassan MH et al (2012) Ameliorative effects of resveratrol on liver injury in streptozotocin-induced diabetic rats. J Biochem Mol Toxicol 26:384–392

Han A, MacDonald A, Ahmed B et al (2017) Butyrate regulates its own metabolic fate as an HDAC inhibitor in colorectal cancer cells. FASEB J 31:300–302

Hao Q, Chen X, Wang X et al (2014) Curcumin attenuates Angiotensin II-induced abdominal aortic aneurysm by inhibition of inflammatory response and ERK signaling pathways. Evid Based Complement Alternat Med 2014, Article ID 270930, 10 pages

Hasan MM, Baek K-H (2013) Induction of resveratrol biosynthesis in grape skins and leaves by ultrasonication treatment. Kor J Hortic Sci Technol 31:496–502

Hatcher H, Planalp R, Cho J et al (2008) Curcumin: from ancient medicine to current clinical trials. Cell Mol Life Sci 65:1631–1652

Hayes JD, Dinkova-Kostova AT, Tew KD (2020) Oxidative stress in cancer. Cancer Cell 38:167–197

He Y, Yue Y, Zheng X et al (2015) Curcumin, inflammation, and chronic diseases: how are they linked? Molecules 20:9183–9213

Hecht F, Pessoa CF, Gentile LB et al (2016) The role of oxidative stress on breast cancer development and therapy. Tumor Biol 37:4281–4291

Heller R, Werner-Felmayer G, Werner ER (2004) Alpha-tocopherol and endothelial nitric oxide synthesis. Ann N Y Acad Sci 1031:74–85

Hemler EC, Hu FB (2019) Plant-based diets for cardiovascular disease prevention: all plant foods are not created equal. Curr Atheroscler Rep 21:18

Hernández L, Afonso D, Rodríguez EM et al (2011) Phenolic compounds in wheat grain cultivars. Plant Foods Hum Nutr 66:408–415

Herzig S, Shaw RJ (2018) AMPK: guardian of metabolism and mitochondrial homeostasis. Nat Rev Mol Cell Biol 19:121–135

Hibino N, Duncan DR, Nalbandian A et al (2012) Evaluation of the use of an induced puripotent stem cell sheet for the construction of tissue-engineered vascular grafts. J Thorac Cardiovasc Surg 143:696–703

Higa R, Roberti SL, Musikant D et al (2014) Effects of maternal dietary olive oil on pathways involved in diabetic embryopathy. Reprod Toxicol 49:185–195

Higgins JA (2012) Whole grains, legumes, and the subsequent meal effect: implications for blood glucose control and the role of fermentation. J Nutr Metab 2012:829238

Hitchon CA, El-Gabalawy HS (2004) Oxidation in rheumatoid arthritis. Arthritis Res Ther 6:265–278

Hoang DV, Pham NM, Lee AH et al (2018) Dietary carotenoid intakes and prostate cancer risk: a case-control study from Vietnam. Nutrients 10:70

Hodge AM, English DR, O'Dea K et al (2006) Alcohol intake, consumption pattern and beverage type, and the risk of type 2 diabetes. Diabet Med 23:690–697

Hodgson JM, Croft KD (2010) Tea flavonoids and cardiovascular health. Mol Asp Med 31:495–502

Hollænder PL, Ross AB, Kristensen M (2015) Whole-grain and blood lipid changes in apparently healthy adults: a systematic review and meta-analysis of randomized controlled studies. Am J Clin Nutr 102:556–572

Holstein A, Beil W (2009) Oral antidiabetic drug metabolism: pharmacogenomics and drug interactions. Expert Opin Drug Metab Toxicol 5:225–241

Honarbakhsh S, Schachter M (2009) Vitamins and cardiovascular disease. Br J Nutr 101:1113–1131

Howard AA, Arnsten JH, Gourevitch MN (2004) Effect of alcohol consumption on diabetes mellitus: a systematic review. Ann Intern Med 140:211–219

Howells LM, Berry DP, Elliott PJ et al (2011) Phase I randomized, double-blind pilot study of micronized resveratrol (SRT501) in patients with hepatic metastases—safety, pharmacokinetics, and pharmacodynamics. Cancer Prev Res 4:1419–1425

Howells LM, Iwuji CO, Irving GR et al (2019) Curcumin combined with FOLFOX chemotherapy is safe and tolerable in patients with metastatic colorectal cancer in a randomized phase IIa trial. J Nutr 149:1133–1139

Hua J, Guerin KI, Chen J et al (2011) Resveratrol inhibits pathologic retinal neovascularization in Vldlr-/- mice. Invest Ophthalmol Vis Sci 52:2809–2816

Huang D (2018) Dietary antioxidants and health promotion. Antioxidants (Basel) 7:9

Huang MT, Newmark HL, Frenkel K (1997) Inhibitory effects of curcumin on tumorigenesis in mice. J Cell Biochem 27:26–34

Huang T-C, Lu K-T, Wo Y-YP et al (2011) Resveratrol protects rats from Aβ-induced neurotoxicity by the reduction of iNOS expression and lipid peroxidation. PLoS One 6:e29102

Huang LH, Liu CY, Wang LY et al (2018) Effects of green tea extract on overweight and obese women with high levels of low density-lipoprotein-cholesterol (LDL-C): a randomised, double-blind, and cross-over placebo-controlled clinical trial. BMC Complement Altern Med 18:294

Huang S-J, Huang J, Yan Y-B et al (2020) The renoprotective effect of curcumin against cisplatin-induced acute kidney injury in mice: involvement of miR-181a/PTEN axis. Ren Fail 42:350–357

Huncharek M, Muscat J, Kupelnick B (2008) Dairy products, dietary calcium and vitamin D intake as risk factors for prostate cancer: a meta-analysis of 26,769 cases from 45 observational studies. Nutr Cancer 60:421–441

Hung PV (2016) Phenolic compounds of cereals and their antioxidant capacity. Crit Rev Food Sci Nutr 56:25–35

Hur KY, Lee M-S (2015) Gut microbiota and metabolic disorders. Diabetes Metab J 39:198–203

Hussein MA, El-Maksoud HA (2013) Biochemical effects of resveratrol and curcumin combination on obese diabetic rats. Mol Clin Pharmacol 4:1–10

Ide K, Yamada H, Takuma N et al (2014) Green tea consumption affects cognitive dysfunction in the elderly: a pilot study. Nutrients 6:4032–4042

Incalza MA, D'Oria R, Natalicchio A et al (2018) Oxidative stress and reactive oxygen species in endothelial dysfunction associated with cardiovascular and metabolic diseases. Vasc Pharmacol 100:1–19

Ingles M, Gambini J, Miguel MG et al (2014) PTEN mediates the antioxidant effect of resveratrol at nutritionally relevant concentrations. Biomed Res Int 2014, Article ID 580852, 6 pages

Islam MT (2017) Oxidative stress and mitochondrial dysfunction-linked neurodegenerative disorders. Neurol Res 39:73–82

Jacobsen C, Sørensen A-DM, Holdt SL et al (2019) Source, extraction, characterization, and applications of novel antioxidants from seaweed. Annu Rev Food Sci Technol 10:541–568

Jamali N, Sorenson CM, Sheibani N (2018) Vitamin D and regulation of vascular cell function. Am J Physiol Heart Circ Physiol 314:H753–H765

James MI, Iwuji C, Irving G et al (2015) Curcumin inhibits cancer stem cell phenotypes in ex vivo models of colorectal liver metastases, and is clinically safe and tolerable in combination with FOLFOX chemotherapy. Cancer Lett 364:135–141

Jang YC, Remmen HV (2009) The mitochondrial theory of aging: insight from transgenic and knockout mouse models. Exp Gerontol 44:256–260

Järvi A, Karlström B, Vessby B et al (2016) Increased intake of fruits and vegetables in overweight subjects: effects on body weight, body composition, metabolic risk factors and dietary intake. Br J Nutr 115:1760–1768

Jędrejek D, Kontek B, Lis B et al (2017) Evaluation of antioxidant activity of phenolic fractions from the leaves and petals of dandelion in human plasma treated with H2O2 and H2O2/Fe. Chem Biol Interact 262:29–37

Ji LL, Gomez-Cabrera MC, Vina J (2006) Exercise and hormesis: activation of cellular antioxidant signaling pathway. Ann N Y Acad Sci 1067:425–435

Jiang Q, Lei Y-H, Krishnadath DC et al (2021) Curcumin regulates EZH2/Wnt/β-catenin pathway in the mandible and femur of ovariectomized osteoporosis rats. Kaohsiung J Med Sci 37:513–519

Jianwen YXWBT, Chenghai G (2006) Determination of the phenolic compounds in pineapple by high performance liquid chromatography. Food Fermentation Industries 2006-02

Jiménez-Osorio AS, García-Niño WR, González-Reyes S et al (2016) The effect of dietary supplementation with curcumin on redox status and Nrf2 activation in patients with nondiabetic or diabetic proteinuric chronic kidney disease: a pilot study. J Ren Nutr 26:237–244

Johnson JJ, Nihal M, Siddiqui IA et al (2011) Enhancing the bioavailability of resveratrol by combining it with piperine. Mol Nutr Food Res 55:1169–1176

Jordan BC, Mock CD, Thilagavathi R et al (2016) Molecular mechanisms of curcumin and its semisynthetic analogues in prostate cancer prevention and treatment. Life Sci 152:135–144

Josse AR, Atkinson SA, Tarnopolsky MA et al (2011) Increased consumption of dairy foods and protein during diet- and exercise induced weight loss promotes fat mass loss and lean mass gain in overweight and obese premenopausal women. J Nutr 141:1626–1634

Jurgoński A, Juśkiewicz J, Zduńczyk Z (2013) An anthocyanin-rich extract from Kamchatka honeysuckle increases enzymatic activity within the gut and ameliorates abnormal lipid and glucose metabolism in rats. Nutrition 29:898–902

Kabtni S, Sdouga D, Bettaib IB et al (2020) Influence of climate variation on phenolic composition and antioxidant capacity of Medicago minima populations. Sci Rep 10:8293

Kanai M, Yoshimura K, Asada M et al (2011) A phase I/II study of gemcitabine-based chemotherapy plus curcumin for patients with gemcitabine-resistant pancreatic cancer. Cancer Chemother Pharmacol 68:157–164

Kapoor MP, Sugita M, Fukuzawa Y et al (2017) Physiological effects of epigallocatechin-3-gallate (EGCG) on energy expenditure for prospective fat oxidation in humans: a systematic review and meta-analysis. J Nutr Biochem 43:1–10

Karimian MS, Pirro M, Johnston TP et al (2017) Curcumin and endothelial function: evidence and mechanisms of protective effects. Curr Pharm Des 23:2462–2473

Karthikeyan A, Senthil N, Min T (2020) Nanocurcumin: a promising candidate for therapeutic applications. Front Pharmacol 11:487

Karuppagounder V, Arumugam S, Giridharan VV et al (2017) Tiny molecule, big power: multi-target approach for curcumin in diabetic cardiomyopathy. Nutrition 34:47–54

Kato M, Nishikawa S, Ikehata A et al (2017) Curcumin improves glucose tolerance via stimulation of glucagon-like peptide-1 secretion. Mol Nutr Food Res 61:1600471

Keck S, Nitsch R, Grune T et al (2003) Proteasome inhibition by paired helical filament-tau in brains of patients with Alzheimer's disease. J Neurochem 85:115–122

Keller JN, Hanni KB, Markesbery WR (2000) Impaired proteasome function in Alzheimer's disease. J Neurochem 75:436–439

Khan H, Ullah H, Nabavi SM (2019) Mechanistic insights of hepatoprotective effects of curcumin: therapeutic updates and future prospects. Food Chem Toxicol 124:182–191

Khanam UKS, Oba S, Yanase E et al (2012) Phenolic acids, flavonoids and total antioxidant capacity of selected leafy vegetables. J Funct Foods 4:979–987

Kim JA, Jung WS, Chun SC et al (2006) A correlation between the level of phenolic compounds and the antioxidant capacity in cooked-with-rice and vegetable soybean (*Glycine max* L.) varieties. Eur Food Res Technol 224:259–270

Kim YH, Kim YS, Roh GS et al (2012) Resveratrol blocks diabetes-induced early vascular lesions and vascular endothelial growth factor induction in mouse retinas. Acta Ophthalmol 90:e31–e37

Kim BH, Lee ES, Choi R et al (2016) Protective effects of curcumin on renal oxidative stress and lipid metabolism in a rat model of type 2 diabetic nephropathy. Yonsei Med J 57:664–673

Kim H, Caulfield LE, Garcia-Larsen V et al (2019) Plant-based diets are associated with a lower risk of incident cardiovascular disease, cardiovascular disease mortality, and all-cause mortality in a general population of middle-aged adults. J Am Heart Assoc 8:e012865

Kirsanov K, Fetisov T, Lesovaya EA et al (2020) Prevention of colorectal carcinogenesis by DNA-binding small-molecule curaxin CBL0137 involves suppression of Wnt signaling. Cancer Prev Res 13:53–64

Kjaer TN, Ornstrup MJ, Poulsen MM et al (2015) Resveratrol reduces the levels of circulating androgen precursors but has no effect on, testosterone, dihydrotestosterone, PSA levels or prostate volume. A 4-month randomised trial in middle-aged men. Prostate 75:1255–1263

Ko J-H, Sethi G, Um J-Y et al (2017) The role of resveratrol in cancer therapy. Int J Mol Sci 18:2589

Kocaadam B, Şanlier N (2017) Curcumin, an active component of turmeric (Curcuma longa), and its effects on health. Crit Rev Food Sci Nutr 57:2889–2895

Koo SI, Noh SK (2007) Green tea as inhibitor of the intestinal absorption of lipids: potential mechanism for its lipid-lowering effect. J Nutr Biochem 18:179–183

Kröger J, Schulze MB (2014) Dietary fish intake and the risk for type 2 diabetes and cardiovascular disease: new insights. Curr Opin Lipidol 25:228–229

Kubota S, Ozawa Y, Kurihara T et al (2012) Roles of AMP-activated protein kinase in diabetes-induced retinal inflammation. Invest Ophthalmol Vis Sci 52:9142–9148

Kulkarni SS, Cantó C (2015) The molecular targets of resveratrol. Biochim Biophys Acta (BBA) Mol Basis Dis 1852:1114–1123

Kumar A, Kaundal RK, Iyer S et al (2007) Effects of resveratrol on nerve functions, oxidative stress and DNA fragmentation in experimental diabetic neuropathy. Life Sci J 80:1236–1244

Kumar AHS, Hynes BG, Mueen AKK (2009) Natural products as regulators of progenitor cells: role in cardiovascular regenerative medicine. J Young Pharm 1:28–35

Kumar A, Negi G, Sharma SS (2013) Neuroprotection by resveratrol in diabetic neuropathy: concepts & mechanisms. Curr Med Chem 20:4640–4645

Kuriyama S, Shimazu T, Ohmori K et al (2006) Green tea consumption and mortality due to cardiovascular disease, cancer, and all causes in Japan: the Ohsaki study. JAMA 296:1255–1265

Kurutas EB (2016) The importance of antioxidants which play the role in cellular response against oxidative/nitrosative stress: current state. Nutr J 15:71

Kwee EM, Niemeyer ED (2011) Variations in phenolic composition and antioxidant properties among 15 basil (*Ocimum basilicum* L.) cultivars. Food Chem 128:1044–1050

La X, Zhang L, Li Z et al (2019) (−)-Epigallocatechin gallate (EGCG) enhances the sensitivity of colorectal cancer cells to 5-FU by inhibiting GRP78/NF-κB/miR-155-5p/MDR1 pathway. J Agric Food Chem 67:2510–2518

Lagouge M, Argmann C, Gerhart-Hines Z et al (2006) Resveratrol improves mitochondrial function and protects against metabolic disease by activating SIRT1 and PGC-1alpha. Cell 127:1109–1122

Lanzotti V (2006) The analysis of onion and garlic. J Chromatogr A 1112:3–22

Lazarovici A, Kalliadasis S, Merkin JH et al (2004) The propagation and inhibition of an exothermic branched-chain flame with an endothermic reaction and radical scavenging. J Eng Math 49:41–55

Lee M-S, Kim C-T, Kim Y (2009) Green tea (−)-epigallocatechin-3-gallate reduces body weight with regulation of multiple genes expression in adipose tissue of diet-induced obese mice. Ann Nutr Metab 54:151–157

Lee SG, Kim B, Yang Y et al (2014) Berry anthocyanins suppress the expression and secretion of proinflammatory mediators in macrophages by inhibiting nuclear translocation of NF-κB independent of NRF2-mediated mechanism. J Nutr Biochem 25:404–411

Lee H, Im SW, Jung CH et al (2016) Tyrosol, an olive oil polyphenol, inhibits ER stress-induced apoptosis in pancreatic β-cell through JNK signaling. Biochem Biophys Res Commun 469:748–752

Lee H, Oh I-N, Kim J et al (2018a) Phenolic compound profiles and their seasonal variations in new red-phenotype head-forming Chinese cabbages. LWT 90:433–439

Lee J, Shin A, Oh JH et al (2018b) The relationship between nut intake and risk of colorectal cancer: a case control study. Nutr J 17:37

Leenders M, Leufkens AM, Siersema PD et al (2014) Plasma and dietary carotenoids and vitamins A, C and E and risk of colon and rectal cancer in the European Prospective Investigation into Cancer and Nutrition. Int J Cancer 135:2930–2939

Leighton T, Ginther C, Fluss L et al (1992) Molecular characterization of quercetin and quercetin glycosides in Allium vegetables: their effects on malignant cell transformation. In: Huang MT, Ho CT, Lee CY (eds) Phenolic compounds from food and their effect on health, ACS Symposium Series, vol 507. American Chemical Society, Washington, DC, pp 220–238

Li Y, Xu Q, Wei P et al (2014a) Chitosan oligosaccharides downregulate the expression of E-selectin and ICAM-1 induced by LPS in endothelial cells by inhibiting MAP kinase signaling. Int J Mol Med 33:392–400

Li M, Fan Y, Zhang X et al (2014b) Fruit and vegetable intake and risk of type 2 diabetes mellitus: meta-analysis of prospective cohort studies. BMJ Open 4:e005497

Li D, Wang P, Luo Y et al (2017a) Health benefits of anthocyanins and molecular mechanisms: update from recent decade. Crit Rev Food Sci Nutr 57:1729–1741

Li X, Fang Q, Tian X et al (2017b) Curcumin attenuates the development of thoracic aortic aneurysm by inhibiting VEGF expression and inflammation. Mol Med Rep 16:4455–4462

Li F, Gao C, Yan P et al (2018) EGCG reduces obesity and white adipose tissue gain partly through AMPK activation in mice. Front Pharmacol 9:1366

Liang L, Liu X, Wang Q et al (2013) Pharmacokinetics, tissue distribution and excretion study of resveratrol and its prodrug 3,5,40-tri-O-acetylresveratrol in rats. Phytomedicine 20:558–563

Liguori I, Russo G, Curcio F et al (2018) Oxidative stress, aging, and diseases. Clin Interv Aging 13:757–772

Limón-Pacheco JH, Hernandez NA, Fanjul-Moles ML et al (2007) Glutathione depletion activates mitogen-activated protein kinase (MAPK) pathways that display organ-specific responses and brain protection in mice. Free Radic Biol Med 43:1335–1347

Lin CL, Chen TF, Chiu MJ et al (2009) Epigallocatechin gallate (EGCG) suppresses beta-amyloid-induced neurotoxicity through inhibiting c-Abl/FE65 nuclear translocation and GSK3 beta activation. Neurobiol Aging 30:81–92

Lin D, Xiao M, Zhao J et al (2016) An overview of plant phenolic compounds and their importance in human nutrition and management of type 2 diabetes. Molecules 21:1374

Lin X, Bai D, Wei Z et al (2019) Curcumin attenuates oxidative stress in RAW264.7 cells by increasing the activity of antioxidant enzymes and activating the Nrf2-Keap1 pathway. PLoS One 14:e0216711

Liu X, Chen K, Zhu L et al (2018) Soyasaponin Ab protects against oxidative stress in HepG2 cells via Nrf2/HO-1/NQO1 signaling pathways. J Funct Foods 45:110–117

Lohan SB, Vitt K, Scholz P et al (2018) ROS production and glutathione response in keratinocytes after application of β-carotene and VIS/NIR irradiation. Chem Biol Interact 280:1–7

Lorenz P, Roychowdhury S, Engelmann M et al (2003) Oxyresveratrol and resveratrol are potent antioxidants and free radical scavengers: effect on nitrosative and oxidative stress derived from microglial cells. Nitric Oxide 9:64–76

Lourenço SC, Moldão-Martins M, Alves VD (2019) Antioxidants of natural plant origins: from sources to food industry applications. Molecules 24:4132

Lu PH, Hsu CH (2016) Does supplementation with green tea extract improve acne in post-adolescent women? A randomized, double-blind, and placebo-controlled clinical trial. Complement Ther Med 25:159–163

Lü J-M, Lin PH, Yao Q et al (2010) Chemical and molecular mechanisms of antioxidants: experimental approaches and model systems. J Cell Mol Med 14:840–860

Ludwig D (2002) The glycemic index: physiological mechanisms relating to obesity, diabetes, and cardiovascular disease. JAMA 287:2414–2423

Ma Y, Zhao L, Gao M et al (2018) Tea polyphenols protect bovine mammary epithelial cells from hydrogen peroxide-induced oxidative damage *in vitro*. J Anim Sci 96:4159–4172

Magyar K, Halmosi R, Palfi A et al (2012) Cardioprotection by resveratrol: a human clinical trial in patients with stable coronary artery disease. Clin Hemorheol Microcirc 50:179–187

Mai F, Glomb MA (2013) Isolation of phenolic compounds from iceberg lettuce and impact on enzymatic browning. J Agric Food Chem 61:2868–2874

Malin SK, Kullman EL, Scelsi AR et al (2018) A whole-grain diet reduces peripheral insulin resistance and improves glucose kinetics in obese adults: a randomized-controlled trial. Metabolism 82:111–117

Man GCW, Wang J, Song Y et al (2020) Therapeutic potential of a novel prodrug of green tea extract in induction of apoptosis via ERK/JNK and Akt signaling pathway in human endometrial cancer. BMC Cancer 20:964

Mancini E, Beglinger C, Drewe J et al (2017) Green tea effects on cognition, mood and human brain function: a systematic review. Phytomedicine 34:26–37

Mancini FR, Affret A, Dow C et al (2018) Dietary antioxidant capacity and risk of type 2 diabetes in the large prospective E3N-EPIC cohort. Diabetologia 61:308–316

Mandair D, Rossi RE, Pericleous M et al (2014) Prostate cancer and the influence of dietary factors and supplements: a systematic review. Nutr Metab 11:30

Mankowski RT, You L, Buford TW et al (2020) Higher dose of resveratrol elevated cardiovascular disease risk biomarker levels in overweight older adults – a pilot study. Exp Gerontol 131:110821

Mansour-Ghanaei F, Pourmasoumu M, Hadi A et al (2019) Efficacy of curcumin/turmeric on liver enzymes in patients with non-alcoholic fatty liver disease: a systematic review of randomized controlled trials. Integr Med Res 8:57–61

Mao X, Gu C, Chen D et al (2017) Oxidative stress-induced diseases and tea polyphenols. Oncotarget 8:81649–81661

Marilo O-V, Enrique B-C, Maria H-L et al (2018) Plant-derived polyphenols in human health: biological activity, metabolites and putative molecular targets. Curr Drug Metab 19:351–369

Maritim AC, Sanders RA, Watkins JB 3rd (2003) Diabetes, oxidative stress, and antioxidants: a review. J Biochem Mol Toxicol 17:24–38

Martini S, Conte A, Tagliazucchi D (2017) Phenolic compounds profile and antioxidant properties of six sweet cherry (*Prunus avium*) cultivars. Food Res Int 97:15–26

Martin-Nizard F, Boullier A, Fruchart JC et al (1998) Alpha-tocopherol but not beta-tocopherol inhibits thrombin-induced PKC activation and endothelin secretion in endothelial cells. J Cardiovasc Risk 5:339–345

Marventano S, Vetrani C, Vitale M et al (2017) Whole grain intake and glycaemic control in healthy subjects: a systematic review and meta-analysis of randomized controlled trials. Nutrients 9:E769

Mattes RD, Dreher ML (2010) Nuts and healthy body weight maintenance mechanisms. Asia Pac J Clin Nutr 19:137–141

Mazidi M, Karimi E, Rezaie P et al (2017) The impact of vitamin D supplement intake on vascular endothelial function; a systematic review and meta-analysis of randomized controlled trials. Food Nutr Res 61:1273574

Miean KH, Mohamed S (2001) Flavonoid (myricetin, quercetin, kaempferol, luteolin, and apigenin) content of edible tropical plants. J Agric Food Chem 49:3106–3112

Mihailović NR, Mihailović VB, Kreft S et al (2018) Analysis of phenolics in the peel and pulp of wild apples (*Malus sylvestris* (L.) Mill.). J Food Compos Anal 67:1–9

Miller K, Feucht W, Schmid M (2019) Bioactive compounds of strawberry and blueberry and their potential health effects based on human intervention studies: a brief overview. Nutrients 11:1510

Milne JC, Lambert PD, Schenk S et al (2007) Small molecule activators of SIRT1 as therapeutics for the treatment of type 2 diabetes. Nature 450:712–716

Mohammadshahi M, Haidari F, Soufi FG (2014) Chronic resveratrol administration improves diabetic cardiomyopathy in part by reducing oxidative stress. J Cardiol 21:39–46

Mohapatra P, Satapathy SR, Das D et al (2014) Resveratrol mediated cell death in cigarette smoke transformed breast epithelial cells is through induction of p21Waf1/Cip1 and inhibition of long patch base excision repair pathway. Toxicol Appl Pharmacol 275:221–231

Moldogazieva NT, Mokhosoev IM, Feldman NB et al (2018) ROS and RNS signalling: adaptive redox switches through oxidative/nitrosative protein modifications. Free Redic Res 52:507–543

Monfoulet LE, Mercier S, Bayle D et al (2017) Curcumin modulates endothelial permeability and monocyte transendothelial migration by affecting endothelial cell dynamics. Free Radic Biol Med 112:109–120

Montazeri M, Sadeghizadeh M, Pilehvar-Soltanahmadi Y et al (2016) Dendrosomal curcumin nanoformulation modulate apoptosis-related genes and protein expression in hepatocarcinoma cell lines. Int J Pharm 509:244–254

Morrison MEW, Hobika EG, Joseph JM et al (2021) Cruciferous vegetable consumption and pancreatic cancer: a case-control study. Cancer Epidemiol 72:101924

Motavaf M, Sadeghizadeh M, Babashah S et al (2020) Protective effects of a nano-formulation of curcumin against cuprizone-induced demyelination in the mouse corpus callosum. Iran J Pharm Res 19:310–320

Moussa C, Hebron M, Huang X et al (2017) Resveratrol regulates neuro-inflammation and induces adaptive immunity in Alzheimer's disease. J Neuroinflammation 14:1

Mozos I, Stoian D, Caraba A et al (2018) Lycopene and vascular health. Front Pharmacol 9:521

Muhammad I, Wang H, Sun X et al (2018) Dual role of dietary curcumin through attenuating AFB_1-induced oxidative stress and liver injury via modulating liver phase-I and phase-II enzymes involved in AFB_1 bioactivation and detoxification. Front Pharmacol 9:554

Musial C, Kuban-Jankowska A, Gorska-Ponikowska M (2020) Beneficial properties of green tea catechins. Int J Mol Sci 21:1744

Mustata GT, Rosca M, Biemel KM et al (2005) Paradoxical effects of green tea (*Camellia sinensis*) and antioxidant vitamins in diabetic rats: improved retinopathy and renal mitochondrial defects but deterioration of collagen matrix glycoxidation and cross-linking. Diabetes 54:517–526

Na LX, Li Y, Pan HZ et al (2013) Curcuminoids exert glucose-lowering effect in type 2 diabetes by decreasing serum free fatty acids: a double-blind, placebo-controlled trial. Mol Nutr Food Res 57:1569–1577

Na LX, Yan BL, Jiang S et al (2014) Curcuminoids target decreasing serum adipocyte-fatty acid binding protein levels in their glucose-lowering effect in patients with type 2 diabetes. Biomed Environ Sci 27:902–906

Nabavi SF, Daglia M, Moghaddam AH et al (2014) Curcumin and liver disease: from chemistry to medicine. Compr Rev Food Sci Food Saf 13:62–77

Nabofa WEE, Alashe OO, Oyeyemi OT et al (2018) Cardioprotective effects of curcumin-nisin based poly lactic acid nanoparticle on myocardial infarction in guinea pigs. Sci Rep 8:16649

Namita P, Mukesh R, Vijay KJ (2012) Camellia Sinensis (Green tea): a review. Global J Pharmacol 6:52–59

Neerati P, Devde R, Gangi AK (2014) Evaluation of the effect of curcumin capsules on glyburide therapy in patients with type-2 diabetes mellitus. Phytother Res 28:1796–1800

Neha K, Haider MR, Pathak A et al (2019) Medicinal prospects of antioxidants: a review. Eur J Med Chem 178:687–704

Nelson KM, Dahlin JL, Bisson J et al (2017) The essential medicinal chemistry of curcumin: mini perspective. J Med Chem 60:1620–1637

Nguyen AV, Martinez M, Stamos MJ et al (2009) Results of a phase I pilot clinical trial examining the effect of plant-derived resveratrol and grape powder on Wnt pathway target gene expression in colonic mucosa and colon cancer. Cancer Manag Res 1:25–37

Nimse SB, Pal D (2015) Free radicals, natural antioxidants, and their reaction mechanisms. RSC Adv 5:27986–28006

Norat T, Riboli E (2003) Dairy products and colorectal cancer: a review of possible mechanisms and epidemiological evidence. Eur J Clin Nutr 57:1–17

Oak MH, Chataigneau M, Keravis T et al (2003) Red wine polyphenolic compounds inhibit vascular endothelial growth factor expression in vascular smooth muscle cells by preventing the activation of the p38 mitogen-activated protein kinase pathway. Arterioscler Thromb Vasc Biol 23:1001–1007

Oh DY, Walenta E, Akiyama TE et al (2014) A Gpr120-selective agonist improves insulin resistance and chronic inflammation in obese mice. Nat Med 20:942–947

Olalere OA, Abdurahman NH, Mohd Yunus R et al (2019) Mineral element determination and phenolic compounds profiling of oleoresin extracts using an accurate mass LC-MS-QTOF and ICP-MS. J King Saud Univ Sci 31:859–863

Oliver JM, Stoner L, Rowlands DS et al (2016) Novel form of curcumin improves endothelial function in young, healthy individuals: a double-blind placebo controlled study. J Nutr Metab 2016, Article ID 1089653, 6 pages

Olsen H, Aaby K, Borge GIA (2009) Characterization and quantification of flavonoids and hydroxycinnamic acids in curly kale (*Brassica oleracea* L. Convar. *acephala* Var. *sabellica*) by HPLC-DAD-ESI-MS^n. J Agric Food Chem 57:2816–2825

Omidian M, Abdolahi M, Daneshzad E et al (2020) The effects of resveratrol on oxidative stress markers: a systematic review and meta-analysis of randomized clinical trials. Endocr Metab Immune Disord Drug Targets 20:718–727

Ostadrahimi A, Taghizadeh A, Mobasseri M et al (2015) Effect of probiotic fermented milk (Kefir) on glycemic control and lipid profile in type 2 diabetic patients: a randomized double-blind placebo-controlled clinical trial. Iran J Public Health 44:228–237

Oyenihi OR, Oyenihi AB, Adeyanju AA et al (2016) Antidiabetic effects of resveratrol: the way forward in its clinical utility. J Diabetes Res 2016, Article ID 9737483, 14 pages

Öztürk E, Arslan AKK, Yerer MB et al (2017) Resveratrol and diabetes: a critical review of clinical studies. Biomed Pharmacother 95:230–234

Pal S, Ellis V, Dhaliwal S (2010) Effects of whey protein isolate on body composition, lipids, insulin and glucose in overweight and obese individuals. Br J Nutr 104:716–723

Pala V, Sieri S, Berrino F et al (2011) Yogurt consumption and risk of colorectal cancer in the Italian EPIC cohort. Int J Cancer 129:2712–2719

Palafox-Carlos H, Yahia EM, González-Aguilar GA (2012) Identification and quantification of major phenolic compounds from mango (*Mangifera indica*, cv. Ataulfo) fruit by HPLC–DAD–MS/MS-ESI and their individual contribution to the antioxidant activity during ripening. Food Chem 135:105–111

Palatty PL, Azmidah A, Rao S et al (2014) Topical application of a sandal wood oil and turmeric based cream prevents radiodermatitis in head and neck cancer patients undergoing external beam radiotherapy: a pilot study. Br J Radiol 87:20130490

Panahi Y, Kianpour P, Mohtashami R et al (2016) Curcumin lowers serum lipids and uric acid in subjects with nonalcoholic fatty liver disease: a randomized controlled trial. J Cardiovasc Pharmacol 68:223–229

Panahi Y, Khalili N, Sahebi E et al (2017a) Antioxidant effects of curcuminoids in patients with type 2 diabetes mellitus: a randomized controlled trial. Inflammopharmacology 25:25–31

Panahi Y, Khalili N, Sahebi E et al (2017b) Curcuminoids modify lipid profile in type 2 diabetes mellitus: a randomized controlled trial. Complement Ther Med 33:1–5

Panahi Y, Ahmadi Y, Teymouri M et al (2018) Curcumin as a potential candidate for treating hyperlipidemia: a review of cellular and metabolic mechanisms. J Cell Physiol 233:141–152

Pannu N, Bhatnagar A (2019) Resveratrol: from enhanced biosynthesis and bioavailability to multitargeting chronic diseases. Biomed Pharmacother 109:2237–2251

Parodi FE, Mao D, Ennis TL et al (2006) Oral administration of diferuloylmethane (curcumin) suppresses proinflammatory cytokines and destructive connective tissue remodeling in experimental abdominal aortic aneurysms. Ann Vasc Surg 20:360–368

Pasinetti GM, Wang J, Ho L et al (2015) Roles of resveratrol and other grape-derived polyphenols in Alzheimer's disease prevention and treatment. Biochim Biophys Acta 1852:1202–1208

Pastorelli D, Fabricio ASC, Giovanis P et al (2018) Phytosome complex of curcumin as complementary therapy of advanced pancreatic cancer improves safety and efficacy of gemcitabine: results of a prospective phase II trial. Pharmacol Res 132:72–79

Patel KR, Brown VA, Jones DJL et al (2010) Clinical pharmacology of resveratrol and its metabolites in colorectal cancer patients. Cancer Res 70:7392–7399

Patel KR, Scott E, Brown VA et al (2011) Clinical trials of resveratrol. Ann N Y Acad Sci 1215:161–169

Pepe G, Calce E, Verdoliva V et al (2020) Curcumin-loaded nanoparticles based on amphiphilic hyaluronan-conjugate explored as targeting delivery system for neurodegenerative disorders. Int J Mol Sci 21:8846

Pereira BLB, Reis PP, Severino FE et al (2017) Tomato (*Lycopersicon esculentum*) or lycopene supplementation attenuates ventricular remodeling after myocardial infarction through different mechanistic pathways. J Nutr Biochem 46:117–124

Perez VI, Bokov A, Remmen HV et al (2009) Is the oxidative stress theory of aging dead? Biochim Biophys Acta 1790:1005–1014

Peschel D, Koerting R, Nass N (2007) Curcumin induces changes in expression of genes involved in cholesterol homeostasis. J Nutr Biochem 18:113–119

Peters U, Poole C, Arab L (2001) Does tea affect cardiovascular disease? A meta-analysis. Am J Epidemiol 154:495–503

Petrella C, Carito V, Carere C et al (2020) Oxidative stress inhibition by resveratrol in alcohol-dependent mice. Nutrition 79–80:110783

Pisoschi AM, Pop A (2015) The role of antioxidants in the chemistry of oxidative stress: a review. Eur J Med Chem 97:55–74

Pohl A, Grabowska A, Kalisz A et al (2019) The eggplant yield and fruit composition as affected by genetic factor and biostimulant application. Notulae Botanicae Horti Agrobotanici Cluj-Napoca 47:929–938

Poljsak B, Šuput D, Milisav I (2013) Achieving the balance between ROS and antioxidants: when to use the synthetic antioxidants. Oxidative Med Cell Longev 2013, Article ID 956792, 11 pages

Popat R, Plesner T, Davies F et al (2013) A phase 2 study of SRT501 (resveratrol) with bortezomib for patients with relapsed and or refractory multiple myeloma. Br J Haematol 160:714–717

Potenza MA, Marasciulo FL, Tarquinio M et al (2007) EGCG, a green tea polyphenol, improves endothelial function and insulin sensitivity, reduces blood pressure, and protects against myocardial I/R injury in SHR. Am J Physiol Endocrinol Metab 292:E1378–E1387

Pourbagher-Shahri AM, Farkhondeh T, Ashrafizadeh M et al (2021) Curcumin and cardiovascular diseases: focus on cellular targets and cascades. Biomed Pharmacother 136:111214

Prentki M, Nolan CJ (2006) Islet β cell failure in type 2 diabetes. J Clin Investig 116:1802–1812

Priebe MG, Wang H, Weening D et al (2010) Factors related to colonic fermentation of nondigestible carbohydrates of a previous evening meal increase tissue glucose uptake and moderate glucose-associated inflammation. Am J Clin Nutr 91:90–97

Priviero FB, Gonçalves TT, Lazaro CM et al (2017) G-hesperidin supplementation impairs the beneficial effects of physical exercise on the body composition, biochemistry profile and oxidative stress in obese rats. FASEB J 31:1013–1019

Priya P, Raj RM, Vasanthakumar V et al (2020) Curcumin-loaded layer-by-layer folic acid and casein coated carboxymethyl cellulose/casein nanogels for treatment of skin cancer. Arab J Chem 13:694–708

Priyadarsini KI (2014) The chemistry of curcumin: from extraction to therapeutic agent. Molecules 19:20091–20112

Przybylska-Balcerek A, Frankowski J, Stuper-Szablewska K (2020) The influence of weather conditions on bioactive compound content in sorghum grain. Eur Food Res Technol 246:13–22

Qin F, Huang X, Zhang HM et al (2009) Pharmacokinetic comparison of puerarin afer oral administration of Jiawei-Xiaoyao-San to healthy volunteers and patients with functional dyspepsia: influence of disease state. J Pharm Pharmacol 61:125–129

Ramírez-Garza SL, Laveriano-Santos EP, Marhuenda-Muñoz M et al (2018) Health effects of resveratrol: results from human intervention trials. Nutrients 10:1892

Rani AJ, Mythili SV (2014) Status in relation to oxidative stress in type 2 diabetes mellitus. J Clin Diagn Res 8:108–110

Rao S, Santhakumar AB, Chinkwo KA et al (2019) Characterization of phenolic compound antioxidant activity in oat varieties using UHPLC–online ABTS and LC Q-TOF. Cereal Chem 96:958–966

Ray A, Rana S, Banerjee D et al (2016) Improved bioavailability of targeted curcumin delivery efficiently regressed cardiac hypertrophy by modulating apoptotic load within cardiac microenvironment. Toxicol Appl Pharmacol 290:54–65

Reynolds A, Mann J, Cummings J et al (2019) Carbohydrate quality and human health: a series of systematic reviews and meta-analyses. Lancet 393:434–445

Ribas-Agustí A, Cáceres R, Gratacós-Cubarsí M et al (2012) A validated HPLC-DAD method for routine determination of ten phenolic compounds in tomato fruits. Food Anal Methods 5:1137–1144

Richheimer SL, Bernart MW, King GA et al (1996) Antioxidant activity of lipid-soluble phenolic diterpenes from rosemary. J Am Oil Chem Soc 73:507–514

Ridker PM (2003) Clinical application of C-reactive protein for cardiovascular disease detection and prevention. Circulation 107:363–369

Rimm EB, Stampfer MJ, Ascherio A et al (1993) Vitamin E consumption and the risk of coronary heart disease in men. New Engl J Med 328:1450–1456

Rivera-Mancía S, Trujillo J, Chaverri JP (2018) Utility of curcumin for the treatment of diabetes mellitus: evidence from preclinical and clinical studies. J Nutr Intermed Metab 14:29–41

Roager HM, Vogt JK, Kristensen M et al (2019) Whole grain-rich diet reduces body weight and systemic low-grade inflammation without inducing major changes of the gut microbiome: a randomised cross-over trial. Gut 68:83–93

Rodrigo R, Guichard C, Charles R (2007) Clinical pharmacology and therapeutic use of antioxidant vitamins. Fundam Clin Pharmacol 21:111–127

Rodriguez C, McCullough ML, Mondul AM et al (2003) Calcium, dairy products, and risk of prostate cancer in a prospective cohort of United States men. Cancer Epidemiol Biomark Prev 12:597–603

Rodriguez-Casado A (2016) The health potential of fruits and vegetables phytochemicals: notable examples. Crit Rev Food Sci Nutr 56:1097–1107

Rodríguez-Enríquez S, Pacheco-Velázquez SC, Marín-Hernández A et al (2019) Resveratrol inhibits cancer cell proliferation by impairing oxidative phosphorylation and inducing oxidative stress. Toxicol Appl Pharmacol 370:65–77

Roke K, Zulyniak M, Gerling C et al (2015) Cardiometabolic profiles and gene expression following fish oil supplementation: is there a role for FADS genotypes? FASEB J 29:915–917

Romiti N, Tongiani R, Cervelli F et al (1998) Effect of curcumin on P-glycoprotein in primary cultures of rat hepatocytes. Life Sci 62:2349–2358

Roselló-Soto E, Koubaa M, Moubarik A et al (2015) Emerging opportunities for the effective valorization of wastes and by-products generated during olive oil production process: non-conventional methods for the recovery of high-added value compounds. Trends Food Sci Technol 45:296–310

Ruiz-López MA, Barrientos-Ramírez L, García-López PM et al (2019) Nutritional and bioactive compounds in Mexican Lupin beans species: a mini-review. Nutrients 11:1785

Saad B, Zaid H, Shanak S et al (2017) Prevention and treatment of obesity-related diseases by diet and medicinal plants. In: Saad B, Zaid H, Shanak S et al (eds) Anti-diabetes and anti-obesity medicinal plants and phytochemicals. Springer, Cham, pp 95–128

Sadeghi N, Mansoori A, Shayesteh A et al (2020) The effect of curcumin supplementation on clinical outcomes and inflammatory markers in patients with ulcerative colitis. Phytother Res 34:1123–1133

Sagoo MK, Gnudi L (2020) Diabetic nephropathy: an overview. In: Gnudi L, Long D (eds) Diabetic nephropathy, Methods in molecular biology, vol 2067. Humana, New York, pp 3–7

Sahebkar A (2014) Curcuminoids for the management of hypertriglyceridaemia. Nat Rev Cardiol 11:123

Sakunthala Devi PR, Gopala Reddy A, Rao GS et al (2015) Pharmacokinetic interaction of curcumin and glibenclamide in diabetic rats. Vet World 8:508–511

Salehi B, Mishra AP, Nigam M et al (2018) Resveratrol: a double-edged sword in health benefits. Biomedicines 6:91

Salim S (2017) Oxidative stress and the central nervous system. J Pharmacol Exp Ther 360:201–205
Samra RA, Wolever TM, Anderson GH (2007) Enhanced food intake regulatory responses after a glucose drink in hyperinsulinemic men. Int J Obes 31:1222–1231
Sanders TA (2012) Role of dairy foods in weight management. Am J Clin Nutr 96:687–688
Santos-Parker JR, Strahler TR, Bassett CJ et al (2017) Curcumin supplementation improves vascular endothelial function in healthy middle-aged and older adults by increasing nitric oxide bioavailability and reducing oxidative stress. Aging 9:187–205
Santos-Sánchez NF, Salas-Coronado R, Villanueva-Cañongo C et al (2019) Antioxidant compounds and their antioxidant mechanism. In: Shalaby E (ed) Antioxidants. IntechOpen, London, United Kingdom
Saraf-Bank S, Ahmadi A, Paknahad Z et al (2019) Effects of curcumin on cardiovascular risk factors in obese and overweight adolescent girls: a randomized clinical trial. Sao Paulo Med J 137:414–422
Saraswati AP, Ali Hussaini SM, Krishna NH et al (2018) Glycogen synthase kinase-3 and its inhibitors: potential target for various therapeutic conditions. Eur J Med Chem 144:843–858
Sattarinezhad A, Roozbeh J, Yeganeh BS et al (2019) Resveratrol reduces albuminuria in diabetic nephropathy: a randomized double-blind placebo-controlled clinical trial. Diabetes Metab 45:53–59
Scheffers FR, Boer JMA, Monique Verschuren WM et al (2019) Pure fruit juice and fruit consumption and the risk of CVD: the European Prospective Investigation into Cancer and Nutrition–Netherlands (EPIC-NL) study. Br J Nutr 121:351–359
Schernthaner G (2011) Kidney disease in diabetology: lessons from 2010. Nephrol Dial Transplant 26:454–457
Scholey A, Downey LA, Ciorciari J et al (2012) Acute neurocognitive effects of epigallocatechin gallate (EGCG). Appetite 58:767–770
Schug ZT, Voorde JV, Gottlieb E (2016) The metabolic fate of acetate in cancer. Nat Rev Cancer 16:708–717
Scicchitano P, Cameli M, Maiello M et al (2014) Nutraceuticals and dyslipidaemia: beyond the common therapeutics. J Funct Foods 6:11–32
Seferidi P, Laverty AA, Pearson-Stuttard J et al (2019) Impacts of Brexit on fruit and vegetable intake and cardiovascular disease in England: a modelling study. BMJ Open 9:e026966
Sellappan S, Akoh CC, Krewer G (2002) Phenolic compounds and antioxidant capacity of Georgia-Grown blueberries and blackberries. J Agric Food Chem 50:2432–2438
Semba RD, Ferrucci L, Bartali B et al (2014) Resveratrol levels and all-cause mortality in older community-dwelling adults. JAMA Intern Med 174:1077–1084
Sena CM, Leandro A, Azul L et al (2018) Oxidative stress: impact and therapeutic approaches. Front Physiol 9:1668
Septembre-Malaterre A, Remize F, Poucheret P (2018) Fruits and vegetables, as a source of nutritional compounds and phytochemicals: changes in bioactive compounds during lactic fermentation. Food Res Int 104:86–99
Serban C, Sahebkar A, Antal D et al (2015) Effects of supplementation with green tea catechins on plasma C-reactive protein concentrations: a systematic review and meta-analysis of randomized controlled trials. Nutrition 31:1061–1071
Seyyedebrahimi S, Khodabandehloo H, Nasli Esfahani E et al (2018) The effects of resveratrol on markers of oxidative stress in patients with type 2 diabetes: a randomized, double-blind, placebo-controlled clinical trial. Acta Diabetol 55:341–353
Shaban S, El-Husseny MWA, Abushouk AI et al (2017) Effects of antioxidant supplements on the survival and differentiation of stem cells. Oxidative Med Cell Longev 2017, Article ID 5032102, 16 pages
Shafi S, Ansari HR, Bahitham W et al (2019) The impact of natural antioxidants on the regenerative potential of vascular cells. Front Cardiovasc Med 6:28
Shahar DR, Abel R, Elhayany A et al (2007) Does dairy calcium intake enhance weight loss among overweight diabetic patients? Diabetes Care 30:485–489

Shaito A, Posadino AM, Younes N et al (2020) Potential adverse effects of resveratrol: a literature review. Int J Mol Sci 21:2084
Shamloo M, Babawale EA, Furtado A et al (2017) Effects of genotype and temperature on accumulation of plant secondary metabolites in Canadian and Australian wheat grown under controlled environments. Sci Rep 7:9133
Shamsi-Goushki A, Mortazavi Z, Mirshekar MA et al (2020) Comparative effects of curcumin versus nano-curcumin on insulin resistance, serum levels of apelin and lipid profile in type 2 diabetic rats. Diabetes Metab Syndr Obes: Targets Ther 13:2337–2346
Shanahan F, van Sinderen D, O'toole PW et al (2017) Feeding the microbiota: transducer of nutrient signals for the host. Gut 66:1709–1717
Sharif Y, Sadeghi O, Benisi-Kohansal S et al (2021) Legume and nuts consumption in relation to odds of breast cancer: a case-control study. Nutr Cancer 73:750–759
Sharifi-Rad J, Rayess YE, Rizk AA et al (2020) Turmeric and its major compound curcumin on health: bioactive effects and safety profiles for food, pharmaceutical, biotechnological and medicinal applications. Front Pharmacol 11:01021
Shen Y, Song SJ, Keum N et al (2014) Olive leaf extract attenuates obesity in high-fat diet-fed mice by modulating the expression of molecules involved in adipogenesis and thermogenesis. Evid Based Complement Alternat Med 2014, Article ID 971890, 12 pages
Shen X, Wang M, Bi X et al (2016) Resveratrol prevents endothelial progenitor cells from senescence and reduces the oxidative reaction via PPARgamma/HO1 pathways. Mol Med Rep 14:5528–5534
Shiratsuchi I, Akagi Y, Kawahara A et al (2011) Expression of IGF-1 and IGF-1R and their relation to clinicopathological factors in colorectal cancer. Anticancer Res 31:2541–2545
Shome S, Talukdar AD, Choudhury MD et al (2016) Curcumin as potential therapeutic natural product: a nanobiotechnological perspective. J Pharm Pharmacol 68:1481–1500
Sies H, Jones DP (2020) Reactive oxygen species (ROS) as pleiotropic physiological signalling agents. Nat Rev Mol Cell Biol 21:363–383
Simitzis PE (2018) Agro-industrial by-products and their bioactive compounds—an ally against oxidative stress and skin aging. Cosmetics 5:58
Singh S (2007) From exotic spice to modern drug? Cell 130:765–768
Singh AK, Vinayak M (2017) Resveratrol alleviates inflammatory hyperalgesia by modulation of reactive oxygen species (ROS), antioxidant enzymes and ERK activation. Inflamm Res 66:911–921
Sinha D, Sarkar N, Biswas J et al (2016) Resveratrol for breast cancer prevention and therapy: preclinical evidence and molecular mechanisms. Semin Cancer Biol 40–41:209–232
Slavin J (2013) Fiber and prebiotics: mechanisms and health benefits. Nutrients 5:1417–1435
Slimestad R, Vangdal E, Brede C (2009) Analysis of phenolic compounds in six Norwegian plum cultivars (*Prunus domestica* L.). J Agric Food Chem 57:11370–11375
Smoliga JM, Blanchard O (2014) Enhancing the delivery of resveratrol in humans: if low bioavailability is the problem, what is the solution? Molecules 19:17154–17172
Sochor J, Zitka O, Skutkova H et al (2010) Content of phenolic compounds and antioxidant capacity in fruits of apricot genotypes. Molecules 15:6285–6305
Socrier L, Quéro A, Verdu M et al (2019) Flax phenolic compounds as inhibitors of lipid oxidation: elucidation of their mechanisms of action. Food Chem 274:651–658
Soleas GJ, Diamandis EP, Goldberg DM (1997) Resveratrol: a molecule whose time has come? And gone? Clin Biochem 30:91–113
Sonawane SK, Chidambaram H, Boral D et al (2020) EGCG impedes human tau aggregation and interacts with tau. Sci Rep 10:12579
Soto-Vaca A, Gutierrez A, Losso JN et al (2012) Evolution of phenolic compounds from color and flavor problems to health benefits. J Agric Food Chem 60:6658–6677
Soufi FG, Mohammad-nejad D, Ahmadieh H (2012) Resveratrol improves diabetic retinopathy possibly through oxidative stress—nuclear factor κB—apoptosis pathway. Pharmacol Rep 64:1505–1514

Sousa GT, Lira FS, Rosa JC et al (2012) Dietary whey protein lessens several risk factors for metabolic diseases: a review. Lipids Health Dis 11:67
Spranger J, Kroke A, Mohlig M et al (2003) Inflammatory cytokines and the risk to develop type 2 diabetes: results of the prospective population-based European Prospective Investigation into Cancer and Nutrition (EPIC)-Potsdam study. Diabetes 52:812–817
Średnicka-Tober D, Barański M, Seal CJ et al (2016) Higher PUFA and n-3 PUFA, conjugated linoleic acid, α-tocopherol and iron, but lower iodine and selenium concentrations in organic milk: a systematic literature review and meta- and redundancy analyses. Br J Nutr 115:1043–1060
Stampfer MJ, Hennekens CH, Manson JE et al (1993) Vitamin E consumption and the risk of coronary disease in women. New Engl J Med 328:1444–1449
Stefanon B, Colitti M (2016) Hydroxytyrosol, an ingredient of olive oil, reduces triglyceride accumulation and promotes lipolysis in human primary visceral adipocytes during differentiation. Exp Biol Med 241:1796–1802
Steigerwalt R, Nebbioso M, Appendino G et al (2012) Meriva®, a lecithinized curcumin delivery system, in diabetic microangiopathy and retinopathy. Panminerva Med 54:11–16
Straub LG, Efthymiou V, Grandl G et al (2019) Antioxidants protect against diabetes by improving glucose homeostasis in mouse models of inducible insulin resistance and obesity. Diabetologia 62:2094–2105
Su Q, Sun Y, Ye Z et al (2018) Pinocembrin protects endothelial cells from oxidized LDL-induced injury. Cytokine 111:475–480
Sun J, Zhao Y, Hu J (2013) Curcumin inhibits imiquimod-induced psoriasis-like inflammation by inhibiting IL-1beta and IL-6 production in mice. PLoS One 8:e67078
Suzuki T, Yoshida S, Hara H (2008) Physiological concentrations of short-chain fatty acids immediately suppress colonic epithelial permeability. Br J Nutr 100:297–305
Szkudelski T, Szkudelska K (2015) Resveratrol and diabetes: from animal to human studies. Mol Basis Dis 1852:1145–1154
Tan BL, Norhaizan ME (2017) Scientific evidence of rice by-products for cancer prevention: chemopreventive properties of waste products from rice milling on carcinogenesis *in vitro* and *in vivo*. Biomed Res Int 2017, Article ID 9017902, 18 pages
Tan BL, Norhaizan ME (2019a) Carotenoids: how effective are they to prevent age-related diseases? Molecules 24:1801
Tan BL, Norhaizan ME (2019b) *Manilkara zapota* (L.) P. Royen leaf water extract triggered apoptosis and activated caspase-dependent pathway in HT-29 human colorectal cancer cell line. Biomed Pharmacother 110:748–757
Tan BL, Norhaizan ME (2019c) Curcumin combination chemotherapy: the implication and efficacy in cancer. Molecules 24:2527
Tan BL, Norhaizan ME, Huynh K et al (2015a) Water extract of brewers' rice induces apoptosis in human colorectal cancer cells via activation of caspase-3 and caspase-8 and downregulates the Wnt/β-catenin downstream signaling pathway in brewers' rice-treated rats with azoxymethane-induced colon carcinogenesis. BMC Complement Altern Med 15:205
Tan BL, Norhaizan ME, Huynh K et al (2015b) Brewers' rice modulates oxidative stress in azoxymethane-mediated colon carcinogenesis in rats. World J Gastroenterol 21:8826–8835
Tan L, Wang W, He G et al (2016) Resveratrol inhibits ovarian tumor growth in an *in vivo* mouse model. Cancer 122:722–729
Tan BL, Norhaizan ME, Liew W-P-P et al (2018a) Antioxidant and oxidative stress: a mutual interplay in age-related diseases. Front Pharmacol 9:1162
Tan BL, Norhaizan ME, Liew W-P-P (2018b) Nutrients and oxidative stress: friend or foe? Oxidative Med Cell Longev 2018, Article ID 9719584, 24 pages
Tan BL, Norhaizan ME, Chan LC (2018c) *Manilkara zapota* (L.) P. Royen leaf water extract induces apoptosis in human hepatocellular carcinoma (HepG2) cells via ERK1/2/Akt1/JNK1 signaling pathways. Evid Based Complement Alternat Med 2018, Article ID 7826576, 17 pages
Tan BL, Norhaizan ME, Chan LC (2018d) ROS-mediated mitochondrial pathway is required for *Manilkara zapota* (L.) P. Royen leaf methanol extract inducing apoptosis in the modulation

of caspase activation and EGFR/NF-κB activities of HeLa human cervical cancer cells. Evid Based Complement Alternat Med 2018, Article ID 6578648, 19 pages

Tang M, Larson-Meyer DE, Liebman M (2008) Effect of cinnamon and turmeric on urinary oxalate excretion, plasma lipids, and plasma glucose in healthy subjects. Am J Clin Nutr 87:1262–1267

Tang KSC, Konczak I, Zhao J (2016) Identification and quantification of phenolics in Australian native mint (*Mentha australis* R. Br.). Food Chem 192:698–705

Tervaert TWC, Mooyaart AL, Amann K et al (2010) Pathologic classification of diabetic nephropathy. J Am Soc Nephrol 21:556–563

Thangapazham RL, Singh AK, Sharma A et al (2007) Green tea polyphenols and its constituent epigallocatechin gallate inhibits proliferation of human breast cancer cells *in vitro* and *in vivo*. Cancer Lett 245:232–241

Thomas R, Williams M, Sharma H et al (2014) A double-blind, placebo-controlled randomised trial evaluating the effect of a polyphenol-rich whole food supplement on PSA progression in men with prostate cancer-the U.K. NCRN Pomi-T study. Prost Cancer Prost Dis 17:180–186

Thordardottir S, Ståhlbom AK, Almkvist O et al (2017) The effects of different familial Alzheimer's disease mutations on APP processing *in vivo*. Alzheimers Res Ther 9:9

Tohma H, Gülçin İ, Bursal E et al (2017) Antioxidant activity and phenolic compounds of ginger (*Zingiber officinale* Rosc.) determined by HPLC-MS/MS. J Food Meas Charact 11:556–566

Tomé-Carneiro J, Gonzálvez M, Larrosa M et al (2012) One-year consumption of a grape nutraceutical containing resveratrol improves the inflammatory and fibrinolytic status of patients in primary prevention of cardiovascular disease. Am J Cardiol 110:356–363

Tsai K-D, Lin J-C, Yang S-M et al (2012) Curcumin protects against UVB-induced skin cancers in SKH-1 hairless mouse: analysis of early molecular markers in carcinogenesis. Evid Based Complement Alternat Med 2012, Article ID 593952, 11 pages

Tsao R (2010) Chemistry and biochemistry of dietary polyphenols. Nutrients 2:1231–1246

Tsuda T, Ueno Y, Aoki H et al (2004) Anthocyanin enhances adipocytokine secretion and adipocyte-specific gene expression in isolated rat adipocytes. Biochem Biophys Res Commun 316:149–157

Tung BT, Rodríguez-Bies E, Ballesteros-Simarro M et al (2014) Modulation of endogenous antioxidant activity by resveratrol and exercise in mouse liver is age dependent. J Gerontol Ser A 69:398–409

Turan B, Tuncay E, Vassort G (2012) Resveratrol and diabetic cardiac function: focus on recent *in vitro* and *in vivo* studies. J Bioenerg Biomembr 44:281–296

Turner RS, Thomas RG, Craft S et al (2015) A randomized, double-blind, placebo-controlled trial of resveratrol for Alzheimer disease. Neurology 85:1383–1391

Ulewicz-Magulska B, Wesolowski M (2019) Total phenolic contents and antioxidant potential of herbs used for medical and culinary purposes. Plant Foods Hum Nutr 74:61–67

Ulrich-Merzenich G, Metzner C, Schiermeyer B et al (2002) Vitamin C and vitamin E antagonistically modulate human vascular endothelial and smooth muscle cell DNA synthesis and proliferation. Eur J Nutr 41:27–34

Ulrich-Merzenich G, Zeitler H, Panek D et al (2007) Vitamin C promotes human endothelial cell growth via the ERK-signaling pathway. Eur J Nutr 46:87–94

Ulrich-Merzenich G, Zeitler H, Vetter H et al (2009) Synergy research: vitamins and secondary plant components in the maintenance of the redox-homeostasis and in cell signaling. Phytomedicine 16:2–16

Ungvari Z, Sonntag WE, de Cabo R et al (2011) Mitochondrial protection by resveratrol. Exerc Sport Sci Rev 39:128–132

USDA (2010) USDA Database for the oxygen radical absorbance capacity (ORAC) of selected foods. Release 2. U.S. Department of Agriculture. ARS. Beltsville, MD. Available from: www.ars.usda.gov/SP2UserFiles/Place/12354500/Data/.../ORAC_R2.pdf. Accessed May 2010

Usharani P, Mateen AA, Naidu MUR et al (2008) Effect of NCB-02, atorvastatin and placebo on endothelial function, oxidative stress and inflammatory markers in patients with type 2 dia-

betes mellitus: a randomized, parallel-group, placebo-controlled, 8-week study. Drugs R&D 9:243–250
Vallejo F, Gil-Izquierdo A, Pérez-Vicente A et al (2004) *In vitro* gastrointestinal digestion study of broccoli inflorescence phenolic compounds, glucosinolates, and vitamin C. J Agric Food Chem 52:135–138
van der Made SM, Plat J, Mensink RP (2015) Resveratrol does not influence metabolic risk markers related to cardiovascular health in overweight and slightly obese subjects: a randomized, placebo-controlled crossover trial. PLoS One 10:e0118393
van der Made SM, Plat J, Mensink RP (2017) Trans-resveratrol supplementation and endothelial function during the fasting and postprandial phase: a randomized placebo-controlled trial in overweight and slightly obese participants. Nutrients 9:596
van der Schaft N, Schoufour JD, Nano J et al (2019) Dietary antioxidant capacity and risk of type 2 diabetes mellitus, prediabetes and insulin resistance: the Rotterdam Study. Eur J Epidemiol 34:853–861
van der Valk FM, Bekkering S, Kroon J et al (2016) Oxidized phospholipids on lipoprotein(a) elicit arterial wall inflammation and an inflammatory monocyte response in humans clinical perspective. Circulation 134:611–624
Vatsavai LK, Kilari EK (2016) Influence of curcumin on the pharmacodynamics and pharmacokinetics of gliclazide in animal models. J Exp Pharmacol 8:69–76
Vella RK, Pullen C, Coulson FR et al (2015) Resveratrol prevents cardiovascular complications in the SHR/STZ rat by reductions in oxidative stress and inflammation. Biomed Res Int 2015, Article ID 918123, 8 pages
Vergnaud AC, Peneau S, Chat-Yung S et al (2008) Dairy consumption and 6-y changes in body weight and waist circumference in middle-aged French adults. Am J Clin Nutr 88:1248–1255
Viitasalo A, Ågren J, Venäläinen T et al (2016) Association of plasma fatty acid composition with plasma irisin levels in normal weight and overweight/obese children. Pediatr Obes 11:299–305
Visioli F, Strata A (2014) Milk, dairy products, and their functional effects in humans: a narrative review of recent evidence. Adv Nutr 5:131–143
Walle T, Hsieh F, DeLegge MH et al (2004) High absorption but very low bioavailability of oral resveratrol in humans. Drug Metab Dispos 32:1377–1382
Waladkhani AR, Clemens MR (1998) Effect of dietary phytochemicals on cancer development. Int J Mol Med 1:747–753
Walsh LA, Damjanovski S (2011) IGF-1 increases invasive potential of MCF 7 breast cancer cells and induces activation of latent TGF-β1 resulting in epithelial to mesenchymal transition. Cell Commun Signal 9:10
Waltner-Law ME, Wang XL, Law BK et al (2002) Epigallocatechin gallate, a constituent of green tea, represses hepatic glucose production. J Biol Chem 277:34933–34940
Wang D, Li LK, Dai T et al (2018a) Adult stem cells in vascular remodeling. Theranostics 8:815–829
Wang J-Y, Wang X, Wang X-J et al (2018b) Curcumin inhibits the growth via Wnt/β-catenin pathway in non-small-cell lung cancer cells. Eur Rev Med Pharmacol Sci 22:7492–7499
Wang W, Zhang ZZ, Wu Y et al (2018c) (−)-Epigallocatechin-3-Gallate ameliorates atherosclerosis and modulates hepatic lipid metabolic gene expression in Apolipoprotein E knockout mice: involvement of TTC39B. Front Pharmacol 9:195
Wang Y, Li L, Liu H et al (2018d) Bioactive compounds and *in vitro* antioxidant activities of peel, flesh and seed powder of kiwi fruit. Int J Food Sci Technol 53:2239–2245
Wang P, Huang C-x, Gao J-j et al (2020) Resveratrol induces SIRT1-dependent autophagy to prevent H2O2-induced oxidative stress and apoptosis in HTR8/SVneo cells. Placenta 91:11–18
Watanabe Y, Nakanishi H, Goto N et al (2010) Antioxidative properties of ascorbic acid and acyl ascorbates in ML/W emulsion. J Am Oil Chem Soc 87:1475–1480
Wei B-b, Liu M-y, Zhong X et al (2019) Increased BBB permeability contributes to EGCG-caused cognitive function improvement in natural aging rats: pharmacokinetic and distribution analyses. Acta Pharmacol Sin 40:1490–1500

White Paper on Consumer Affairs (2015) Consumer Affairs Agency, Government of Japan 2015; A76: EGCG. http://www.caa.go.jp/foods/todoke_76-100.html
Wichmann A, Allahyar A, Greiner TU et al (2013) Microbial modulation of energy availability in the colon regulates intestinal transit. Cell Host Microbe 14:582–590
Williamson G (2013) Possible effects of dietary polyphenols on sugar absorption and digestion. Mol Nutr Food Res 57:48–57
Wojtunik-Kulesza KA, Oniszczuk A, Oniszczuk T et al (2016) The influence of common free radicals and antioxidants on development of Alzheimer's Disease. Biomed Pharmacother 78:39–49
Wolfram S, Raederstorff D, Preller M et al (2006) Epigallocatechin gallate supplementation alleviates diabetes in rodents. J Nutr 136:2512–2518
Wong MS, Leisegang MS, Kruse C et al (2014) Vitamin D promotes vascular regeneration. Circulation 130:976–986
Wong KE, Ngai SC, Chan K-G et al (2019) Curcumin nanoformulations for colorectal cancer: a review. Front Pharmacol 10:152
Wood LG, Williams EJ, Berthon BS et al (2017) Effects of an encapsulated fruit and vegetable juice concentrate on obesity-induced systemic inflammation. FASEB J 31:161–166
Wu L, Zhang Y, Ma X et al (2012) The effect of resveratrol on FoxO1 expression in kidneys of diabetic nephropathy rats. Mol Biol Rep 39:9085–9093
Wu X, Xu Y, Zhu B et al (2018) Resveratrol induces apoptosis in SGC-7901 gastric cancer cells. Oncol Lett 16:2949–2956
Xiao Q, Zhu W, Feng W et al (2019) A review of resveratrol as a potent chemoprotective and synergistic agent in cancer chemotherapy. Front Pharmacol 9:1534
Xu F, Wang Y, Cui W et al (2014) Resveratrol prevention of diabetic nephropathy is associated with the suppression of renal inflammation and mesangial cell proliferation: possible roles of Akt/NF-κB pathway. Int J Endocrinol 2014, Article ID 289327, 9 pages
Xu D-P, Li Y, Meng X et al (2017) Natural antioxidants in foods and medicinal plants: extraction, assessment and resources. Int J Mol Sci 18:96
Xu R, Bai Y, Yang K et al (2020a) Effects of green tea consumption on glycemic control: a systematic review and meta-analysis of randomized controlled trials. Nutr Metab 17:56
Xu R, Yang K, Li S et al (2020b) Effect of green tea consumption on blood lipids: a systematic review and meta-analysis of randomized controlled trials. Nutr J 19:48
Yadav A, Kumari R, Yadav A et al (2016) Antioxidants and its functions in human body-A review. Res Environ Life Sci 9:1328–1331
Yamagata K (2020) Protective effect of epigallocatechin gallate on endothelial disorders in atherosclerosis. J Cardiovasc Pharmacol 75:292–298
Yang WS, Wang WY, Fan WY et al (2014) Tea consumption and risk of type 2 diabetes: a dose-response meta-analysis of cohort studies. Br J Nutr 111:1329–1339
Yang H, Xu W, Zhou Z et al (2015) Curcumin attenuates urinary excretion of albumin in type II diabetic patients with enhancing nuclear factor erythroid-derived 2-like 2 (Nrf2) system and repressing inflammatory signaling efficacies. Exp Clin Endocrinol Diabetes 123:360–367
Yang CS, Zhang J, Zhang L et al (2016) Mechanisms of body weight reduction and metabolic syndrome alleviation by tea. Mol Nutr Food Res 60:160–174
Yang Y, Nirmagustina DE, Kumrungsee T et al (2017) Feeding of the water extract from Ganoderma lingzhi to rats modulates secondary bile acids, intestinal microflora, mucins, and propionate important to colon cancer. Biosci Biotechnol Biochem 81:1796–1804
Yang C, Zhu K, Yuan X et al (2020) Curcumin has immunomodulatory effects on RANKL-stimulated osteoclastogenesis *in vitro* and titanium nanoparticle-induced bone loss *in vivo*. J Cell Mol Med 24:1553–1567
Yar AS, Menevse S, Alp E (2011) The effects of resveratrol on cyclooxygenase-1 and -2, nuclear factor kappa beta, matrix metalloproteinase-9, and sirtuin 1 mRNA expression in hearts of streptozotocin-induced diabetic rats. Genet Mol Res 10:2962–2975
Yin H, Xu L, Porter NA (2011) Free radical lipid peroxidation: mechanisms and analysis. Chem Rev 11:5944–5972

Yoshimura H, Yoshida H, Matsuda S et al (2019) The therapeutic potential of epigallocatechin-3-gallate against human oral squamous cell carcinoma through inhibition of cell proliferation and induction of apoptosis: *in vitro* and *in vivo* murine xenograft study. Mol Med Rep 20:1139–1148

Yun S-Y, Kim S-P, Song D-K (2006) Effects of (−)-epigallocatechin-3-gallate on pancreatic beta-cell damage in streptozotocin-induced diabetic rats. Eur J Pharmacol 541:115–121

Yun J-M, Chien A, Jialal I et al (2012) Resveratrol upregulates SIRT1 and inhibits cellular oxidative stress in the diabetic milieu: mechanistic insights. J Nutr Biochem 23:699–705

Zalazar L, Pagola P, Miró MV et al (2019) Bacterioruberin extracts from a genetically modified hyperpigmented *Haloferax volcanii* strain: antioxidant activity and bioactive properties on sperm cells. J Appl Microbiol 126:796–810

Zanotti I, Dall'Asta M, Mena P et al (2015) Atheroprotective effects of (poly)phenols: a focus on cell cholesterol metabolism. Food Funct 6:13–31

Zenebe W, Pechánová O, Bernátová I (2001) Protective effects of red wine polyphenolic compounds on the cardiovascular system. Exp Clin Cardiol 6:153–158

Zeng YC, Peng LS, Zou L et al (2017) Protective effect and mechanism of lycopene on endothelial progenitor cells (EPCs) from type 2 diabetes mellitus rats. Biomed Pharmacother 92:86–94

Zhang D, Hamauzu Y (2004) Phenolic compounds and their antioxidant properties in different tissues of carrots (*Daucus carota* L.). Int J Food Agric Environ 2:95–100

Zhang H, Tsao R (2016) Dietary polyphenols, oxidative stress and antioxidant and anti-inflammatory effects. Curr Opin Food Sci 8:33–42

Zhang CX, Ho SC, Fu JH et al (2011) Dairy products, calcium intake, and breast cancer risk: a case-control study in China. Nutr Cancer 63:12–20

Zhang N, Li H, Jia J et al (2015) Anti-inflammatory effect of curcumin on mast cell-mediated allergic responses in ovalbumin-induced allergic rhinitis mouse. Cell Immunol 298:88–95

Zhang B, Zhao Q, Guo W et al (2018) Association of whole grain intake with all-cause, cardiovascular, and cancer mortality: a systematic review and dose–response meta-analysis from prospective cohort studies. Eur J Clin Nutr 72:57–65

Zhang S, Zhu Q, Chen J-Y et al (2020) The pharmacological activity of epigallocatechin-3-gallate (EGCG) on Alzheimer's disease animal model: a systematic review. Phytomedicine 79:153316

Zhang S, Willett SA, Hyatt JR et al (2021) Phenolic compounds as antioxidants to improve oxidative stability of menhaden oil-based structured lipid as butterfat analog. Food Chem 334:127584

Zhao J, Sun XB, Ye F et al (2011) Suppression of fatty acid synthase, differentiation and lipid accumulation in adipocytes by curcumin. Mol Cell Biochem 351:19–28

Zhao C-N, Meng X, Li Y et al (2017) Fruits for prevention and treatment of cardiovascular diseases. Nutrients 9:598

Zhen P, Zhao Q, Hou D et al (2012) Genistein attenuates vascular endothelial impairment in ovariectomized hyperhomocysteinemic rats. J Biomed Biotechnol 2012, Article ID 730462, 10 pages

Zheng J, Yang B, Ruusunen V et al (2012) Compositional differences of phenolic compounds between black currant (*Ribes nigrum* L.) cultivars and their response to latitude and weather conditions. J Agric Food Chem 60:6581–6593

Zheng Z, Sun Y, Liu Z et al (2015) The effect of curcumin and its nanoformulation on adjuvant-induced arthritis in rats. Drug Design Dev Ther 9:4931–4942

Zhou X, Yang J, Zhou M et al (2019) Resveratrol attenuates endothelial oxidative injury by inducing autophagy via the activation of transcription factor EB. Nutr Metab 16:42

Zhu L, Ding X, Zhang D et al (2015) Curcumin inhibits bovine herpesvirus type 1 entry into MDBK cells. Acta Virol 59:221–227

Zhu X, Lei X, Wang J et al (2020) Protective effects of resveratrol on hyperoxia-induced lung injury in neonatal rats by alleviating apoptosis and ROS production. J Matern Fetal Neonatal Med 33:4150–4158

Chapter 7
The Role of Antioxidant on Health and Age-Related Diseases in Aging

7.1 Low Molecular Weight Antioxidant

Low molecular weight antioxidant (LMWA), for instance, phenolic compounds, glutathione, carotenoids, vitamins E and C, lipoic acid, and uric acid are critical antioxidative defense mechanisms of organisms and cells (Grune et al. 2005; Tan et al. 2018). LMWA is a small molecule (˂900 daltons) biological compound that mediates the physiological process in the body (Macielag 2012; Veber et al. 2002). LMWA exerts preventive ability toward oxidative damage by stabilizing the redox potential of transition metals and scavenging free radicals (Kohen 1999). In particular, tocopherol (vitamin E) and ascorbic acid (vitamin C) are crucial LMWA cannot be synthesized by the human (Podda and Grundmann-Kollmann 2001). Many compounds are synthesized in the body which exerts an antioxidant potential such as coenzyme Q, keto acids, uric acid, glutathione, melanins, melatonin, taurine, and lipoic acid (Sifuentes-Franco et al. 2017).

LMWA is one of the efficient agents for the skin defense mechanism toward environmental factors, for instance, pollutants, pesticides, cigarette smoke, ozone, and ultraviolet light (Kohen and Gati 2000). The LMWA protects skin tissues by neutralizing the free radicals or chelating the metallic ions (Martins et al. 2020). In particular, vitamin E, ascorbic acid, ubiquinol, uric acid, and glutathione are detectable in the stratum corneum. The gradient capacity of LMWA in keratinocytes is inversely associated with carcinogenesis risk (Martins et al. 2020). Grammenandi et al. (2016) explored the concentration of LMWA in both nonmelanoma skin carcinomas and normal-looking adjacent. The study found that normal-looking skin showed higher ascorbic acid and uric acid concentration compared to nonmelanoma skin carcinomas. The deficiency of hydrophilic LMWA such as uric and ascorbic acids in tumor samples is likely due to the decrease of the water levels in a cancer area and thereby leading to the reduction of the solubility of hydrophilic antioxidants. Ultimately, their activities are further decreased (Grammenandi et al. 2016).

B. L. Tan, M. E. Norhaizan, *The Role of Antioxidants in Longevity and Age-Related Diseases*, https://doi.org/10.1007/978-3-030-82859-2_7

Nonetheless, Grammenandi et al. (2016) did not identify an association of lipophilic LMWA (α-tocopherol, ubiquinol, and β-carotene) between nonmelanoma skin carcinomas and normal-looking skin. Further, aging and exposure to oxidative stress caused a significant reduction in skin water-soluble LMWA activity and levels (Kohen and Gati 2000). The skin is the outermost barrier of the body, which is susceptible to exposure to oxidative stress such as UV-irradiation (Strozyk and Kulms 2013). Oxidative stress has been implicated in the development of skin aging, namely photo-aging (Gu et al. 2020). In this regard, direct topical application or through high intake of vegetables and fruits are thought to increase amounts of LMWA and thereby may protect against oxidative stress and modulate the development of skin aging (Podda and Grundmann-Kollmann 2001).

Besides its effects on cancer and skin aging, a relatively low LMWA was also found in the rat embryos and yolk sacs cultured for 28 h under the diabetic condition *in vitro* (Ornoy et al. 1999). It has been demonstrated that diabetes caused an increase in uric acid and a decrease in vitamin E and ascorbic acid *in vitro*, indicates that dysregulation of LMWA may increase diabetes and anomalies (Ornoy et al. 1999). Moreover, research evidence also indicates that patients with type 2 diabetes mellitus had a relatively low level of LMWA (Farhood et al. 2019). Of all LMWA, glutathione is one of the predominant cellular antioxidants (Narayanankutty et al. 2019).

7.2 Glutathione

Glutathione (GSH) (Fig. 7.1) is a pleiotropic tripeptide, comprised of glutamic acid, cysteine, and glycine (Huang and Yin 2020). GSH is a crucial antioxidant present in plants, microorganisms, and animals, which is involved in the modulation of cellular homeostasis (Lushchak 2012). The synthesis of GSH involved two

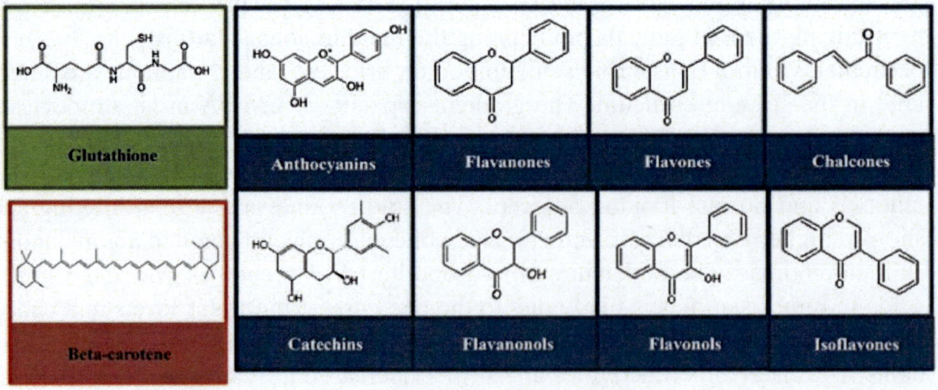

Fig. 7.1 Molecular structures of glutathione, polyphenols (anthocyanins, flavanones, flavones, chalcones, catechins, flavanonols, flavonols, and isoflavones), and beta-carotene

ATP-dependent enzymatic steps catalyzed by glutathione synthase (GS) and glutamylcysteine synthase (GCS) (Halliwell and Gutteridge 1989). GCS forms a peptide bond between cysteine and glutamate, the glycine is added with the facilitation of GS (Lu 2013). Despite GS is not crucial in the mediation of GSH synthesis, numerous evidence indicates that GS may play a pivotal role in specific tissues and/or under stressful circumstances (Luo et al. 1998). GSH contains non-protein intracellular thiol and reaching a concentration of millimolar in the intracellular region. The active thiol group exists as part of the cysteine residue and allows the GSH to serve as an antioxidant.

The function and biological role of GSH depends on the chemical structure (Calabrese et al. 2017). GSH plays a predominant role in the detoxifying of endogenous compounds and xenobiotics (Desari et al. 2018). These compounds are electrophiles and formed a conjugate with GSH catalyzed by GSH-S-transferases (GST), either enzymatically or spontaneously (Meister 1988). GSH prevents the cellular damage induced by ROS and RNS via detoxification or through GSH-dependent peroxidase-catalyzed reactions (Circu and Aw 2012). Elevation of GSH appears to be a universal cellular response against oxidative stress. By contrast, reduced GSH seems to be aggravated the development of many diseases (Ballatori et al. 2009). Nearly 90% of intracellular GSH is detected in the cytosol, while about 10–15% of intracellular GSH is found within mitochondria and the endoplasmic reticulum contains a small percentage of GSH (Lu 2009). However, impairment of the mitochondrial ETC is linked to the deficiency of GSH (Hargreavas et al. 2005). In this regard, mitochondrial depletion of GSH may lead to an elevation of ROS and depletion of ATP. Alteration of GSH levels may cause a detrimental health impact, for instance, dysregulation of transcription of detoxification enzymes, cell growth, and apoptosis (Townsend et al. 2003). Importantly, compromised GSH concentration could be a crucial indicator of various diseases due to its ability in modulating the redox balance in the body (Bhattacharjee et al. 2017). When ROS levels increased during immune response and metabolism, antioxidant including GSH concentration is increased to compensate (Nita and Grzybowski 2016; Mak et al. 2017). Subsequently, the upstream regulators are needed to mediate the recycling molecules and *de novo* synthesis to maintain optimal GSH concentration (Vomhof-Dekrey and Picklo 2012; Habib et al. 2015; Lin et al. 2016a). Nuclear factor erythroid 2-related factor 2 (Nrf2) is one of the pivotal antioxidant regulators (Gong and Yang 2020). Nrf2 is a transcription factor linked to the antioxidant response element (ARE), a promoter region of several genes coding for antioxidant-related enzymes (Vomhof-Dekrey and Picklo 2012). During redox equilibrium, Nrf2 is restricted to the cell cytoplasm by Kelch-like ECH-associated protein 1 (Keap1). Therefore, this mechanism prevents Nrf2 from increase the transcription of antioxidant-related enzymes. By contrast, Nrf2 is dissociated to Keap1 is favored by Keap1 inhibitors. Subsequently, Nrf2 is translocated into the cell nucleus and bound with the ARE of the antioxidant-related gene promoters (Li et al. 2013; Steele et al. 2013; Pekovic-Vaughan et al. 2014; Vomund et al. 2017). The elevation of these transcriptions promotes the upregulation of GSH to mediate the redox equilibrium.

In human study, a significant decrease in blood GSH concentration was found in the healthy elderly, particularly in individuals aged 60–79 years (Lang et al. 1992). Low GSH in the elderly may increase the risk of metabolic ailments due to the reduced capacity to maintain various detoxification and metabolic reactions modulated by GSH (Lang et al. 1992). Compared with young individuals, elderly people had a relatively low GSH concentration and high plasma oxidative stress (Sekhar et al. 2011). Deficiency of GSH was linked to the increased oxidative stress in the elderly, which indicates that the primary contributor to low GSH and high oxidative stress in aging is a decreased rate of GSH synthesis, which could be due to the low availability of its precursor amino acids (Sekhar et al. 2011; Jahoor et al. 2019). Dietary supplementation of GSH precursors glycine (1.33 mmol/kg/day) and cysteine (0.81 mmol/kg/day) for 14 days is effective at restoring GSH in the elderly to the levels similar to young subjects, suggesting that replenish the supply of glycine and cysteine can restore the GSH synthesis and decrease the oxidant damage (Sekhar et al. 2011). GSH is a tripeptide of glycine, cysteine, and glutamate (Minich and Brown 2019). Elderly people have adequate glutamate concentration because amino acid nitrogen is recycled by regulating the glutamate that serves as an intermediary (Sekhar et al. 2011). Importantly, the flux of a nonessential amino acid comprises its release from *de novo* synthesis and protein breakdown (Zhang et al. 2017a). Reduced intracellular glycine and cysteine levels in the elderly could be due to the reduced *de novo* synthesis or delayed body protein turnover (Sekhar et al. 2011). Research evidence indicates the aging promotes the reduction of protein turnover, suggesting that glycine and cysteine are reduced with aging (Young et al. 1975; Golden and Waterlow 1977). Several studies revealed that feeding animals with a diet low in GSH precursor amino acids, particularly cysteine, increased the development of GSH deficiency (Grimble et al. 1992; Jahoor et al. 1995; Bella et al. 1999). In this regard, these findings showed that deficiency of intracellular GSH in aging is likely due to the reduced *in vivo* synthesis secondary to a reduced supply of the GSH precursors cysteine and glycine (Sekhar et al. 2011). Table 7.1 summarizes the effects of GSH on age-related diseases in *in vivo* and human studies.

In malignant and normal cells, the elevation of GSH concentration is linked to the proliferative response, which is required for cell cycle progression (Messina and Lawrence 1989; Lu and Ge 1992). Holmgren (1981) demonstrated that the mechanisms underlying GSH in DNA synthesis are complex and might relate to the regulation of thioredoxin and reduced glutaredoxin, which is needed for the ribonucleotide reductase activity. As opposed to the beneficial effects of the other antioxidants, GSH has a complex role in relation to cancers. GSH exhibits a dual role in cancer by inducing cancer proliferation or through safeguarding cell homeostasis from proneoplastic oxidative damage (Desideri et al. 2019). The concentration of GSH was associated with cellular proliferation and metastasis of melanoma cells and liver cancer (Gamcsik et al. 2012; Carretero et al. 1999; Huang et al. 2001; Marengo et al. 2010). An animal study has demonstrated that intrasplenic inoculation of B16 melanoma (B16M) cells into C57BL/6J mice triggered metastatic foci formation (Carretero et al. 1999). The size and number of metastatic foci were markedly increased when B16M cells with high GSH concentration were inoculated *in vivo*

Table 7.1 Effects of glutathione on age-related diseases in *in vivo* and human studies

Age-related diseases	Experimental model	Outcomes	References
Breast cancer	Breast cancer women aged 30–99 years (mean 58 years)	Associated with the clinical progression of the breast cancer	Jardim et al. (2013)
Ovarian cancer	Women with ovarian cancer	Combination of GSH and cisplatin increased the quality of life	Smyth et al. (1997)
Cancer	Systematic review of dietary supplementation of GSH in patients receiving chemotherapy with or without radiation for various malignancies	Counteract detrimental effects of cancer-induced antioxidant depletion	Ladas et al. (2004)
Type 2 diabetes mellitus	Type 2 diabetic, prediabetic, and normoglycemic subjects	Prediabetic and type 2 diabetes mellitus patients had higher erythrocyte GPx	Gunawardena et al. (2019)
Type 2 diabetes mellitus	Patients with type 2 diabetes mellitus (7 without and 9 with microvascular complications)	Diabetic patients with microvascular complications had lower erythrocyte GSH levels	Lutchmansingh et al. (2018)
CVD	Apolipoprotein E-deficient mice fed with 50 mg/kg/day of liposomal coated GSH for 2 months	Upregulated GSH concentration and decreased lipid peroxides and ox-LDL levels	Rosenblat et al. (2007)
CVD	A prospective cohort study involving 909 atrial fibrillation patients during a mean follow-up of 43.4 months	Patients with atrial fibrillation had relatively low amounts of serum GPx3	Pastori et al. (2016)
Atherosclerosis	245 type 2 diabetes mellitus patients	High concentration of GPx3 decreased the carotid plaque and carotid intima-media thickness	Ling et al. (2020)

CVD cardiovascular disease, *GPx* glutathione peroxidase, *GSH* glutathione, *ox-LDL* oxidized-low density lipoprotein

(Carretero et al. 1999). Intriguingly, high percentages of tumor cells with high GSH levels were able to proliferate in the presence of oxidative and nitrosative stress (Carretero et al. 2001). Hence, maintenance of a high intracellular concentration of GSH is crucial for the extravascular proliferation of metastatic cells (Carretero et al. 2001). Godwin et al. (1992) and Mulcahy et al. (1994) found that elevation of GSH concentration is related to the drug resistance in tumor cells. GSH is a predominant contributor to chemoresistance by participating in DNA repair processes, preventing the DNA or protein damage, interacting with ROS, and reacting with drugs (Traverso et al. 2013). A retrospective study evaluated the breast cancer tumor fragments in women aged 30 to 99 years revealed that estrogen receptor (ER)-positive had a relatively higher GSH protein concentration compared to ER-negative (Jardim et al. 2013). Further, GPx and GSH were found to be associated with the clinical progression of breast cancer. Breast cancer patients with high GPx levels were significantly correlated with a high rate of mortality (Jardim et al. 2013). Elevation of GSH is characterized as an indicator of low response toward chemotherapy, which may further lead to the development of metastasis. It has been suggested that high amounts of ROS are produced upon exposure to doxorubicin that is responsible for the antioxidant synthesis including GPx and GSH, and thus causing the cells to be more resistant to oxidative damages (Gaudiano et al. 2000), indicated that high expression of GPx in breast cancer patients may link to the development of chemotherapy resistance (Jardim et al. 2013). Although research has revealed the detrimental impact of GSH in cancer, not all studies showed such a link. Ladas et al. (2004) found that dietary supplementation of GSH during chemotherapy was effective to counteract the detrimental effects of cancer-induced antioxidant depletion. Several studies by Nunes and Serpa (2018) have suggested that GSH could have a potential co-adjuvant of chemotherapy in ovarian cancer patients. In support of this, a clinical trial has demonstrated that the combination of GSH with cisplatin increased the quality of life in patients with advanced-stage ovarian cancer (Smyth et al. 1997).

Besides its effects on cancer, GSH also plays a crucial role in type 2 diabetes mellitus. A study by Lutchmansingh et al. (2018) evaluated the GSH in relation to type 2 diabetes mellitus with microvascular complications. The data showed that patients with type 2 diabetes mellitus showed reduced absolute synthesis rates and lower erythrocyte GSH levels, particularly to those diabetic patients with microvascular complications, compared to non-diabetic controls (Lutchmansingh et al. 2018). Circulating concentration of proinflammatory cytokines and ROS are increased in type 2 diabetes mellitus with macro- and microvascular complications, which may further decrease the GSH levels (Brownlee 2001; Nguyen et al. 2014). This finding could be attributed to the irreversible loss of GSH in diminishing the rate of synthesis and decreasing GSH levels (Lutchmansingh et al. 2018). Indeed, an irreversible utilization of GSH in diabetes may also occur with activated activity and high oxidative stress via the polyol pathway (Lutchmansingh et al. 2018). In addition, prediabetic and type 2 diabetes mellitus patients had higher erythrocyte GPx activity compared to an individual with normoglycemic (Gunawardena et al. 2019). Further, high GPx activity was also found in patients with type 2 diabetes mellitus with poor glycemic control, suggesting a compensatory adaptive response for poor glycemic control concomitantly with markedly increased lipid peroxidation (Gunawardena et al. 2019). GPx has been identified as an initial protective response to control H_2O_2 levels after oxidative damage and under normal physiological conditions (Lee et al. 2008). GPx plays a critical role in mediating the H_2O_2 balance. When the level of H_2O_2 is markedly increased through the lipid peroxidation pathway, it may elevate GPx activity (Gunawardena et al. 2019). Moreover, the study also showed that GPx is positively correlated with HbA1c of type 2 diabetes mellitus (Gunawardena et al. 2019).

GSH plays a crucial etiological role in the development of CVD and cardiometabolic disease (Bajic et al. 2019). The progression and development of CVD are characterized by substantial changes in the GSH levels and oxidation state (Espinosa-Díez et al. 2018). Rosenblat et al. (2007) exploring the impact of GSH on atherosclerosis using liposomes. The data showed that feeding atherosclerotic apolipoprotein E-deficient mice with 50 mg/kg/day of liposomal coated GSH for 2 months decreased lipid peroxides and ox-LDL levels and upregulated the GSH concentration. This favorable effect is likely due to the ability of GSH to increase the 7 alpha-hydroxylase cholesterol levels, and subsequently promotes the biosynthesis of bile acids from cholesterol (Hassan et al. 1993; Lin et al. 2004). A 43.4-month follow-up prospective cohort study conducted from September 2007 to October 2015 involving 909 patients with atrial fibrillation showed that patients with atrial fibrillation had relatively low amounts of GPx3 in serum compared to those without cardiovascular disease (Pastori et al. 2016). The study further demonstrated that GPx3 activity progressively declines with aging, with a marked reduction in elderly aged ≥70 years (Pastori et al. 2016). These findings suggest that decreasing natural antioxidants could be a predisposing factor to cardiovascular events in the elderly (Pastori et al. 2016). Of all glutathione peroxidase isoforms, GPx3 is the predominant form of antioxidant in plasma and only identified in the extracellular space (Holley et al. 2016). Data from the animal study further revealed that deficiency of GPx3 increased occluded vessels, enhanced platelet-rich thrombi, and stimulated platelet activation to a greater extent compared to wild-type (Jin et al. 2011). The association between thrombosis and GPx3 levels could be attributed to H_2O_2, which is a stimulus for thromboxane A_2 production (Pignatelli et al. 1998). Thromboxane A_2 is a potent aggregating molecule derived from COX-1 stimulation (Smyth 2010). Indeed, high serum GPx3 levels are associated with the decrease in carotid plaque and carotid intima-media thickness, suggesting that reduced GPx3 concentration may act as an independent predictor for carotid atherosclerosis in type 2 diabetes mellitus (Ling et al. 2020). Taken together, GSH plays a crucial role in cell redox homeostatic mode of action. It was evident that GSH involved in many intracellular homeostatic functions, hence there has been a tremendous potential to be used as a biomarker in diseases suffered by the elderly. The primary role played by GSH in pathophysiology conditions underlying age-related diseases worth further elucidation in long-term clinical trials.

7.3 Polyphenol

Polyphenol or known as polyhydroxyphenols, are characterized by several phenol structural units (Nascimento-Souza et al. 2018). Polyphenols are non-nutrient (Goñi and Hernández-Galiot 2019) or secondary metabolites (Pandey and Rizvi 2009) produced by plants, which are usually found in vegetables and fruits (Lima et al. 2014). In general, the features and number of these phenol structures contribute to the unique characteristics of the polyphenol compounds in terms of physical, chemical, and biological (Estrela et al. 2017). Apart from that, polyphenols can affect the color, odor, and flavor which may contribute to the sensory perception of food (Pandey and Rizvi 2009). These phenolics are responsible for the attractive color of flowers, fruits, and leaves (Mikulic-Petkovsek et al. 2015). Among the polyphenolic compounds, the flavonoid is the most common class of polyphenol (Kumar and Pandey 2013). Phenolics comprised of around 8000 naturally occurring components, in which all of them exert one common structural feature, namely a phenol (Leopoldini et al. 2011). Further classification divides them into simple phenols and polyphenols, based on the number of phenol subunits. Polyphenols exert at least two phenol subunits such as stilbenes, flavonoids, and compounds that possess at least three phenol subunits, for instance, tannins. Phenolic acids are phenols that exert one carboxylic acid. It contains two distinctive constitutive carbon, namely hydroxybenzoic and hydroxycinnamic acids (Kumar and Goel 2019).

Flavonoids are the most studied group of polyphenol. The diphenylpropane skeleton is the basic structure of flavonoid, consist of two benzene rings (rings B and A) bound with three-carbon chains and formed a closed pyran ring (a heterocyclic ring containing oxygen, ring C) (de Castro Peixoto et al. 2019). The structures of flavonoids are denoted as C6-C3-C6. The B ring is connected to position 3/4. Flavonoids, for instance, genistein, fisetin, isorhamnetin, kaempferol, and quercetin are widely found in vegetables and fruits (Dou et al. 2014; Ganai and Farooqi 2015; Chin et al. 2018; Kanberoglu et al. 2019; Imran et al. 2021). In nature, flavonoid components are products derived from plants, which are found in several parts of the plants (Stankovic et al. 2011). It belongs to the category of low-molecular-weight phenolic compounds that are extensively found in the plant kingdom. Flavonoids are widely found in beverages and foods of plant origin, for instance, wine, cocoa, tea, vegetables, and fruits; hence they are also known as dietary flavonoids (Panche et al. 2016). Flavonoids are further classified into several subgroups (anthocyanins, flavanones, flavones, chalcones, catechins, flavanonols, and flavonols) (Fig. 7.1) according to the degree of unsaturation, the carbon of the ring C bound to the ring B, and the oxidation of the ring C (Panche et al. 2016). Indeed, the physiological function of the flavonoid is depends on the pattern of hydroxylation and glycosylation of the three rings and the structural characteristics (Halbwirth 2010). This subgroup has unique major sources. For instance, tea and onions are the predominant dietary sources of flavones and flavonols (Maiti et al. 2019; Li et al. 2020).

Isoflavones (Fig. 7.1) are one of the flavonoid in which the B ring is bound to the position 3 of the C ring. The B ring is linked to position 2 of the C ring, which can be further categorized into several subgroups, namely chalcones, flavones, flavonols, flavanones, and anthocyanins (Fig. 7.1 and Table 7.2).

There are nearly 6000 flavonoids that form the colorful pigments of vegetables, fruits, medicinal plants, and herbs (Kumar 2017; Li et al. 2018a). Flavonoids have long been recognized to be synthesized in certain parts and are responsible for the aroma and color of flowers (Samanta et al. 2011). The health-promoting potentials of flavonoids have been demonstrated in preclinical and clinical studies (Revathy et al. 2018; Rodríguez-García et al. 2019; Dükel et al. 2021). Substantial evidence revealed an inverse relationship between plasma or serum total antioxidant capacity

Table 7.2 Subgroups of flavonoids (Hano and Tungmunnithum 2020)

Subgroups	Characteristics
Chalcones	The absence of ring C, or known as open-chain flavonoids
Flavones	Containing a ketone on the ring C at position 4 and a double bond between positions 2 and 3. Majority of them comprised of a hydroxyl group in position 5 of ring A; whereas hydroxylation in other positions, usually at positions 3′ and 4′ of the B ring or position 7 of the A ring
Flavonols	-The largest and the most common flavonoid subgroup, comprised of a hydroxyl and a ketone group in position 3 of the ring C
	-Exhibit different patterns of glycosylation, methylation, and hydroxylation
	-Constitute the building blocks of proanthocyanin
Flavanones	Fully saturated ring C
Anthocyanins	Pigments responsible for the coloring of fruits, flowers, and plants, varying depends on the acylation or methylation of the hydroxyl groups on the rings B and A as well as the pH

and both the progression and onset of age-related diseases (Wu et al. 2017; Mullan et al. 2018). It has been suggested that antioxidant supplementation could be served as a promising therapy based on the Free Radical Theory of Aging (FRTA) (Harman 1956, 2006).

The biochemical, medicinal, and pharmacological properties of phenolics have been widely reported (Semaming et al. 2015; Kiokias et al. 2020). The anti-inflammatory, antimutagenic, anticarcinogenic, and antioxidative properties as well as their abilities to mediate prominent cellular enzyme function have drawn interest from the medicinal, pharmaceutical, cosmetic, and nutraceutical industries. Indeed, most interest has been attributed to their antioxidant activity of flavonoids by scavenging free radicals and reducing free radical formation (Treml and Šmejkal 2016).

Flavonoids are also known as potent inhibitors for enzymes including phosphoinositide 3-kinase, lipoxygenase, COX, and XO (Tronina et al. 2017; Ibrahim et al. 2018; Santi et al. 2018; Zhang et al. 2019a). A study reported by Cos et al. (1998) evaluated the structure-function activity of flavonoids. The data showed that flavonoid luteolin (tetrahydroxyflavone) is the most potent inhibitor of XO (Cos et al. 1998). XO is a source of oxygen free radicals (Dowell et al. 1993). In this regard, the ability of flavonoids in chelating trace metals plays a crucial role in oxygen metabolism (Malešev and Kuntić 2007).

Flavonoids have beneficial antioxidant activity and biochemical effects in relation to oxidative stress-induced diseases in the elderly such as neurodegenerative disease (Devi et al. 2021), diabetes mellitus (Hussain et al. 2020), CVD (Yamagata and Yamori 2020), and cancer (Liskova et al. 2020) (Tables 7.3 and 7.4). Some flavonoids such as tea catechins (Rahman et al. 2020), apigenin (Adham et al. 2021), and quercetin (Raja et al. 2017) have been found to have anti-inflammatory activity via suppressing iNOS and COX-2. The effectiveness of flavonoids in the prevention of neurodegenerative disease has been reported in recent decades (Cirmi et al. 2016; Flanagan et al. 2018; Borowiec and Michalak 2021). In particular, the study is concerned about Alzheimer's disease, Parkinson's disease, and dementia (Jung and

Table 7.3 *In vitro* and *in vivo* studies conducted in polyphenols and their effects on age-related diseases

Age-related diseases	Cell lines/ animal models	Source of polyphenols	Dosages	Findings	References
Parkinson disease	6-OHDA lesion rat model	Tangeretin, a citrus flavonoid	20 mg/kg/day for 4 days	Protect nigrostriatal pathway induced by neurotoxic substance 6-OHDA	Datla et al. (2001)
Colon and breast cancers	LoVo and MCF-7 cell lines	Peel-flavonoids (Peel-F) and flesh flavonoids (Flesh-F) extracted from Pink Lady apples	LoVo cells (Peel-F = 110.33 ± 2.52 mg/mL and Flesh-F = 378.14 ± 1.64 mg/mL)	↓ Proliferation of LoVo and MCF-7 cells	Yang et al. (2015)
			MCF-7 cells (Peel-F = 58.42 ± 1.39 mg/mL and Flesh-F = 296.06 ± 3.71 mg/mL)		
Colon cancer	Pirc rats (F344/NTac-Apc^{am1137})	Flavonoid-rich extract from bergamot juice	35 mg/kg and 70 mg/kg body weight for 12 weeks	↓ Colon preneoplastic lesions MDF	Navarra et al. (2020)
CVD	Male Wistar rats	EGCG	10 mg/kg intravenously	↓ Myocardial damage by reducing NF-κB activity and AP-1 pathway	Aneja et al. (2004)
CVD	Sprague-Dawley rats	Quercetin	5 μM for 30 min	↓ Vasoconstrictor sensitivity in rat aortic ring	Khoo et al. (2010)
Diabetes mellitus	Systematic review and meta-analysis of animal studies	Quercetin	10, 25, and 50 mg/kg	↓ Serum glucose levels	Bule et al. (2019)
Diabetes mellitus	32 male albino Wistar rats	Apple peel extract or apple juice	Apple peel extract (1 g/kg) or cloudy apple juice (15 mL/kg) for 21 days	Improved antioxidant enzyme activity	Fathy and Drees (2016)
				↓ Lipid peroxidation levels	
Osteoporosis	Three-month-old female Sprague Dawley rats	Total flavonoids of *Herba Taxilli*	400, 200, and 100 mg/kg/day for 28 days	↑ Serum calcium (Ca^{2+}) level	Li et al. (2019a)

(continued)

Table 7.3 (continued)

Age-related diseases	Cell lines/ animal models	Source of polyphenols	Dosages	Findings	References
Osteoporosis	Systematic review (including 11 animal studies and 1 human study) (between January 2016 to October 2019)	Flavonoids	–	Exert a positive effect on osteoporotic fracture healing	Chiavarini et al. (2020)

AP-1 activator protein-1, *CVD* cardiovascular disease, *EGCG* (−)-epigallocatechin-3-gallate, *LoVo* human colon cancer cell line, *MCF-7* human breast cancer cell line, *MDF* mucin-depleted foci, *NF-κB* nuclear factor-kappa B, *6-OHDA* 6-hydroxydopamine

Kim 2018; Shishtar et al. 2020; Uddin et al. 2020). Flavonoids appear to mediate neuronal function (Jaeger et al. 2018). Diets high in flavonoids were found to be beneficially affecting the human neuronal function, probably via enhancement of their regeneration and function (Cirmi et al. 2021). RNS and ROS are involved in the progression and development of neurodegenerative diseases; while dietary flavonoids have been demonstrated to combat oxidative neuronal damage (Khan et al. 2020). A study reported by Panche et al. (2015) and Jager and Saaby (2011) evaluated different plant metabolites in relation to neurodegenerative diseases. The data showed that flavonoids may perform a crucial role in receptor and key systems in the brain as well as exert a significant effect on the central nervous system. Substantial studies have been conducted on flavonoids in Alzheimer's disease using molecular docking. The study found that flavonoid is a potent acetylcholinesterase (AChE) inhibitor (Hu et al. 2009). Interestingly, flavonoids showed more potent inhibitory activity against AChE compared to rivastigmine, an Alzheimer's disease drug (Hu et al. 2009). AChE is a crucial enzyme in the central nervous system and suppression of this enzyme could result in the elevation of neural acetylcholine levels. In this regard, suppression of cholinesterases is one of the predominant focuses for drug development to combat Alzheimer's disease (Sharma et al. 2020). It has been demonstrated that a strong relationship between flavonoids and suppression of NF-κB pathway (Spagnuolo et al. 2018). The extract from *Ginkgo biloba* plant, rich in flavonoids, may beneficially affect Alzheimer's disease and dementia (Singh et al. 2019). A study in the animal model study has suggested that tangeretin (20 mg/kg/day for 4 days), a citrus flavonoid, could protect the nigrostriatal pathway induced by neurotoxic substance 6-hydroxydopamine (Datla et al. 2001). In support of this, the Personnes Age'es QUID (PAQUID study) involving 1640 subjects aged 65 years and above suggested that regular consumption of dietary

Table 7.4 The clinical studies of polyphenols on age-related diseases

Age-related diseases	Study conditions	Source of polyphenols	Durations	Outcomes	References
Dementia	Prospective epidemiologic study (1640 nondemented participants aged 65 years and above)	Flavonoids (Fruit juice, soup, coffee, tea, chocolate, oats flakes, sweet pepper, asparagus, French beans, spinach, cabbage, dried fruits, other fruits, kiwi, and/or citrus fruits)	14.33 mg/day over a 10-year period	Improved cognitive function	Letenneur et al. (2007)
Lung cancer	Case-control study (649 incident cases of primary lung cancer among women diagnosed from February 1992 to January 1994 and control group of 675 women)	Green tea	>1500 g/year	↓ Lung cancer risk in non-smoking women	Zhong et al. (2001)
Lung cancer	Meta-analysis of epidemiological studies (5 cohort and 12 case-control studies) involving 12,276 cases and 102,516 controls	Coffee	≤1 cup/day, 2–3 cups per day, or ≥3 cups/day	↑ Lung cancer risk	Xie et al. (2016)

Lung cancer	Prospective cohort study (34,708 post-menopausal women)	Flavonoids-containing foods including red wine, chocolate, tea, soy, citrus juices and fruits (orange juice, oranges, grapefruit juice, grapefruits), bran added to foods, broccoli, grapes and raisins, berries (blueberries and strawberries), and fresh pears and apples	Median intake of total flavonoids = 239.2 mg/day	↓ Lung cancer risk, particularly in women who had stopped smoking and current smokers	Cutler et al. (2008)
Lung cancer	Prospective cohort study (63,257 Singaporean Chinese men and women)	Black tea	At least 2 cups/day, with an average of 17.7 years of follow-up	↓ 33% of lung cancer risk in men compared to those who never or less-than-weekly black tea drinkers	Seow et al. (2020)
Lung cancer	Prospective cohort study (63,257 Singaporean Chinese men and women)	Green tea	More than 2 cups/day, with an average of 17.7 years of follow-up	No effect	Seow et al. (2020)
Lung cancer	Prospective cohort study (63,257 Singaporean Chinese men and women)	Coffee	One or more cups/day	↑ Risk of lung cancer compared to rarely or non-daily coffee drinkers	Seow et al. (2020)

(continued)

Table 7.4 (continued)

Age-related diseases	Study conditions	Source of polyphenols	Durations	Outcomes	References
Cancer	Cohort study (A total of 3234 incident cancer cases among 38,408 women aged 45 years and above)	Foods rich in flavonoids including tofu, onion, broccoli, apple, and tea	Median intake = 8.88–47.44 mg/day during 11.5 years of follow-up	No significant association	Wang et al. (2009)
Cancer	Cohort study (1063 women aged 75 years old and above)	Total flavonoids (from black tea, onions, pears, apples, tea, fruit juice, and oranges)	≥813 mg/day over the 5 years follow-up period	↓ Risk of cancer mortality	Ivey et al. (2015)
Gastric cancer	European Prospective Investigation into Cancer and Nutrition (EPIC) study (477,312 subjects (29.8% men) aged 35–70 years old from 10 European countries)	Total flavonoid, flavanols, flavones, flavonols, and anthocyanidins	Total flavonoid = (433.8 ± 330.6 mg/day), flavanols (350.8 ± 304.1 mg/day), flavones (3.46 ± 3.92 mg/day), flavonols (26.70 ± 17.38 mg/day), and anthocyanidins (29.51 ± 22.75 mg/day) during an average follow-up of 11 years	↓ Gastric adenocarcinoma risk in women	Zamora-Ros et al. (2012)
Gastric cancer	Prospective cohort study (1297 gastric cancer cases)	Flavonoids from 124 food items	≤84.1 mg/day or 84.2–4211.2 mg/day over a mean of 12 years follow-up	No effect	Sun et al. (2017)
Gastric cancer	Case-control study using 334 cases and 334 matched controls aged 35–75 years	Total flavonoid from black soybeans, tofu, and green tea	105.2 ± 77.9 mg/day	↓ Gastric cancer risk in women but not in men	Woo et al. (2014)
Breast cancer	A meta-analysis of case-control and prospective cohort studies (9513 cases and 181,906 controls)	Total flavonoids, flavanones, or flavan-3-ols	Total flavonoids (median intake = 19.13–29.1 mg/day), flavanones (median intake = 31.2–33.7 mg/day), or flavan-3-ols (median intake = 7.9–162 mg/day)	No significant association	Hui et al. (2013)

Breast cancer	A meta-analysis of case-control and prospective cohort studies (9513 cases and 181,906 controls)	Flavones and flavonols	Flavones (median intake = 0.13–2.5 mg/day) and flavonols (median intake = 9.8–27.8 mg/day) intake	↓ Breast cancer risk in postmenopausal women	Hui et al. (2013)
Colorectal cancer	Spanish case-control study (424 cases with incident colorectal cancer and 401 hospital-based controls)	Total flavonoid intake from vegetables, legumes, wine, and fruits	Median = 196.4 mg/day	↓ Colorectal cancer risk	Zamora-Ros et al. (2013)
Colorectal cancer	Case-control study in a Chinese population (1632 colorectal cancer cases and 1632 frequency-matched controls)	Flavonoids from fruits and vegetables	Flavones (3.67 mg/day) from lettuce, peppers, pumpkin, leaf mustard, and Chinese celery, flavanones (3.76 mg/day) from tomato, white wine, citrus juice, citrus fruits, and leaf mustard, and anthocyanidins (20.64 mg/day) from banana, apple, grapes, and vegetables (radish and eggplant)	↓ Colorectal cancer risk	Xu et al. (2016)
Colorectal cancer	Prospective cohort study from 2519 colorectal cancer cases (1061 in men, 1458 in women)	Flavonoid subclasses (anthocyanins, flavan-3-ols, flavanones, flavones, and flavonols) from onions, tea, red wine, oranges, blueberries, strawberries, pears, and apples	116–769 mg/day and 107–808 mg/day during 26 years of follow-up	No association	Nimptsch et al. (2016)
Prostate cancer	Case-control study (118 histopathological-verified prostate cancer cases and 222 controls)	Catechins and flavonols	Catechins (36.18 mg/day) and flavonols (37.14 mg/day)	↓ Prostate cancer risk	Reale et al. (2018)

(continued)

Table 7.4 (continued)

Age-related diseases	Study conditions	Source of polyphenols	Durations	Outcomes	References
Prostate cancer	Case-control study (118 histopathological-verified prostate cancer cases and 222 controls)	Flavanones	81.32 mg/day	↑ Prostate cancer risk	Reale et al. (2018)
Thyroid cancer	Prospective study of European Prospective Investigation into Cancer and Nutrition (EPIC) study	Coffee or tea	100 mL/day	No association with the total differentiated thyroid cancer	Zamora-Ros et al. (2019)
Coronary heart disease	Zutphen Elderly Study (805 males aged 65–84 years)	Flavonoids-containing foods such as apples, onions, or tea	Followed-up for 5 years	↓ Mortality of coronary heart disease	Hertog et al. (1993)
Stroke	Cohort study from 42 cases of first fatal or nonfatal stroke (552 men aged 50–69 years)	Flavonoids	≥28.6 mg/day	↓ Stroke incidence in elderly	Keli et al. (1996)
Stroke	Cohort study from 42 cases of first fatal or nonfatal stroke (552 men aged 50–69 years)	Flavonoid from tea	≥4.7 cups	Protect against stroke	Keli et al. (1996)
CVD	Meta-analysis of randomized controlled trials	Catechin supplements	20–300 mg/day	↑ Flow-mediated dilation	Shafabakhsh et al. (2020)
				↓ Pulse wave velocity	
				No changes in endothelial function markers	

CVD	Randomized, double-blind, controlled, cross-over design, 19 healthy men	Black tea	Twice daily containing 800 mg tea flavonoids/day for 1 week	Improved flow-mediated dilatation in healthy men	Grassi et al. (2009)
CVD	33 male and female patients with coronary artery disease submitted to elective PCI (between 18 and 75 years)	Flavonoid-based antioxidant-rich diet such as grapes, cherry tomatoes, onions, and cruciferous vegetables	40 mg/day antioxidant compounds for 6 months	↓ LDL-C Not related to the decreasing of inflammatory and oxidative stress markers	Cammerer et al. (2018)
Diabetes mellitus	Meta-analysis of 6 prospective cohort studies involving 18,146 cases and 284,806 participants	Total flavonoids	An increased in 500 mg/day	↓ 5% in diabetes mellitus risk	Liu et al. (2014a)
Diabetes mellitus	Meta-analysis involving 8 prospective studies (312,015 subjects with 19,953 developed type 2 diabetes mellitus during the follow-up periods of 4–28 years)	Total flavonoids	≥550 mg/day	↓ Incident of type 2 diabetes mellitus	Xu et al. (2018)

(continued)

Table 7.4 (continued)

Age-related diseases	Study conditions	Source of polyphenols	Durations	Outcomes	References
Osteoporosis	Meta-analysis of 16 studies (7 cohort and 9 case-control studies) involving 772,707 participants with 37,166 fracture cases	Tea	≥3 cups/day	↓ Fractures risk	Xiang et al. (2019)
Osteoporosis	Systematic review and meta-analysis of 52 randomized controlled trials	Soy isoflavones	˂90 mg/day	Effective in the prevention of osteoporosis-related bone loss in overweight/obese individuals	Akhlaghi et al. (2020)
Osteoporosis	Systematic review and meta-analysis of 63 randomized controlled trials involving 6427 postmenopausal women	Ipriflavone (600 mg/day) and genistein (54 mg/day)	Ipriflavone (600 mg/day) and genistein (54 mg/day)	Improved bone mineral density	Sansai et al. (2020)

CVD cardiovascular disease, *LDL-C* low-density lipoprotein cholesterol, *PCI* percutaneous coronary intervention flavonoids (mean flavonoid intake is 14.33 mg/day for 10 years) in fruit juice, soup, coffee, tea, chocolate, oats flakes, sweet pepper, asparagus, French beans, spinach, cabbage, dried fruits, other fruits, kiwi, and/or citrus fruits could be beneficial on cognitive function during aging (Letenneur et al. 2007). Flavonoids are believed to be mediated by regulating the neuronal signal cascade through suppression of cell apoptosis, and thereby promote neuronal differentiation and survival (Vauzour et al. 2008). Flavonoids also seem to possess beneficial effects on the central and peripheral nervous system by changing the central blood flow (Socci et al. 2017). Folch et al. (2018) demonstrated that flavonoids may reduce the progression and development of Alzheimer's disease-like pathophysiology by inducing α-secretase (A disintegrin and metalloproteinase 10 (ADAM10)), interfering amyloid β protein production, and suppressing of β-secretase (β active site cleavage enzyme-1 (BACE-1)) (Fig. 7.2). Despite regular consumption of foods rich in flavonoids seems to decrease the risk of neurodegenerative diseases and combat the onset of age-related cognitive disorders, the mechanisms of actions remain obscure. Overall, it remains unknown which of the substances could produce the strongest protection of the nervous system and when to use it to produce the optimal effectiveness. Therefore, further study is necessary to conclusively resolve the question behind these effects.

Notably, data from epidemiological and clinical studies found that flavonoids may reduce the risk of cancer (Storelli et al. 2019; Sun et al. 2021). In the promotion

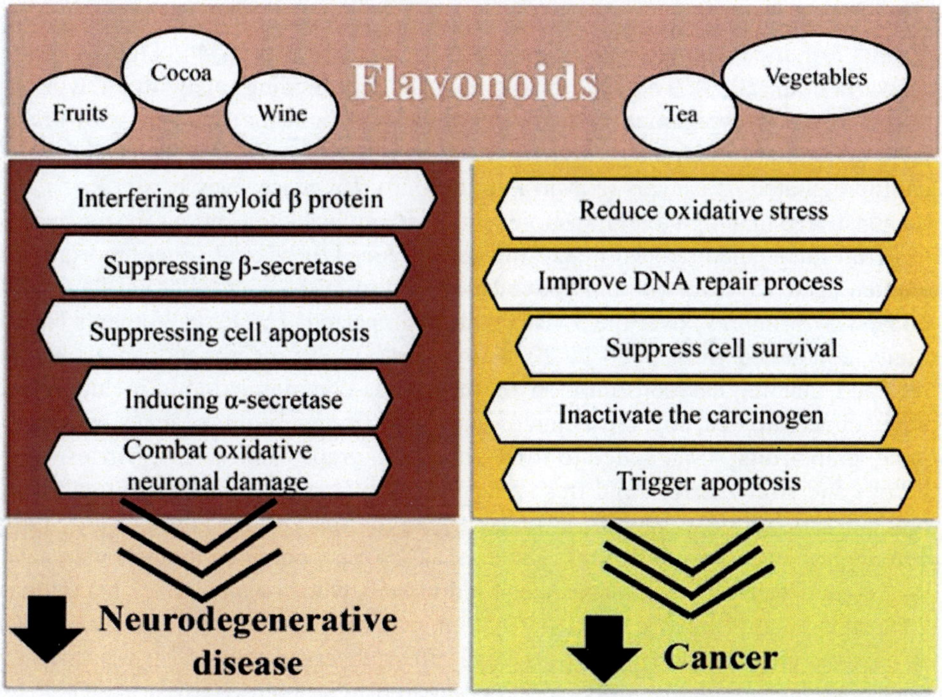

Fig. 7.2 Mechanisms of flavonoids in the modulation of neurodegenerative disease and cancer

and initiation stages, flavonoids decrease the proliferation of cancerous cells through several mechanisms including reduction of oxidative stress, improvement of DNA repair process, suppression of cell survival, and inactivation of the carcinogen (Zaidun et al. 2018; Zhang et al. 2018a) (Fig. 7.2). While in the progression phase, flavonoids may show cytostatic or cytotoxic activity against cancer cells, exhibit antioxidant activity, suppress angiogenesis, and trigger apoptosis (Sp et al. 2017; Loung et al. 2019; Elhady et al. 2020). Prevention of metabolic stimulation of pro-carcinogens is associated with flavonoids interaction with Phase I enzymes, which is responsible for the metabolism of several exogenous and endogenous substrates (Batra and Sharma 2013). Such findings could be attributed to the suppression of the cytochrome P450 enzymes, for instance, CYP1A2 and CYP1A1. In addition, flavonoids may protect against cellular damage arising from the stimulation of carcinogenic factors through reinforcement of mutagen detoxification by inducing phase II enzymes, for instance, UDP-glucuronyl transferase (UDP-GT) and glutathione S-transferase (GST) (Shih et al. 2007; Jiang and Hu 2012). Flavonoids were shown to negatively associate with several types of cancers in animal and human studies. Apples are a rich source of flavanoids (Tu et al. 2017). It has been demonstrated that peel-flavonoids (Peel-F) and flesh flavonoids (Flesh-F) extracted from Pink Lady apples suppressed the colon cancer (LoVo) (Peel-F = 110.33 ± 2.52 mg/mL and Flesh-F = 378.14 ± 1.64 mg/mL) and breast cancer (MCF-7) (Peel-F = 58.42 ± 1.39 mg/mL and Flesh-F = 296.06 ± 3.71 mg/mL) cells (Yang et al. 2015). Further, consumption of at least one apple/day decreased the risk of colorectal cancer (Jaganathan et al. 2014). An animal study has shown that a flavonoid-rich extract from bergamot juice (35 mg/kg and 70 mg/kg body weight for 12 weeks) reduced colon preneoplastic lesions mucin-depleted foci (MDF) in Pirc rats (F344/NTac-Apc^{am1137}) (Navarra et al. 2020). The data further revealed that feeding rats with a flavonoid-rich extract from bergamot juice (70 mg/kg body weight for 12 weeks) downregulated inflammation-related genes (*Arginase 1*, *IL-10*, *IL-6*, *IL-1β*, *iNOS*, and *COX-2*) and upregulated *p53* levels (Navarra et al. 2020). Likewise, Ivey et al. (2015) also demonstrated that high total flavonoids intake (from black tea, onions, pears, apples, tea, fruit juice, and oranges) (≥813 mg/day) reduced the risk of cancer mortality in women aged 75 years old and above compared to those with low total flavonoids intake (˂525 mg/day) over the 5 years follow-up period. The Iowa Women's Health Study conducted from 1986 to 2004 involving 34,708 post-menopausal women revealed that regular consumption of flavonoids-containing foods including red wine, chocolate, tea, soy, citrus juices and fruits (orange juice, oranges, grapefruit juice, grapefruits), bran added to foods, broccoli, grapes and raisins, berries (blueberries and strawberries), and fresh pears and apples (median intake of total flavonoids = 239.2 mg/day) significantly decreased the risk of lung cancer, particularly among past and current smokers (Cutler et al. 2008). Compared to those who never or less-than-weekly black tea drinkers, individuals who consume black tea (at least 2 cups/day) significantly reduced 33% of lung cancer risk in men (Seow et al. 2020). A possible explanation for the protective effect observed only in men is that they may have high oxidative stress levels compared to women (Barp et al. 2002; Ide et al. 2002; Matarrese et al. 2011), and thus the antioxidant effect of black tea to restore balance may be more significant among men. Although several meta-analyses found that green tea reduced the risk of lung cancer (Zhong et al. 2001; Wang et al. 2014), the recent study failed to show any significant association between lung cancer risk and green tea consumption (more than 2 cups/day) (Seow et al. 2020). In another study involving 38,408 women aged 45 years and above further revealed that no significant association between the risk of cancer and consumption of foods rich in flavonoids including tofu, onion, broccoli, apple, and tea over 11.5 years of follow-up (median intake = 8.88–47.44 mg/day) (Wang et al. 2009). The same effect was also observed in the prospective study. European Prospective Investigation into Cancer and Nutrition (EPIC) study showed that coffee or tea consumption (100 mL/day) was not associated with total differentiated thyroid cancer (Zamora-Ros et al. 2019).

Compared to those who rarely or non-daily coffee drinkers, individuals who consume one or more cups of coffee per day have a relatively high risk of lung cancer (Seow et al. 2020). Similarly, three recent meta-analyses found that coffee consumption (≤1 cup/day, 2–3 cups per day, or ≥3 cups/day) was positively associated with the risk of lung cancer (Xie et al. 2016; Galarraga and Boffetta 2016; Wang et al. 2016a). The increased cancer risk from drinking coffee is possibly modulated by intermediate compounds and mechanisms, for instance, pyrogallol and gallic acid, and thus leading to DNA damage (Hossain et al. 2013). In the context of breast cancer, a meta-analysis of epidemiological studies involving 9513 cases and 181,906 controls revealed that no significant association between total flavonoids (median intake = 19.13–29.1 mg/day), flavanones (median intake = 31.2–33.7 mg/day), or flavan-3-ols (median intake = 7.9–162 mg/day) intake and breast cancer risk (Hui et al. 2013). Nonetheless, the study found that flavones (median intake = 0.13–2.5 mg/day) and flavonols (median intake = 9.8–27.8 mg/day) intake is negatively associated with breast cancer risk in postmenopausal women (Hui et al. 2013). This finding could be attributed to the flavonoids that have a similar chemical composition as oestrogens, and thus flavonoids possess an ability to decrease menopause symptoms (Gardeazabal et al. 2019). However, further study is necessary to investigate the effects of flavonoid consumption in post- and pre-menopause breast cancer risk. Furthermore, a population-based case-control study reported by Reale et al. (2018) evaluated flavonoids in relation to prostate cancer risk in Sicilian men from January 2015 to December 2016. The data found that a high intake of catechins (36.18 mg/day) and flavonols (37.14 mg/day) reduced the prostate cancer risk. Notably, the study found that high consumption of flavanones (81.32 mg/day) increased prostate cancer risk (Reale et al. 2018). In addition to the effects mentioned above, total flavonoid intake (105.2 ± 77.9 mg/day) from black soybeans, tofu, and green tea reduced the gastric cancer risk in women but not in men (Woo et al. 2014). A similar finding was also observed in the EPIC study, who reported that total flavonoid intake (433.8 ± 330.6 mg/day during an average follow-up of 11 years) is negatively associated with gastric adenocarcinoma risk in women (Zamora-Ros et al. 2012). Such relationship was also observed in several flavonoid subgroups, for instance, flavanols (350.8 ± 304.1 mg/day), flavones (3.46 ± 3.92 mg/day), flavonols (26.70 ± 17.38 mg/day), and anthocyanidins (29.51 ± 22.75 mg/day) (Zamora-Ros et al. 2012). However, Sun et al. (2017) failed to show any protective effect between flavonoid intake (≤84.1 mg/day or 84.2–4211.2 mg/day over 12 years of follow-up) and gastric cancer risk. In the context of colorectal cancer, several studies reported by Hazafa et al. (2020) have shown that flavonoids exert their efficacy in the inhibition of colon cancer *in vitro*. In support of this, Xu et al. (2016) showed that flavonoid consumption such as flavones (3.67 mg/day) from lettuce, peppers, pumpkin, leaf mustard, and Chinese celery, flavanones (3.76 mg/day) (tomato, white wine, citrus juice, citrus fruits, and leaf mustard), and anthocyanidins (20.64 mg/day) (banana, apple, grapes, and vegetables (radish and eggplant)), is negatively linked to the colorectal cancer risk. Zamora-Ros et al. (2013) further revealed that total flavonoid intake (median = 196.4 mg/day) from vegetables, legumes, wine, and fruits is inversely associated with colorectal cancer risk in a Spanish population. Nonetheless, a prospective study evaluated the association of daily flavonoids consumption (116–769 mg/day and 107–808 mg/day during 26 years of follow-up) (from onions, tea, red wine, oranges, blueberries, strawberries, pears, and apples) and colorectal cancer has demonstrated that no association between flavonoid intake and colorectal cancer risk (Nimptsch et al. 2016). Overall, the protective effects of flavonoids on cancer in epidemiological studies remain controversial. Indeed, these findings may vary from diet and geographical location. Therefore, further studies are required to confirm the effects of flavonoids on certain types of cancers.

Flavonoids not only modulate neurodegenerative diseases and cancers but also decrease CVD (Mahmoud et al. 2019). Inflammation plays a prominent role in atherosclerosis (Moriya 2019). Flavonoids exert bioactivity that beneficially influence cardiovascular risk factors, for instance, blood platelet aggregation, endothelial dysfunction, dyslipidemia, and lipoprotein oxidation (Mulvihill et al. 2016; Malakul et al. 2018; Rolnik et al. 2020) (Fig. 7.3). The cardioprotective effect of flavonoids could be due to its lipid-lowering properties, anti-thrombogenic properties, and antioxidant activities (Nijveldt et al. 2001; Zeka et al. 2017). Several epidemiological studies reported by Ponzo et al. (2015) and Dalgaard et al. (2019) examined flavonoids in relation to CVD risk. Considerable studies showed that flavonoid intake can decrease total mortality and the risk of CVD (Kim and Je 2017). In Zutphen Elderly Study involving 805 males aged 65–84 years and followed-up for 5 years revealed that the mortality of coronary heart disease was reduced after consuming flavonoids-containing foods such as apples, onions, or tea (Hertog et al. 1993), suggesting that regularly consuming food rich in flavonoids may decrease the risk of mortality from coronary heart disease in elderly men. A 15-year follow-up study further reported that high consumption of flavonoids (≥28.6 mg/day) such as black tea can decrease the stroke incidence in the elderly (Keli et al. 1996). Notably, data from a meta-analysis of randomized controlled trials included a study from inception to March 2019 that evaluated catechin supplementation in relation to endothelial function (Shafabakhsh et al. 2020). The study showed that catechin supplementation (20–300 mg/day) significantly increase flow-mediated dilation and reduce pulse wave velocity, but did not change endothelial function markers (Shafabakhsh et al. 2020). By contrast, some research has emerged to suggest that flavonoids can modulate endothelial function (Bondonno et al. 2018; Zhang et al. 2018b).

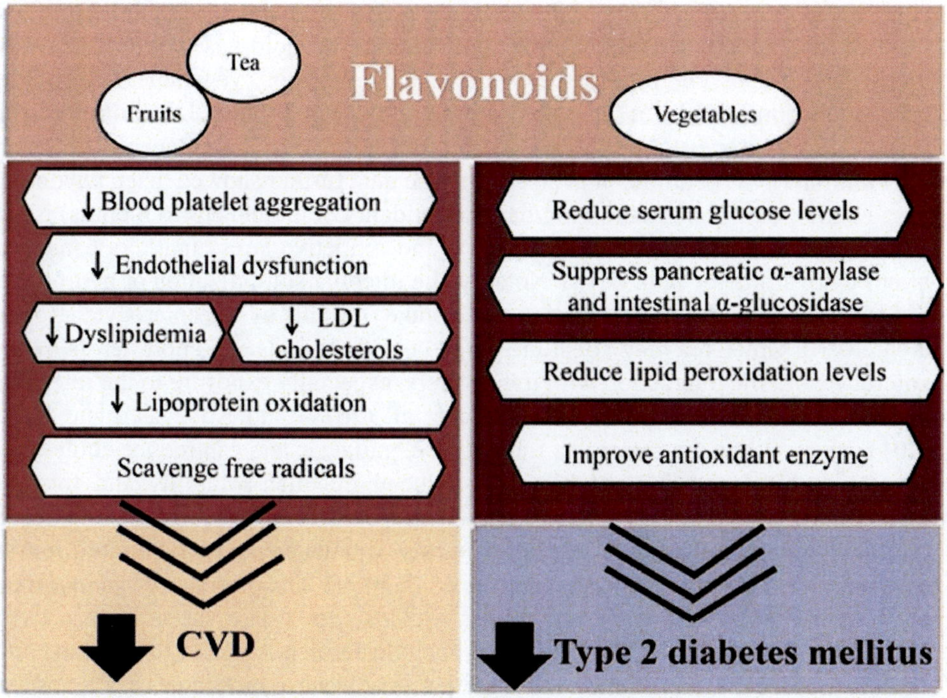

Fig. 7.3 Mechanisms of flavonoids in the modulation of cardiovascular disease and type 2 diabetes mellitus. *CVD* cardiovascular disease, *LDL* low-density lipoprotein

Endothelial-derived NO via the activity of constitutive NO synthase is crucial for maintaining the dilation of blood vessels (Sandoo et al. 2010). However, a high dosage of NO produced by iNOS in macrophages can result in oxidative damage (Xue et al. 2018). It has been suggested that flavonoids can scavenge the free radicals by decreasing the formation of peroxynitrite, and thereby leading to less damage (Nugroho et al. 2017). Furthermore, flavonoids decreased the ischemia-reperfusion injury by disrupting iNOS activity (Cebova and Pechanova 2020). In this regard, NO scavenging is hypothesized to play a crucial role in the therapeutic effects of tea flavonoids (Panat et al. 2016). A study reported by Grassi et al. (2009) evaluated the effects of tea flavonoids in relation to flow-mediated dilatation of the brachial artery. The data showed that black tea (twice daily containing 800 mg tea flavonoids/day for 1 week) improved flow-mediated dilatation in healthy men (Grassi et al. 2009). The endothelium is the inner lining of the blood vessels (Favero et al. 2014). It plays an important role in selectively permeable barriers between tissues and blood. Tea catechins have been shown to mediate the redox-sensitive transcriptional activity, for instance, AP-1 and NF-κB in cardiovascular cells (Aneja et al. 2004). NF-κB has been demonstrated to be involved in the mediation of several oxidative-stress related vascular activities (Pierce et al. 2009). The previous study found that EGCG administration (10 mg/kg intravenously) is beneficial for the treatment of reperfusion-induced myocardial damage by reducing NF-κB activity and AP-1 pathway (Aneja et al. 2004). A beneficial role of flavonoids has also been observed on lipid profiles. Data from a randomized clinical trial showed that flavonoid-based antioxidant-rich diet such as grapes, cherry tomatoes, onions, and cruciferous vegetables (40 mg/day antioxidant compounds for 6 months) significantly decrease the LDL cholesterols in patients with coronary artery disease submitted to elective percutaneous coronary intervention (PCI) (Cammerer et al. 2018). The data further showed that a flavonoid-based antioxidant-rich diet (40 mg/day antioxidant compounds for 6 months) is not related to the decreasing of inflammatory and oxidative stress markers 6 months after PCI (Cammerer et al. 2018). Notably, the micellar solubilization of hydrolyzed lipids is important for the absorption and uptake of lipids by the enterocyte (Robert et al. 2020). Green tea may affect the uptake of other lipids and cholesterols by the enterocyte via the interaction with transporters, especially exposure to the intestinal lumen (Koo and Noh 2007). Sakakibara et al. (2019) and Kobayashi and Ikeda (2017) showed that tea catechins, for instance, gallate esters can reduce cholesterol absorption by decreasing the bile acid-induced micellar solubility and forming insoluble co-precipitates of cholesterols. It has been reported that green tea catechins are potent and selective inhibitors in rats' squalene epoxidase, a rate-limiting enzyme for cholesterol biosynthesis (Abe et al. 2000). The presence of galloyl moiety has been suggested to be crucial for squalene epoxidase suppression activity (Abe et al. 2000). In this regard, the hypocholesterolemic effect of tea catechins could be attributed to the inhibitory activity on cholesterol biosynthesis and absorption to improve cholesterol excretion (Kobayashi and Ikeda 2017). Taken together, the experimental studies suggest that flavonoids may beneficial to the cardiovascular system and decrease the CVD risk. Despite substantial evidence on flavonoids are promising, many studies are still needed to prove the efficacy of CVD treatment and prevention. The potential implication of flavonoids on CVD worth further investigation in long-term clinical studies.

Diabetes mellitus is a chronic disorder characterized by increased levels of blood glucose, and thereby resulted in the damage of nerves, kidneys, eyes, blood vessels, and heart (World Health Organization 2020). Nearly 422 million people have diabetes worldwide, with 1.6 million deaths are attributed to diabetes annually (World Health Organization 2020). Data from systematic review and meta-analysis of animal studies included a study between July 2018 and August 2018 have demonstrated that quercetin with a dose of 10, 25, and 50 mg/kg could reduce the serum glucose levels (Bule et al. 2019) (Fig. 7.3). Another common flavonoid, anthocyanidins are commonly found in radishes, red cabbage, and apples (Rodriguez-Amaya 2019; Yuste et al. 2019). Cyanidin and its glycosides, which belong to anthocyanins, have been shown to suppress pancreatic α-amylase and intestinal α-glucosidase *in vitro* (Akkarachiyasit et al. 2010) (Fig. 7.3). Further, the animal study showed that feeding rats with apple peel extract (1 g/kg) or cloudy apple juice (15 mL/kg) for 21 days improved antioxidant enzyme activity and reduced lipid peroxidation levels, suggesting that apple juice/extract may have protective effects against the deleterious complication of diabetes mellitus (Fathy and Drees 2016). Notably, emerging epidemiological studies in Europe and United States suggested that dietary flavonoids can protect against diabetes mellitus (Mondal and Rahaman

2020). Several prospective cohort studies in a meta-analysis have corroborated this finding and found that an increase in the total flavonoid consumption of 500 mg/day was linked to a 5% reduction in type 2 diabetes mellitus (Liu et al. 2014a). Consistent with the study reported by Liu et al. (2014a), a meta-analysis involving 8 prospective studies (312,015 subjects with 19,953 developed type 2 diabetes mellitus during the follow-up periods of 4–28 years) revealed that high consumption of total flavonoids (≥550 mg/day) is inversely associated with the incident of type 2 diabetes mellitus (Xu et al. 2018). The study further demonstrated that the type 2 diabetes mellitus risk was reduced by 5% for every 300 mg/day increment in total flavonoid consumption (Xu et al. 2018). These improvements could be due to the antioxidant properties of flavonoids that protect the body against deleterious effects of hyperglycemia in type 2 diabetes mellitus via modulation of aldose reductase, glucose co-transporter, or α-glucosidase (Fang et al. 2019).

Substantial studies have suggested that flavonoids could decrease the risk of osteoporosis (Qi et al. 2019; Liu et al. 2020). Feeding rats with total flavonoids of *Herba Taxilli* (400, 200, and 100 mg/kg/day for 28 days) was shown an ability in treating osteoporotic rats induced by retinoic acid (Li et al. 2019a). This efficacy could be attributed to the reduction of tartrate-resistant acid phosphatase (TRAP) and alkaline phosphatase (ALP) and elevation of serum Ca^{2+} level (Li et al. 2019a) (Fig. 7.4). Data from a systematic review reported by Chiavarini et al. (2020) involving 11 animal studies and 1 human study (between January 2016 to October 2019)

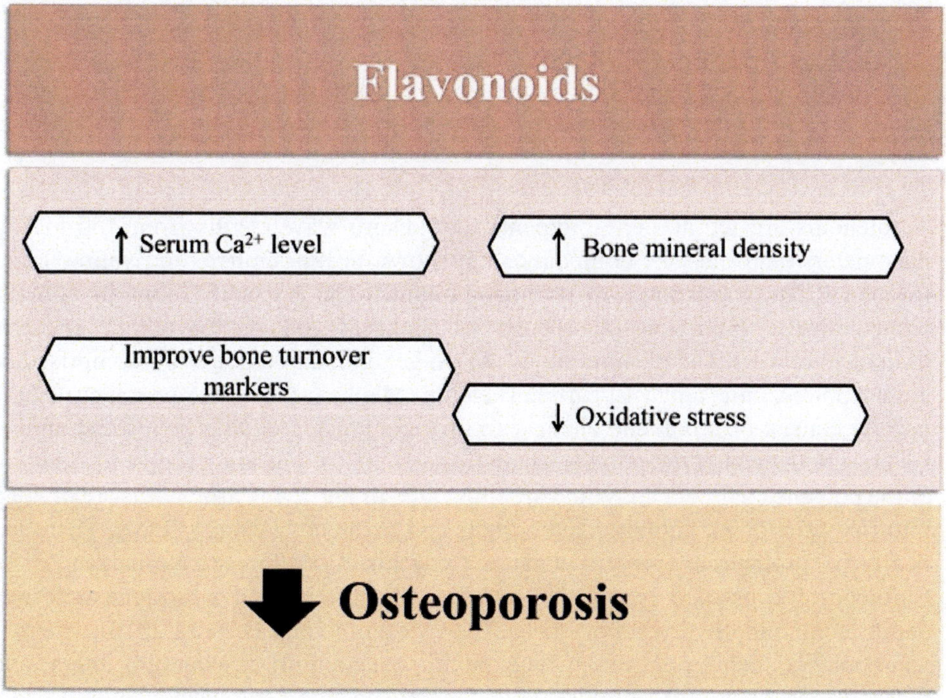

Fig. 7.4 Biological mechanism of flavonoids in relation to osteoporosis

further revealed that flavonoids exert a positive effect on osteoporotic fracture healing. This finding indicates that flavonoids could be used as a therapeutic agent for fracture healing. Likewise, a meta-analysis of 772,707 participants with 37,166 fracture cases demonstrated that tea consumption (≥3 cups/day) may be linked to the reduced risk of fractures (Xiang et al. 2019). In another systematic review and meta-analysis including 8 studies from inception to 9th January 2019 demonstrated that puerarin could enhance the bone mineral density in ovariectomy-induced postmenopausal osteoporosis in the murine model (Yang et al. 2020a) (Fig. 7.4). Puerarin is an isoflavone that exerts health-promoting properties (Zhang et al. 2019b; Jeon et al. 2020; Chen et al. 2021). Similarly, a systematic review and meta-analysis of randomized controlled trials further revealed that bone turnover markers (C-telopeptides, pyridinoline, and osteoprotegerin) were favorably affected by soy isoflavones (Akhlaghi et al. 2020). Notably, the data from bone markers analysis further demonstrated that isoflavones at dosage $<$90 mg/day show a better preventive effect in osteoporosis-associated bone loss in overweight/obese individuals (Akhlaghi et al. 2020). In the context of bone mineral density in postmenopausal women, a systematic review and meta-analysis of 63 randomized controlled trials involving 6427 postmenopausal women revealed that ipriflavone (600 mg/day) and genistein (54 mg/day) shows a beneficial effect on bone mineral density (Sansai et al. 2020). In this regard, these findings imply that flavonoids may improve bone mineral density and reducing the symptoms of osteoporosis. Taken together, flavonoids might be a promising approach for the prevention of age-related diseases and amelioration of oxidative stress. The crucial role played by flavonoids and their subclasses is worth attention to and further evaluated in randomized clinical trials.

7.4 Carotenoids

Carotenoids are tetraterpene pigments that absorb wavelengths from 400 to 550 nanometers, and thus the compounds are shown in red, orange, and yellow color (Maoka 2020). It is a naturally occurring pigment and is usually found in animals, plants, algae, fungi, several species of archaea, and photosynthetic bacteria (Ahrazem et al. 2016; Liang et al. 2018). In general, carotenoids are comprised of eight isoprene units and a 40-carbon skeleton (Maoka 2020). The general structures usually consist of a polyene chain with nine conjugated double bonds and an end group at both ends of the polyene chain (Maoka 2020). The end group and structure of the polyene chain of carotenoid are illustrated in Fig. 7.1. Carotenoids are divided into two categories, namely xanthophylls and carotenes (Maoka 2020). There are nearly 50 carotenes are present in nature (Khachik 2006; Tan and Norhaizan 2019). Carotenes, for instance, lycopene, γ-carotene, β-carotene, and α-carotene are carotenoids containing hydrocarbons (Fig. 7.1) (Mezzomo and Ferreira 2016); whereas xanthophylls such as peridinin, fucoxanthin, astaxanthin, zeaxanthin, lutein, and β-cryptoxanthin, are carotenoids containing oxygen atoms, for instance, furanoxide, epoxide, carboxylic, aldehyde, and carbonyl groups (Maoka 2020). Several xanthophylls are present as protein complexes, sulfates, glycosides, and fatty acid esters. There are nearly 800 xanthophylls that have been naturally discovered up until 2018 (Maoka 2009).

Carotenoids are essential components found in chlorophylls of photosynthetic plants, algae, and bacteria that are involved in photoprotection and photosynthesis (Maoka 2020). It produces light energy and converts this energy to chlorophylls via singlet-singlet excitation transfer (Kvíčalová et al. 2016). This conversion is a low energy state transfer used during photosynthesis. Conversely, carotenoids absorb unrestricted energy from chlorophylls via triplet-triplet conversion and generate excessive energy through polyene vibration (Maoka 2020). This triplet-triplet transfer is a high-energy state that is involved in photoprotection. During photosynthesis, ROS including superoxide anion radicals, hydroxyl radicals, and singlet oxygen are released from light and oxygen (Phaniendra et al. 2015). Notably, carotenoids with more than 11 conjugated double bonds have demonstrated an ability to scavenge singlet oxygen (Fiedor and Burda 2014). This mechanism is a physical reaction, in which carotenoids take up the thermal energy from singlet oxygen and produce this energy through polyene vibration (Maoka 2020). During the xanthophyll cycle, it involves the enzymatic removal of epoxy groups from xanthophylls (lutein epoxide, antheraxanthin, and violaxanthin) to produce de-epoxy xanthophylls (lutein and zeaxanthin) (Maoka 2020). This enzymatic cycle plays a crucial role in activating the energy liberation within light-harvesting antenna proteins through non-photochemical scavenging by decreasing the energy levels that reach the photosynthetic reaction centers (Polívka and Frank 2010).

Besides photosynthetic, carotenoids also appeared in non-photosynthetic organs of plants including flowers, roots, seeds, pericarps, and fruits (Maoka 2020). Carotenoids in non-photosynthetic organs serve as precursors of plant hormones, color attractants, antioxidants, and photo-protectors (Maoka 2020). During the ripening stage, seeds and fruits turn to purple or red color. This color changes could be due to the formation of anthocyanins and/or carotenoids (Fernández-López et al. 2020). For instance, the color of the pericarp of tomato changed to deep red from greenish-yellow color during the ripening process (Carrillo-López and Yahia 2014). This change is due to the conversion of phytoene to lycopene in the pericarp of the tomato (Fraser et al. 2007). Phytoene is a colorless compound found predominantly in greenish-yellow tomato and is converted to neurosporene (orange) and lycopene (red) by phytoene desaturase (Britton et al. 1998, 2008). Lycopene is a carotenoid that exerts some photoprotection and singlet oxygen scavenging activities (Britton et al. 1998, 2008).

Carotenes are one of the first intermediates in the synthesis of carotenoids, and thus these components exist in all photosynthetic organisms and are abundantly found in our diets (Maoka 2020). There are nearly 50 carotenoids found in human foods, in which 20 of them ingested from food are found in the serum or plasma (Maoka 2020). Of all of these carotenoids, zeaxanthin, lutein, β-cryptoxanthin, lycopene, α-carotene, and β-carotene are the major compounds that make up more than 90% of the total carotenoids (Maoka 2020). Carotenoids ingested from the diet are absorbed by the small intestine (Reboul 2019); whereas xanthophyll esters are hydrolyzed by esterase or lipase and absorbed (Breithaupt et al. 2007). The provitamin A carotenoids are converted by β-carotene-15,15′-dioxygenase into the retinal in the mucous of the small intestine (von Lintig 2012). Subsequently, the absorbed carotenoids are incorporated into chylomicrons and transported to the liver and other organs in the bloodstream (Reboul 2013).

Many studies have shown that carotenoids exert beneficial antioxidant and biochemical effects on animals and humans (Sajjadi and Bathaie 2017; Cui et al. 2019; Deding et al. 2020; Lin et al. 2020). Data from epidemiological studies revealed a negative link between high consumption of carotenoids and the risk of chronic diseases (Tan and Norhaizan 2019; Umigai et al. 2020). These favorable effects could be mediated via several mechanisms, for instance, regulation of cell growth (Moccia et al. 2020), enhancement of immune system (Liu et al. 2019), and modulation of cell differentiation, apoptosis, and cell cycle (Sheng et al. 2020). Carotenoids are highly lipophilic molecules that reside intracellularly to protect the membranes from oxidative stress (Fiedor and Burda 2014). Carotenoids are well-known as an eye-sight protecting agent (Bungau et al. 2019). These carotenoids are categorized as pro-vitamin A which comprised of the unsubstituted β-ionone ring such as γ-carotene, β-cryptoxanthin, β-carotene, and α-carotene, and thereby converted into retinal (Grune et al. 2010). Substantial evidence highlights that zeaxanthin and lutein are crucial dietary carotenoids in decreasing and preventing age-related macular degeneration (AMD) and cataracts (Maci and Santos 2015; Eisenhauer et al. 2017). AMD is the most common cause of blindness in adults aged 65 years old and above (Wong et al. 2014). Oxidative stress within the retina is implicated in the development and progression of AMD (Abokyi et al. 2020). Zeaxanthin and lutein have been identified as the major components of the macular pigments (Ma et al. 2012a). Studies have demonstrated a significant correlation between high concentrations of lutein in serum or ocular tissues and decreased risk of AMD (Korobelnik et al. 2017; Arslan et al. 2019; Rinninella et al. 2018). The preventive role of lutein toward AMD is more likely due to the antioxidant effect. Lutein can inhibit ROS and inflammatory mediator production, and thereby reduce NADPH oxidase subunit Nox4 (Tuzcu et al. 2017; Yanai et al. 2018). Data from the meta-analysis showed that serum lutein and zeaxanthin levels were inversely linked to the risk of nuclear cataracts (Liu et al. 2014b) (Table 7.5). The preventive role of lutein and zeaxanthin is more likely due to their blue-light filtering activities and antioxidant properties (Chitchumroonchokchai et al. 2004). The lens of the eye is prone to oxidative damage which subsequently promotes high-energy short-wavelength light exposure and increases the production of ROS. The ROS produced leads to aggregation or cross-linking of the crystalline proteins in the lens of epithelial cells, and ultimately caused the generation of cataracts (Gao et al. 2011). In particular, isomer zeaxanthin and lutein are concentrated in the lens, indicates that they may have a preventive role in the vital ocular tissues (Mares 2016). A systematic review and meta-analysis of 6 cohort studies comprising a total of 2477 incident AMD cases and follow-up from 5 to 18 years revealed that dietary intake of lutein and zeaxanthin can affect the risk of AMD (Ma et al. 2012b). The data showed that increased dietary intake of lutein and zeaxanthin may protect against late AMD (Ma et al.

Table 7.5 The clinical studies of carotenoids on age-related diseases

Age-related diseases	Study conditions	Source of carotenoids	Durations	Outcomes	References
Nuclear cataract	Meta-analysis (7 cross-sectional studies and 1 cohort study)	Serum lutein and zeaxanthin	–	↓ Nuclear cataract risk	Liu et al. (2014b)
AMD	Systematic review and meta-analysis of 6 cohort studies comprising a total of 2477 incident AMD cases and follow-up from 5 to 18 years	Dietary intake of lutein and zeaxanthin	–	Protect against late AMD	Ma et al. (2012b)
				Not associated with a reduced risk of early AMD	
Age-related cataract	Dose-response meta-analysis (6 prospective cohort studies involving 4416 cases and 41,999 participants)	Zeaxanthin/lutein	Each 300 μg/day increment	↓ Risk of posterior subcapsular cataract and nuclear cataract by 3%	Ma et al. (2014)
AMD	The multicenter Eye Disease Case-Control Study (356 case subjects who were diagnosed with the advanced stage of AMD within 1 year prior to their enrollment, aged 55 to 80 years old, and 520 control subjects)	Lutein from dark green, leafy vegetables	~6 mg/day	↓ Risk of AMD	Seddon et al. (1994)
AMD	Meta-analysis of 8 randomized controlled trials involving 1176 AMD patients	Zeaxanthin/lutein supplements	Each 1 mg/day increment (ranged from 6 to 36 months)	↓ 0.003 of visual acuity scale	Liu et al. (2015)
AMD	Meta-analysis (9 prospective, randomized controlled trials involving 855 patients diagnosed with AMD)	Lutein supplements	10 or 20 mg/day for more than 6 months	Improved visual acuity and MPOD in AMD patients	Feng et al. (2019a)
CVD	Systematic review and meta-analysis including from inception to August 2016, involving 634 participants	Lycopene-containing food such as tomato	70–400 g/day, followed-up for 2 months on average, range from 1 to 180 days	↓ IL-6 levels and LDL-cholesterol	Cheng et al. (2017)
				Improved flow-mediated dilation	
CVD	Placebo-controlled, randomized, double-blind, cross-over trial with 16 postmenopausal women	100% watermelon juice	Daily dose of 14.4 ± 0.34 mg lycopene for 4 weeks	No effect on serum lipids or antioxidant capacity	Crowe-White et al. (2020)
				↑ Circulating lycopene levels	
CVD	Diet-control-led, repeated measures cross-over design with 10 healthy subjects	Tomato or watermelon juice	Tomato juice (122 g each at dinner and breakfast, 18.4 mg lycopene/day) or watermelon juice (260 g each at dinner, lunch, and breakfast, 20.1 mg lycopene/day) for 3 weeks	No effect on plasma cholesterol levels or antioxidant status	Collins et al. (2004)

(continued)

Table 7.5 (continued)

Age-related diseases	Study conditions	Source of carotenoids	Durations	Outcomes	References
CVD	Prospective cohort of 39,876 middle-aged and older women	Tomato-based products (pizza, tomato sauce, tomato juice, and tomatoes)	≥7 servings/week during a median follow-up of 7.2 years	↓ CVD risk by 30%	Sesso et al. (2003)
Hypertension	Quasi-experimental study with pretest-posttest control group design, he Community Health Center of Magelang in July 2016–January 2017, pregnant women in trimester I, II, and III with high blood pressure	Tomato juice	250 mL/day	↓ 2.81 mmHg diastolic pressure	Anita et al. (2017)
				↓ 8.26 mmHg systolic pressure	
Atherosclerosis	Case-control study involving 125 subjects with early atherosclerosis and 107 controls aged 45–68 years old	Serum levels of lutein and β-carotene	–	Serum lutein level showed a greater early protection against atherosclerosis compared to serum β-carotene level	Zou et al. (2011)
Coronary artery disease	PBMC from patients with stable angina	Lutein	25 µM for 48 h	Inhibits LPS-induced IL-6, TNF, and IL-1β mRNA levels	Chung et al. (2017)
Colorectal cancer	Case-control study with 1846 controls and 923 colorectal cancer patients	Carrot, persimmon, peach, pumpkin, and ginger	49.4 ± 45.4 g/day	↑ Risk of colorectal cancer in men	Lee et al. (2017)
Lung cancer/prostate cancer	Randomized, double-blind, placebo-controlled primary-prevention trial involving 29,133 male smokers 50–69 years old	Alpha-tocopherol or β-carotene	Alpha-tocopherol (50 mg/day) alone, beta carotene (20 mg/day) alone and follow-up for 5–8 years	No reduction in the incidence of lung cancer among male smokers	The Alpha-Tocopherol Beta-Carotene Cancer Prevention Study Group (1994)
				High incidence of lung cancer in men who received β-carotene compared those who did not received	
				Induce prostate cancer among individuals who received α-tocopherol than those who did not	

Bladder cancer	Meta-analysis involving 516,740 adults included a study from inception to April 2019	Circulating concentration of zeaxanthin, lutein, β-carotene, and α-carotene	–	↓ Bladder cancer risk by 56% for every 1 μmol/L increase in daily circulating concentration of zeaxanthin and lutein	Wu et al. (2020)
		Dietary β-cryptoxanthin intake		↓ Bladder cancer risk by 27% for every 1 μmol/L elevation in circulating concentration of β-carotene	
				↓ Bladder cancer risk by 76% for every 1 μmol/L increase in circulating concentration of α-carotene	
				↓ Bladder cancer risk by 42% for every 1 mg increase in β-cryptoxanthin intake	
Alzheimer's disease	Case-control study performed from August 2006 to 2008	Serum carotenoids (zeaxanthin, lycopene, lutein, β-cryptoxanthin, β-carotene, and α-carotene)	–	Patients with Alzheimer's disease showed lower levels of serum carotenoids compared to cognitively normal participants	Mullan et al. (2017)
Parkinson's disease	Cohort study involving a total of 1036 cases (482 Nurses' Health Study and 554 Health Professionals Follow-up Study)	Dietary carotenoids (nuts, vegetables, and fruits)	Median intake of dietary carotenoids = 7502–26,940 mg/day over 12 years of follow-up	No effect	Hughes et al. (2016)
Alzheimer's disease and dementia	Community-based cohort of older adults (927 participants free from Alzheimer' disease and dementia at baseline)	Total carotenoids	6.7–24.8 mg/day during the follow-up for a mean of 7 years	Less global Alzheimer's disease and dementia pathology	Yuan et al. (2021)

AMD age-related macular degeneration, *CVD* cardiovascular disease, *IL-1β* interleukin-1beta, *IL-6* interleukin-6, *LDL-C* low-density lipoprotein cholesterol, *LPS* lipopolysaccharide, *MPOD* macular pigment optical density, *PBMC* peripheral blood mononuclear cells, *TNF-α* tumor necrosis factor-α

2012b). However, Ma et al. (2012b) did not identify an association between dietary lutein and zeaxanthin and early AMD. Ma et al. (2014) showed that every 300 μg/day increments of zeaxanthin/lutein decreased posterior subcapsular cataract and nuclear cataract risk by 3%. In addition, zeaxanthin/lutein supplementation was also significantly improved contrast sensitivity and visual acuity (Liu et al. 2015). The study showed that each 1 mg/day increment in zeaxanthin/lutein is linked to the 0.003 reductions on a visual acuity scale (Liu et al. 2015). Of all carotenoids, zeaxanthin and lutein, which are predominantly obtained from dark green leafy vegetables, were strongly related to the decreased risk of AMD (Seddon et al. 1994). The intake of lutein ~6 mg/day is sufficient to reduce the risk of AMD (Seddon et al. 1994). In support of this, a meta-analysis conducted by Feng et al. (2019a) showed that dietary intake of lutein (10 or 20 mg/day) for more than 6 months can significantly improve visual acuity and macular pigment optical density (MPOD) in AMD patients. In the context of toxicity, a systematic risk assessment of lutein supplements is safe up to 20 mg/day (Shao and Hathcock 2006). The AREDS2 trial showed that no adverse effects were reported (except skin yellowing) after supplementation of zeaxanthin and lutein (2 and 10 mg/day, respectively) for 5 years in patients with intermediate AMD (Age-Related Eye Disease Study 2 Research G 2013). Despite the data for safety is available for lutein, future studies are needed to evaluate the chronic intake of high lutein-containing supplements in a certain population. Table 7.5 summarizes the clinical studies of carotenoids on age-related diseases.

Carotenoids have been reported to prevent several chronic diseases induced by oxidative stress such as coronary artery disease (Chung et al. 2017) and CVD (Costa-Rodrigues et al. 2018). Oxidative stress can activate signaling pathways, predominantly through NF-κB, and thus leading to stimulation of IL-6 and TNF-α (Brasier 2010; Fischer and Maier 2015; Panahi et al. 2018). The implication of carotenoids in relation to the pathogenesis of CVD has been widely evaluated in both *in vivo* and *in vitro* models (Hajizadeh-Sharafabad et al. 2019). Increased plasma low-density lipoprotein (LDL) levels and oxidation state of LDL fraction increased the prevalence of CVD (Zhang et al. 2019c). Despite many therapeutic methods such as angioplasty or thrombolysis are available today, special attention needs to be paid to the diet (Tong et al. 2016). The prevalence of cardiovascular disorders is unevenly distributed in developed countries; while Southern Europe seems to be protected by having less incidence of CVD. This effect could be attributed to the dietary factors, for instance, the Mediterranean diet, which is containing olive oil and tomatoes (Krasinska et al. 2017). Watermelon, tomato sauce, and tomatoes are crucial sources of lycopene and may surrogate for the Mediterranean diet to a certain degree (Naz et al. 2014). Lycopene is a carotenoid with high antioxidant activity, which can modulate inflammation and apoptosis (Thies et al. 2017). Many studies evaluated lycopene in relation to cardiovascular risk. Although the findings are inconsistent, most of the studies showed that consumption of lycopene-containing food such as tomato (70–400 g/day for 1–180 days) significantly improved flow-mediated dilation and decreased IL-6 and LDL-cholesterol (Cheng et al. 2017), suggesting the tomatoes may protect against the development of CVD. Despite lycopene shows a high antioxidant activity but in certain circumstances, its cardiovascular preventive role appears to be likely linked to the anti-inflammatory properties than to the suppression of the LDL oxidation (Muller et al. 2016). This phenomenon could be attributed to the elevation of hydrophobicity in lycopene; hence it is more likely to be found in the nuclear hydrophobic core of the lipoprotein (Muller et al. 2016). Research evidence has suggested that intakes of one or more servings/day of tomato-rich products, such as pizza, tomato sauce, tomato juice, and tomatoes, reduced the CVD risk by 30% in middle-aged and older women (Sesso et al. 2003). The effects were not only observed in lycopene consumption, the serum concentration may also affect the cardiovascular risk (Mozos et al. 2018). Low adipose tissue and serum lycopene levels were correlated with major acute cerebrovascular and acute coronary diseases and early atherosclerosis and demonstrated to be more reliable in risk assessment compared to daily consumption of lycopene (Kim et al. 2011). A placebo-controlled, randomized, double-blind, cross-over trial involving 16 postmenopausal women demonstrated that watermelon juice (14.4 ± 0.34 mg lycopene for 4 weeks), one of the rich sources of dietary lycopene, was significantly increased circulating lycopene (Crowe-White et al. 2020). However, the study showed that consumption of watermelon juice did not improve antioxidant capacity or serum lipids (Crowe-White et al. 2020). In support of this, intakes of lycopene from tomato juice (122 g each at dinner and breakfast, 18.4 mg lycopene/day) or watermelon juice (260 g each at dinner, lunch, and breakfast, 20.1 mg lycopene/day) for 3 weeks did not affect the plasma cholesterol levels or antioxidant status among middle-aged adults (Collins et al. 2004). Oxidative stress causes endothelial dysfunction due to oxidative injury of the endothelial cells and uncoupling of the NO synthase (Mozos and Luca 2017), in which both of these are linked to the inflammation. By decreasing ROS and oxidative stress, lycopene decreased lipids, protein, and mitochondrial damage, enhances endothelium-dependent vasodilation, and promotes the bioavailability of NO (Nakamura et al. 2017). For instance, watermelon promotes plasma L-arginine and enabling NO production due to the presence of L-citrulline (Figueroa et al. 2017). The previous study stated that lycopene supplementation enhanced the endothelial-modulated vasodilation in CVD patients, but not in healthy individuals, implied the crucial role of lycopene in secondary cardiovascular prevention (Costa-Rodrigues et al. 2018). These favorable effects of lycopene could be attributed to the suppression of intracellular adhesion molecule-1 expression, TNF-α-induced NF-κB activation, and interaction of endothelial cells and monocytes (Hung et al. 2008). Further, lycopene can also suppress T lymphocyte activation and decrease the secretion of metalloproteinases by macrophages (Thies et al. 2017). The animal study showed that lycopene (30 mg/kg/day for 8 weeks) prevents transplant vasculopathy and reduces the smooth muscle cell proliferation and intimal hyperplasia in allograft vessels via modulating the NO/cyclic guanosine monophosphate (cGMP) pathways and decreasing Rho-associated kinases (He et al. 2016). Importantly, lycopene is a regulator for cholesterol levels by modulating a few mechanisms including down-regulation of kexin type 9/proprotein convertase subtilisin mRNA levels and suppression of HMG-CoA reductase, suggesting that lycopene supplementation could be beneficial for patients with statin intolerance (Alvi et al. 2017). In addition, lycopene possesses an antihypertensive effect by enhancing the production of NO in the endothelium, decreasing oxidative stress induced by angiotensin-II, and suppressing an angiotensin-converting enzyme in the endothelium (Belovic et al. 2016). This finding is supported by the study reported by Anita et al. (2017), who found that women who drank 250 mL of tomato juice/day can result in diastolic pressure drop by 2.81 mmHg and systolic pressure to decrease by 8.26 mmHg. Besides lycopene, lutein has also emerged as a potential atheroprotective agent in recent decades (Chung et al. 2017). In a study by Zou et al. (2011) focusing on serum lutein and β-carotene in relation to atherosclerosis, serum lutein level was shown to have a greater likelihood of early protection against atherosclerosis compared to serum β-carotene. The protective role of the cardiovascular system is more likely due to the ability to mediate cholesterol metabolism and antioxidative activity (Palloza et al. 2012), and thus lead to the prevention of LDL cholesterol oxidation (Hu et al. 2008). Further, the data from *ex-vivo* have also been demonstrated that lutein exerts anti-inflammatory effects on human peripheral blood mononuclear cells (PBMC) (Chung et al. 2017). Administration of lutein (25 μM for 48 h) in PBMC inhibits the LPS-induced IL-6, TNF, and IL-1β mRNA levels (Chung et al. 2017). Several studies have also reached a similar finding, in which high intake of carotenoids reduced the incidence of hypertension, stroke, myocardial infarction, and all-cause mortality (Yang et al. 2011). Although epidemiological studies have demonstrated an inverse relationship between β-carotene and the risk of CVD, several large randomized trials failed to observe any negative association between β-carotene and CVD. For instance, Women's Antioxidant Cardiovascular Study (WACS) (Cook et al. 2007), α-tocopherol and β-carotene (ATBC) study (Rapola et al. 1997), and MRC/BHF Heart Protection Study (Heart Protection Study Collaborative Group 2002) failed to show any beneficial effects. Although some controversy was reported in certain effects of carotenoids on cardiovascular health, most of the evidence showed the unequivocal benefits of lycopene consumption on vascular, cardiac, and endothelial health and function.

In addition to the effects observed on vision health and CVD, a beneficial effect of carotenoid supplementation has also been reported on cancer, demonstrating the enormous functional potential of carotenoids (Saini et al. 2020). Lycopene has gained attention for its health benefits including the treatment and prevention of cancer (Cha et al. 2017; Lim and Wang 2020). It has been suggested that the intake of lycopene from food or by itself may decrease cancer risk (da Costa Pereira Soares et al. 2019). Carotenoids were found to negatively link to several types of cancer (Gong et al. 2018; Kim et al. 2019; Mirahmadi et al. 2020). The anticancer ability of carotenoids is modulated via a few mechanisms including modulation of phase II detoxification enzyme and anti-inflammatory activities, regulation of angiogenesis and metastasis, and mediation of apoptosis induction, cell cycle arrest, and growth factor signaling (Trejo-Solís et al. 2013). These findings indicate the unique chemical structure of the carotenoids, which confers a strong antioxidant property (Mezzomo and Ferreira 2016). A meta-analysis involving 644 animals showed that lycopene (1.1–15 mg/kg/day or 20 mg/kg for 3x a week, range from 4 to 70 weeks via diet) significantly decreased the growth, number, and incidence of hepatocellular carcinoma (HCC) (Mekuria et al. 2020) (Table 7.6). Lycopene has been demonstrated to suppress liver tumor initiation through several mechanisms including stimulating antioxidant defense system, scavenging oxygen free radicals, and inhibiting the cytochrome P450 2E1 enzymes (Aizawa et al. 2016; Wang et al. 2010; Stice et al. 2018). The former could be due to its ability to activate the Nrf2 which is known to be an important regulator for the cellular response to oxidative

Table 7.6 Effects of carotenoids on age-related diseases *in vitro* and *in vivo*

Age-related diseases	Cell lines/Animal models	Source of carotenoids	Dosages	Findings	References
CVD	Male Brow-Norway (BN) rats (n = 16) and male Lewis rats (n = 32)	Lycopene	30 mg/kg/day for 8 weeks	Prevents transplant vasculopathy	He et al. (2016)
Cancer	Systematic review and meta-analysis involving 644 animals (Long-Evans Cinnamon rats, ferrets, C57BL/6J mice, C3H/HeN mice, WT/BCO2-KO mice, Sprague-Dawley rats, Wistar rats, and BALB/c mice), studies between 2001 and 2018	Lycopene	1.1–15 mg/kg/day or 20 mg/kg for 3x a week, range from 4 to 70 weeks via diet	↓ Growth, number, and incidence of HCC	Mekuria et al. (2020)
Dementia	Female Sprague-Dawley rats (n = 60)	Lycopene	100 mg/kg for 2 months	↑ SOD activity in the hippocampus ↓ NRSF levels	Zhu et al. (2020)
Alzheimer's disease	P301L transgenic mice	Lycopene	5 mg/kg for 8 weeks	↓ The increase of tau phosphorylation at Ser396, Ser262, Ser235, and Thr231 in brain tissues	Yu et al. (2017)
Alzheimer's and Parkinson's diseases	Microglial cells	Crocetin and crocin	Crocetin (10 and 20 μM) and crocin (10 and 20 μM)	Inhibit the generation of NO and proinflammatory cytokines	Nam et al. (2010)
Cognition	Male Wistar rats	Lycopene	4 mg/kg/day for 90 days	↓ Cognitive impairment of the $AlCl_3$-induced hippocampal lesions	Cao et al. (2019)

HCC hepatocellular carcinoma, *NO* nitric oxide, *NRSF* neuron-restrictive silencer factor, *SOD* superoxide dismutase stress (Bartolini et al. 2018). Nrf2 can stimulate detoxifying or antioxidant enzymes, for instance, catalase, HO-1, GPx, glutathione-S-transferase (GST), and SOD (Gupta et al. 2013; Bhatia et al. 2018). Further, lycopene exerts an anti-inflammatory effect by suppressing HCC progression and development (Bhatia et al. 2018). This is largely modulated by suppression of the oncogenic transcriptional NF-κB and downstream cascades (Ip et al. 2013; Bhatia et al. 2018), and thus resulting in the suppression of the NF-κB induced expression of inflammatory cytokines, for instance, IL-6 and TNF-α, that influence the proliferation and survival of premalignant cells (Vendrell et al. 2015; Todoric et al. 2016). The beneficial effect of lycopene was not only observed in liver cancer, the anticancer effect was also reported in bladder cancer. Data from a meta-analysis involving 516,740 adults included a study from inception to April 2019 demonstrated that bladder cancer risk was reduced by 56% for every 1 μmol/L increase in daily circulating concentration of zeaxanthin and lutein, by 27% for every 1 μmol/L elevation in circulating concentration of β-carotene, by 76% for every 1 μmol/L increase in circulating concentration of α-carotene, and by 42% for every 1 mg increase in β-cryptoxanthin intake (Wu et al. 2020). Orange/yellow fruits and vegetables are thought to be an excellent source of carotene, which can act as provitamin A (Liu 2004). Carotenoids may exert a positive effect against cancer by inhibiting the metastasis and proliferation of cancer cells (Reczek and Chandel 2015). However, intake of orange/yellow fruits and vegetables including carrot, persimmon, peach, pumpkin, and ginger (49.4 ± 45.4 g/day) are linked to an increased risk of colorectal cancer in men (Lee et al. 2017). The possible mechanisms underlying the increased risk of colorectal cancer in men should be further elucidated to confirm these relationships.

Notably, high blood concentration and dietary consumption of carotenoids were linked to the reduced risk of total cancer and all-cause mortality (Aune et al. 2018). However, the negative association between blood concentration and dietary intake of antioxidants is not likely due to the single antioxidants but due to the combination of bioactive compounds present in vegetables and fruits (Aune et al. 2018). Such findings support the recommendation to increase consumption of vegetables and fruits, but not antioxidant supplements for chronic disease prevention (Aune et al. 2018). However, randomized clinical trials are consistently shown no clear benefits of antioxidant supplements in relation to chronic disease risk as well as potential harms of supplementation of β-carotene and vitamin E on lung cancer risk and mortality, respectively (The Alpha-Tocopherol Beta-Carotene Cancer Prevention Study Group 1994; Bjelakovic et al. 2012, 2013). The study evaluated the use of supplements with a single of several antioxidants; whereas the observational studies have assessed these nutrients based on the dietary intake of food or blood concentrations. The dietary sources of these antioxidants found in vegetables and fruits also exert a broad spectrum of bioactive constituents that may have synergistic effects. The findings from the observational studies could reflect a small number of bioactive constituents that have not been assessed in randomized trials. In support of this, the randomized intervention study evaluated the intake of kiwi fruit (3 kiwi fruits/day added to the regular diet) and using gene expression profiles as a proxy outcome, the study showed that phytochemicals work in concert rather than individually (Bøhn et al. 2010).

An emerging role of carotenoids in relation to neuroinflammation and the occurrence of its positive effects on Alzheimer's disease pathophysiology has been extensively evaluated in both *in vivo* and *in vitro* models (Nazem et al. 2015; Ardura-Fabregat et al. 2017). Neuroinflammation is a response of the nervous system during autoimmune disorders, trauma, and neurodegeneration (Stephenson et al. 2018). Carotenoids have been shown to decrease proinflammatory cytokine release and lipid peroxidation by inhibiting the stimulation of the NF-κB pathway in the presence of oxidative stressors (Kim et al. 2012; Wu et al. 2015; Liu et al. 2017a). The animal study showed that feeding rats with lycopene (100 mg/kg for 2 months) increased SOD activity in the hippocampus and reduced the neuron-restrictive silencer factor (NRSF) levels in the vascular dementia rats (Zhu et al. 2020), indicating that the memory and learning impairment of vascular dementia rats could be alleviated by lycopene. NRSF plays a crucial role in numerous cellular processes in the nervous systems (Warburton et al. 2015), in which the NRSF levels are markedly increased under stress-associated conditions of Alzheimer's disease and aging. A study reported by Yu et al. (2017) further demonstrated that administration of lycopene (5 mg/kg for 8 weeks) can reduce the increase of tau phosphorylation at Ser396, Ser262, Ser235, and Thr231 in brain tissues of P301L transgenic mice, implied that suppression of tau protein phosphorylation may modulate the Alzheimer's disease. In addition to lycopene, the beneficial effects of dietary carotenoids, for instance, crocetin, crocin, astaxanthin, and fucoxanthin have been investigated in recent decades (Lin et al. 2016b; Finley and Gao 2017; Tribuzi et al. 2017; Yuan et al. 2020; Hafez et al. 2021). The previous study has demonstrated that crocetin (10 and 20 μM) and crocin (10 and 20 μM) can inhibit the generation of NO and proinflammatory cytokines stimulated by β-amyloid (Aβ), interferon-γ, and LPS in microglial cells (Nam et al. 2010). Astaxanthin has also been demonstrated to protect neurons in Parkinson's disease (Lee et al. 2011; Ye et al. 2012, 2013) and Alzheimer's disease (Lobos et al. 2016). Consistent with the beneficial effects of carotenoids treatment on the neurodegenerative study on cell culture and animal studies, the cohort study has also shown the correlation between the intakes of total carotenoids (6.7–24.8 mg/day during the follow-up for a mean of 7 years) and had less global Alzheimer's disease and dementia pathology (Tan and Norhaizan 2019; Yuan et al. 2021). A study evaluated zeaxanthin and lutein in relation to cognitive performance among 2011–2014 National Health and Nutrition Examination Survey (NHANES) among US adults aged ≥60 years old (Christensen et al. 2020). The data showed that a high intake of zeaxanthin and lutein from foods and supplements may prevent or delay cognitive decline (Christensen et al. 2020). In support of this, a case-control study further revealed that patients with Alzheimer's disease showed lower serum levels of carotenoids (zeaxanthin, lycopene, lutein, β-cryptoxanthin, β-carotene, and α-carotene) compared to cognitively normal subjects (Mullan et al. 2017). Despite experimental and human studies have proposed a possible relationship between increased consumption of both nonprovitamin A carotenoid and provitamin A species and the decreased risk of Parkinson disease; however, the risk reduction was small and did not always reach statistical significance (Guest and Grant 2016). A study reported by Hughes et al. (2016) showed that intake of carotenoids from nuts, vegetables, and fruits (median intake of dietary carotenoids = 7502–26,940 mg/day over 12 years of follow-up) was not related to the risk of Parkinson's disease. Carotenoids may suppress the onset of neurodegenerative diseases via a few mechanisms. It has been suggested that carotenoids may alleviate oxidative stress and mitochondrial dysfunction by regulating amyloid oligomer-induced signaling, modulating anti-amyloid aggregation activity and anti-neuroinflammatory effects, upregulating of antioxidant enzyme, and scavenging of ROS (Lakey-Beitia et al. 2017; Lin et al. 2017; Xiang et al. 2017; Zhao et al. 2017). Notably, the mechanisms of action of carotenoids are likely to be regulated simultaneously. For instance, feeding with a diet containing lycopene (4 mg/kg/day for 90 days) ameliorated cognitive impairment of the hippocampus in aluminium chloride ($AlCl_3$)-induced rats by inhibiting oxidative stress-mediated apoptosis and inflammation (Cao et al. 2019). Such an effect was accompanied by an elevation of SOD and GSH activity and decreased 8-hydroxy-2′-deoxyguanosine and MDA levels (Cao et al. 2019). The study further revealed that administration of lycopene resulted in the concomitant upregulation of *Nrf2* and *HO-1*, downregulation of NF-κB protein and TNF-α mRNA levels, implies that lycopene ameliorates $AlCl_3$-induced hippocampal lesions by suppressing oxidative stress-mediated inflammation in rats (Cao et al. 2019). Although many studies showed the beneficial effects of carotenoids on neurodegenerative diseases, further studies are required to elucidate whether the anti-apoptotic and antioxidative activity of carotenoids in relation to Alzheimer's disease or Parkinson's disease is interconnected and whether modulation of neuroinflammatory response could result in the therapeutic effect of carotenoids. Taken together, carotenoids are a good antioxidant that is best obtained by consuming specific foods rich with carotenoids. High consumption of carotenoids should be included in the diet of the elderly through carotenoids-rich foods.

7.5 Dietary Minerals

Dietary minerals are essential components required by all living organisms for proper body metabolism (Gharibzahedi and Jafari 2017). Minerals are naturally occurring elements with definite chemical formulas and universal structures. Minerals play a key role in transmitting nerve impulses, secreting specific hormones, regulating heartbeat, and facilitating bone formation (Pravina et al. 2013). Minerals are not only can be obtained from drinking water and food, it can also found in mineral supplements (Akram and Rehman 2018; Ershow et al. 2018). Indeed, most of the minerals can be consumed through animal sources, fruits, and vegetables (Tasić et al. 2017; Wong et al. 2019; Czech et al. 2020). However, individuals who did not meet the daily dietary mineral intake can obtain the minerals through supplementation (Wan et al. 2019).

Minerals are classified into two classes, namely, trace minerals and macrominerals. Macrominerals are dietary minerals that are needed in a large amount, such as calcium, magnesium, chloride, phosphorus, sodium, and potassium; whereas trace minerals including fluoride, zinc, copper, iron, iodine, and selenium (Sousa et al. 2019). These trace elements are usually served as a functional part of the enzyme

(Yatoo et al. 2013). Nevertheless, consumption of a high amount of trace elements is noxious to both animals and humans (Counotte et al. 2019; Mehri 2020). For example, selenium is a crucial trace element that should be present in the diet to obtain an adequate intake. Selenium exerts anti-inflammatory and antioxidant actions and is responsible for immune response regulation (Ibrahim et al. 2019). However, long-term excessive selenium (2 and 6 mg/L for 85 days) supplementation significantly elevates the systolic blood pressure in male Wistar rats (Grotto et al. 2018) (Table 7.7). Selenium (200 μg/day) supplementation was also significantly increased the risk of type 2 diabetes mellitus in the elderly, suggesting that

Table 7.7 *In vivo* studies conducted in dietary minerals and their effects on age-related diseases

Age-related diseases	Animal models	Source of minerals	Dosages	Findings	References
Parkinson's disease	Male Wistar rats	Selenium	11.18 μg/L in drinking water	Effectively prevented the harmful effects of the toxin in locomotor activity, reducing bradykinesia, and DNA damage in leukocytes	Ellwanger et al. (2015)
Alzheimer's disease	Wistar rats	Selenium	1 mg/kg for 21 days	Attenuate lipid peroxidation and protect against aluminium chloride-induced Alzheimer's disease	Lakshmi et al. (2015)
Osteoporosis	Male Wistar rats	Zinc supplements	300 mg zinc/kg diet or 3000 mg zinc/kg diet for 4 weeks	Reduced bone mineral density	Suzuki et al. (2016)
Diabetic cardiomyopathy	Female Wistar rats	Normal basal diet (AIN-76) with the addition of zinc salt (zinc carbonate)	Five times (0.19 g/kg diet) and ten times (0.38 g/kg diet) for 6 weeks	Reduced hypercholesterolemia and hyperlipidemia, ameliorated hyperglycemia-induced oxidative stress, and showed cardioprotective effects	Barman and Srinivasan (2017)
Diabetic nephropathy	Male Wistar rats	Zinc supplemented diet	5 mg/kg in drinking tap water	Protect diabetic-induced renal damage in rats	Elsaed and Mohamed (2017)
Hypertension	Male Wistar rats	Selenium	2 and 6 mg/L for 85 days	Elevate systolic blood pressure	Grotto et al. (2018)
Huntington's disease	R6/1 mice	Dietary zinc	Zinc-deficient diet (chow: Specialty Feeds) for 15 weeks	Caused cognitive impairment and exacerbated hippocampal LTP deficit	Ayton et al. (2020)

AIN American Institute of Nutrition, *LTP* long-term potentiation selenium supplementation may increase the development of diabetes (Thompson et al. 2016).

They are several minerals showed their antioxidant properties such as magnesium, selenium, zinc, and copper (Salmonowicz et al. 2014). For instance, zinc plays a crucial role in the body by suppressing the nicotinamide adenine dinucleotide phosphate-oxidase (NADPH-oxidase) enzyme (do Nascimento Marreiro et al. 2017), acts as a cofactor for SOD enzymes (Cruz et al. 2015), regulating the GSH metabolism (Omata et al. 2013), and modulating the metallothionein expression (Ganger et al. 2016). Table 7.8 summarizes the roles and primary sources of macrominerals and trace minerals.

The deficiency of selenium, iodine, zinc, and iron affects about 15, 30, 30, and 60%, respectively, of the world population (Gharibzahedi and Jafari 2017). The mineral malnutrition in developing and industrial countries has drawn attention from the researchers, which attempts to design a technological and scientific study to overcome the minerals deficiency (Gharibzahedi and Jafari 2017). The deficiency of trace minerals can affect organs in the body (Zemrani and Bines 2020). In children and infants, the predominant causes of zinc deficiency are low zinc levels or malnutrition in breast milk, undernourishment, and parenteral nutrition (Livingstone 2015). Further, eating disorders, for instance, bulimia and anorexia nervosa may also cause zinc deficiency in both adults and children (Humphries et al. 1989). Malabsorption disorders also caused zinc deficiency (Prasad 2020). For example, inherited diseases such as cystic fibrosis and acrodermatitis enteropathica, as well as chronic inflammatory bowel diseases including ulcerative colitis and Crohn's disease may also lead to zinc malabsorption. Apart from that, high consumption of phytic acid, iron, or copper can also lead to dietary zinc malabsorption (Glutsch et al. 2019).

Dietary minerals have beneficial antioxidant and biochemical effects on several age-related diseases in the elderly (Tan et al. 2018; Skrajnowska and Bobrowska-Korczak 2019; Squitti et al. 2020) (Table 7.9). Zinc was found to negatively associate with CVD in humans based on several studies. Research evidence has revealed that zinc decreases TG, LDL-C, and TC and increases HDL-C levels (Li et al. 2010; El-Ashmony et al. 2012; Olechnowicz et al. 2018). A systematic review of prospective cohort studies comprising of 91,708 participants revealed that high serum zinc level was found to be associated with the reduced risk of cardiometabolic diseases, particularly in CVD (Chu et al. 2016). From the study reviewed, it showed that the protective effect of zinc was more pronounced in vulnerable individuals, specifically for those with coronary angiography or patients with type 2 diabetes mellitus (Chu et al. 2016). In support of this, low serum zinc levels ($<$14.1 μmol/L) increased the incidence of myocardial infarction by 37% (Soinio et al. 2007). Notably, data from a prospective cohort study involving 20,460 Australian women aged 50–61 years revealed that consumption of food rich in zinc such as meat, dairy products, cereals, poultry, and fish, has a greater likelihood to increase the incidence of CVD in women (Milton et al. 2018). A recent systematic review and meta-analysis of 16 randomized controlled trials and 26 prospective cohort studies demonstrated that dietary calcium intake (72.2–4405 mg/day and follow-up for

Table 7.8 The roles and primary sources of macrominerals and trace minerals

Groups	Minerals	Main functions	Examples of food sources		
			Vegetables/Fruits	Animals	Others
Macrominerals	Calcium (Ca)	Blood pressure, blood clotting, immune system modulation, nerve functioning, muscles contraction and relaxation, and healthy teeth and bones maintenance	Legumes, mustard greens, and broccoli	Canned fish with bones (sardines and salmon), fortified soy milk and tofu, dairy products, and milk	–
	Magnesium (Mg)	Prevents constipation, nerve transmission, immune system health modulation, muscle contraction, and protein formation	Green and leafy vegetables, artichokes, legumes, seeds, and nuts	Seafood, dairy products, and milk	Drinking water and chocolate
	Chloride (Cl)	Essential for digestive (stomach) juices and body fluids maintenance	Olives, celery, lettuce, tomatoes, rye, and seaweed	Meat and trace amounts in milk	Breads, processed foods, soy sauce, and table salt
	Potassium (K)	Waste elimination, blood pressure maintenance, muscle contraction, nerve transmission, and fluid balance regulation	Legumes, whole grains, and dried and fresh vegetables and fruits	Yogurt (non-fat/skim, plain), milk, salmon fish, and meats	Acorn squash and baked potato
	Phosphorus (P)	Modulate nerve signaling and heartbeat, maintains acid-base balance, healthy teeth and bones maintenance, energy and ATP production, and required for protein synthesis	Legumes (lentils and beans), Brazil nuts, and squash and pumpkin seeds	Cheese, non-fat yogurt, milk, eggs, poultry, shellfish (scallops), salmon fish, and meats	Soya foods (tofu) and processed foods
	Sulfur (S)	Maintain structural integrity of skin, required for connective tissues development, and protect against toxic substances	Nuts, legumes, chives, leeks, onions, garlic, turnips, Brussels sprouts, kale, cabbage, cauliflower, and broccoli	Milk, eggs, fish, poultry, and meats	–
	Sodium (Na)	Nerve transition, muscle contraction, modulates some metabolic activities, heart function, and fluid and electrolyte balance	Canned vegetables (sweet peppers) and pickles (cucumber)	Cheese, trace amounts in milk, and cured fish and meat (bacon)	Fast foods, snacks, salted and roasted seeds and nuts (pumpkin seeds), instant soups, soy sauce, breads, and table salt

(continued)

Groups	Minerals	Main functions	Examples of food sources		
			Vegetables/Fruits	Animals	Others
Trace minerals	Chromium (Cr)	Regulates blood glucose levels, stimulates cholesterol and fatty acid synthesis, modulates body processes and brain function, and carbohydrates and fats metabolism	Molasses, black pepper, spinach, bananas, apples, green peppers, nuts, wheat germ, and whole grains	Butter, cheese, oysters, eggs chicken, liver, and beef	Unrefined foods and brewer's yeast
	Boron (B)	Important in modulating organ and cellular membrane functions, embryonic development, and improving the estrogen levels	Oat and wheat bran, legumes (beans and lentils), peanut butter, nuts, dried fruits (raisings and prunes), olives, avocados, apples, pears, peaches, red grapes, banana, potatoes, onions, carrots, and broccoli	–	Bee pollen, honey, and fruit juices
	Cobalt (Co)	Reduces neuromuscular, digestive, and fatigue	Figs, mushrooms (especially shitake), nuts, green leafy vegetables (spinach, turnip, lettuce, cabbage, and broccoli), and cereals (oats)	Shellfish, fish, mussels, oysters, milk, kidneys, liver, and meat	–
	Fluoride (F)	Delays bone density loss and maintain bone structure, decreases cavities in children, prevents tooth decay, teeth and bones development	Teas	Deboned chicken/meat, seafood (shellfish and fish)	Canned shellfish and fish, wine, beer, processed cereals, and drinking water
	Iodine (I)	Involved in several crucial component of hormones produced by the thyroid gland such as production of blood cells, muscle and nerve function, reproduction, metabolism, development, and growth	Organic potatoes, navy beans, strawberries, dried prunes, cranberries, and bananas	Boiled eggs, baked turkey breast, dairy products and milk, and seafood (lobster, shrimp, and cod fish)	Canned corn, tuna, baked potatoes, bread, and iodized salt
	Copper (Cu)	Neutralizes free radicals, promotes healing, repairs injured tissues, promotes immune system, required for proper functioning of organs and metabolic processes, and involved in protein metabolism	Avocados, dried fruit (prunes), mushroom, kale, whole grains, sesame seeds, cashew nuts, and legumes (chickpeas and cooked beans)	Goat cheese, oysters, and organ meats	Fermented soy foods (tempeh) and drinking water
	Iron (Fe)	An integrated part for crucial enzyme systems in several tissues, a transport medium for electrons within cells, needed for energy metabolism, and needed for the formation of hemoglobin in red blood cells	Dried fruits, dark leafy green vegetables, bran, whole grains, white beans, lentils, nuts, and pumpkin and squash seeds	Poultry, eggs, lamb, beef, clams, mussels, oysters, and liver (chicken)	Tofu, cereals, breads, cocoa powder, and dark chocolate
	Manganese (Mn)	Crucial for the nervous system activity and normal functioning of the brain, important for the growth of human bone structure, and preventing osteoporosis	Tea (black, brewed), whole grains (brown rice), spinach, beans, pumpkin seeds, and nuts	Fish and seafood	Tofu and bread (whole-wheat)
	Molybdenum (Mo)	Cell protection, produce energy within the cells, and produce enzymes that are required to remove waste from the body	Cucumber, wheat flour, sunflower seeds, nuts, leafy vegetables (spinach), whole grains, and legumes (lentils, peas, and beans)	Organ meats (lambs), cheese, milk, liver, and eggs	Breads and pasta
	Selenium (Se)	Assist in detoxification process, protect the organism from various viruses, stimulate the immune system, and protect the body from free radicals, heavy metals, and other harmful substances	Mushrooms, brown rice, lima/pinto beans, whole grains (rye), Brazil nuts, seeds (flaxseed, sesame, sunflower, and chia), and green and leafy vegetables (spinach, cabbage, and broccoli)	Pork, turkey and chicken, lamb and beef, fish, and seafood (oysters)	Selenium-enriched yeast and whole-wheat bread
	Zinc (Zn)	Important for protein synthesis and genetic materials, exert a key role in modulating the immune system, normal growth, fetal development, wound healing, and taste perception and improve digestion	White mushrooms, beans, nuts, squash and pumpkin seeds, spinach, wheat germ, and leavened whole grains	Pork, poultry, fish, lam and beef, and seafood (oyster)	Cocoa powder and chocolate

Table 7.9 The clinical studies of dietary minerals on age-related diseases

Age-related diseases	Study conditions	Source of minerals	Durations	Outcomes	References
Type 2 diabetes mellitus	Phase III, randomized, placebo-controlled study	Selenium supplements	200 μg/day for 33.0 months (range from 0 to 82.6 months)	↑ Risk of type 2 diabetes mellitus in elderly	Thompson et al. (2016)
Type 2 diabetes mellitus	Systematic review of prospective cohort studies comprising of 334,387 participants	Zinc intake (dietary and/or supplements)	Median intake = 4.9–18.0 mg/day and follow-up for 4.8–24 years	No association	Chu et al. (2016)
Type 2 diabetes mellitus	Prospective cohort study involving 8921 women aged 45–50 years	Dietary zinc consumption from sugar, fat spreads, eggs, dairy products, bread, vegetables, and fruits	5.94–17.35 mg/day during 6 years of follow-up	↓ Risk of type 2 diabetes mellitus in women	Vashum et al. (2013)
CVD	Systematic review of prospective cohort studies comprising of 91,708 participants	Zinc (serum level)	≤14.1 μmol/L	↓ CVD risk	Chu et al. (2016)
CVD	Cohort study (20,460 Australian women aged 50–61 years)	Zinc-containing foods (meat, dairy products, cereals, poultry, and fish)	5.94–17.35 mg/day, with a mean intake of 10.66 mg/day and follow-up for 6 years	↑ Incidence of CVD in women aged 50 and above	Milton et al. (2018)
CVD	Systematic review and meta-analysis of 16 randomized controlled trials and 26 prospective cohort studies	Dietary calcium from dairy foods and nonalcoholic beverages	72.2–4405 mg/day and follow-up for 5.5–65 years	Do not increase the risk of CVD such as stroke and coronary heart disease	Yang et al. (2020b)
CVD	Systematic review and meta-analysis of 16 randomized controlled trials and 26 prospective cohort studies	Calcium supplements	1000–1500 mg/day during 6 months-7.1 years of follow-up	↑ Risk of myocardial infarction	Yang et al. (2020b)

(continued)

Age-related diseases	Study conditions	Source of minerals	Durations	Outcomes	References
Alzheimer's disease	10 patients (aged 54–93 years) with early or moderate AD, as well as 60 healthy people (aged 52–83 years)	Serum zinc and selenium	–	Low concentrations in Alzheimer's disease patients compared to healthy individuals	Socha et al. (2021)
Cognition	5435 community-dwelling men aged ≥65 years, a cohort study with a median follow-up for cognitive function of 4.6 years	Serum sodium	126–140 mmol/L	Low serum sodium increased risk of cognitive decline	Nowak et al. (2018)
Osteoporosis	12,794 community-dwelling individuals (6301 men and 6493 women) aged 40–74 years and follow-up for 5 years	Dietary calcium (fish, beans, vegetables, dairy products and milk, fruits, and cereals)	Women = 537 mg/day and men = 442 mg/day	Low calcium intake increased vertebral fractures and total fractures in women	Platonova et al. (2021)
Osteoporosis	Multicenter, hospital-based, and cross-sectional study involving 277 women with osteoporotic fractures.	Dietary calcium	Mean intake = 503.7 ± 274.7 mg/day, ranged from 36.4 to 2090.8 mg/day	Positive correlation on bone mineral density of femoral neck	Yoon et al. (2016)
Osteoporosis	Randomized controlled trial comprising a total of 1994 osteopenic postmenopausal women aged 65 years old and above during 6 years of follow-up	Dietary calcium	886 mg/day	No effect	Bristow et al. (2019)

CVD cardiovascular disease

5.5–65 years) from dairy foods and nonalcoholic beverages do not increase the risk of CVD such as stroke and coronary heart disease; whereas calcium supplementation (1000–1500 mg/day during 6 months-7.1 years of follow-up) may potentially lead to the development of coronary heart disease, particularly in myocardial infarction (Yang et al. 2020b). These findings suggest that intake of calcium should be included in the diet of the elderly via calcium-rich containing food. Table 7.9 summarizes the clinical studies of dietary minerals on age-related diseases.

Zinc is thought to play a crucial role in insulin secretion in peripheral tissues (Ruz et al. 2019). The binding of zinc to insulin is crucial for the maturation, crystallization, and biosynthesis of the hormone (Chabosseau and Rutter 2016). Zinc is transported to the insulin secretory granules of β-cells via ZnT8 (Chu et al. 2016). The previous study found that deletion or lack of ZnT8 reduces insulin secretion and crystallization (Wijesekara et al. 2010). Suppressing the stimulation of NF-κB, zinc leads to the inhibition of IL-6, TNF-α, and IL-1β secretion from macrophages and monocytes (Chasapis et al. 2020). A study reported by Vashum et al. (2013) evaluated dietary zinc in relation to type 2 diabetes mellitus. The data showed that high dietary zinc consumption (5.94–17.35 mg/day during 6 years of follow-up) from sugar, fat spreads, eggs, dairy products, bread, vegetables, and fruits is negatively associated with the type 2 diabetes mellitus in women (Vashum et al. 2013). Zinc intake has been linked to the decrease of oxidative stress through a few mechanisms. Indeed, zinc is involved in the insulin secretion, production, and action that served as a catalytic factor for the carboxypeptidase H enzyme. This enzyme facilitates the conversion from an inactive form of proinsulin into an active form of insulin. Moreover, zinc is also caused the phosphorylation of the insulin receptor by releasing more glucose into the cells (Norouzi et al. 2018). In a systematic review, zinc was found to improve insulin resistance in obese individuals (Cruz et al. 2017). Despite several studies have found an inverse association between dietary zinc and diabetes mellitus risk, not all findings showed such a link. Chu et al. (2016) did not identify an association between type 2 diabetes mellitus risk and zinc dietary/supplements intake (4.9–18.0 mg/day and follow-up for 4.8–24 years). Collectively, zinc plays a crucial role as an antioxidant nutrient that mediates insulin levels in type 2 diabetes mellitus.

In addition to the effects mentioned above, zinc has also shown a significant impact on brain function (Qi and Liu 2019). In particular, the hippocampus contains the highest zinc levels in the body (Frederickson 1989). The loss of zinc may compromise neuronal activity (Ayton et al. 2020). Low cortical zinc was shown to trigger cognitive impairment (Ayton et al. 2020). An animal study further demonstrated that dietary restriction of zinc exacerbated hippocampal long-term potentiation (LTP) deficit, suggesting that zinc may play a crucial role in maintaining the brain function of Huntington's disease (Ayton et al. 2020) (Table 7.7). Huntington's disease is characterized by cognitive decline and chorea motor impairment (Paulsen 2011), and typically occurred in mid-life, suggesting that aging may play a key role in the pathogenesis of Huntington's disease (Machiela and Southwell 2020). In the brain, abnormal metabolism of trace elements could modulate the mitochondrial functions and synaptic signaling pathways, induce inflammation, oxidative stress, protein aggregation, and lead to synaptic dysfunction in the brain of Alzheimer's disease patients (De Benedictis et al. 2019). The previous study stated that patients with Alzheimer's disease had a relatively low concentration of serum zinc and selenium levels compared to healthy individuals (Socha et al. 2021). It has been found that proper zinc homeostasis is crucial in Alzheimer's disease because extremely high levels of zinc may lead to tau aggregation-induced toxicity and apoptosis and trigger the formation of granular tau aggregates in neuronal cells, which subsequently contributes to the hyper-phosphorylation of tau. By contrast, very low amounts of zinc can contribute to the development of amyloid fibril (Cristóvão et al. 2016; Hu et al. 2017). Frequent consumption of tea, jam, meat, poultry, honey, and flour products may positively correlated with serum zinc levels; whereas frequent intakes of cottage cheese, raw vegetables, sausages, and cakes may negatively link to the serum zinc concentration (Socha et al. 2021). A study reported by Nowak et al. (2018) evaluated the serum sodium level in relation to cognitive function in the community-dwelling older man. The data showed that low serum sodium levels (126–140 mmol/L) were related to the risk of cognitive decline in community-dwelling older men (Nowak et al. 2018). This finding could be attributed to the reducing extracellular sodium which may lead to the increases in oxidative stress markers (Barsony et al. 2011). Indeed, oxidative stress is one of the predominant factors that contribute to brain senescence, which can lead to cognitive decline (Forster et al. 1996).

Calcium is an essential nutrient crucial in modulating bone metabolism, and deficiency of calcium has been associated with osteoporosis (Park et al. 2017), particularly in East Asians (Balk et al. 2017). In a study by Platonova et al. (2021) focusing on dietary calcium-containing foods such as fish, beans, vegetables, dairy products and milk, fruits, and cereals (women = 537 mg/day and men = 442 mg/day) on osteoporotic fracture risk in middle-aged and elderly Japanese women and men, women were shown to have a greater likelihood of vertebral fractures and total fractures with low calcium intake compared to the men during a 5-year follow-up, suggesting that high intakes of dairy products and milk may be useful in women with low calcium consumption. In another cross-sectional study comprising 277 women with osteoporotic fractures showed that dietary intake of calcium (503.7 ± 274.7 mg/day) showed a positive effect on bone mineral density (Yoon et al. 2016). The beneficial role of calcium against osteoporosis is mediated through two main mechanisms, namely paracellular and transcellular pathways (de Barboza et al. 2015). Calcitriol ($1,25(OH)_2D_3$) is the major regulating hormone responsible for the intestinal Ca^{2+} absorption (de Barboza et al. 2015), which can change the function and structure of enterocytes (Tolosa de Talamoni et al. 1989; Alisio et al. 1997), and subsequently improve the Ca^{2+} transport across the intestine (de Barboza et al. 2015). Intriguingly, the data further revealed that nearly 80% of elderly Korean women with osteoporotic fractures had calcium intake below the recommended dietary intake (˂700 mg/day) (Yoon et al. 2016). Despite research has demonstrated a beneficial effect of calcium intake and osteoporosis, not all studies showed such a relationship. A recent study by Bristow et al. (2019) did not identify an association between dietary calcium (886 mg/day) intake and postmenopausal bone loss during

6 years of follow-up in osteopenic postmenopausal women aged 65 years old and above. This finding suggests that an increase in calcium consumption is unlikely to reduce bone loss in the elderly. Overall, minerals are a good antioxidant to be consumed by ingesting specific food high with their chemical elements of interest. The crucial role played by minerals required further elucidation. Nonetheless, an overdose of mineral consumption is not recommended and may lead to an adverse outcome to health.

7.6 Ascorbic Acid

Ascorbic acid ($AscH_2$) or known as vitamin C, is a water-soluble ketolactone comprised of two ionizable hydroxyl groups (Du et al. 2012) (Fig. 7.5). Under physiological pH, the ascorbate monoanion, $AscH^-$, is the predominant form of vitamin C (Du et al. 2012). Ascorbate serves as a reducing agent and is readily to undergo two consecutive reactions, namely one-electron oxidation to form ascorbate radical ($ASc^{\bullet-}$) and dehydroascorbic acid (DHA) (Zhitkovich 2020). Due to the resonance stabilization of the unpaired electron, the ascorbate radical is relatively unreactive and is readily dismute to ascorbate and DHA (Du et al. 2012). In this regard, these properties make ascorbate an effective donor antioxidant (Padayatty et al. 2003). In general, there are three primary biological functions of ascorbic acid including acceptor/donor in electron transport either at the chloroplasts or plasma membrane, radical scavenger, and enzyme cofactor (Davey et al. 2000). Ascorbic acid is comprised of 4 –OH groups that can donate hydrogen to an oxidizing agent. Ascorbic acid is prone to chelate metal ions (Fe^{++}) due to the presence of –OH groups toward the carbon atoms (Timoshnikov et al. 2020). Pure ascorbic acid is a white crystalline powder, which is highly soluble in water and thus contributes to a colorless solution (Du et al. 2012). At high amounts of ascorbic acid (>1000 mg/kg), it tends to change the balance between ferrous (Fe^{2+}) and ferric iron (Fe^{3+}) and thus suppress oxidation and scavenge the oxygen (Timoshnikov et al. 2020). Ascorbic acid serves as a cofactor for several enzyme-catalyzed reactions, for example, mediating the hematopoiesis and leukocyte activity, hydroxylation of proline and lysine, neuroprotection, improving the iron absorption and collagen biosynthesis, and modulating the vascular and connective tissue's integrity (Kishimoto et al. 2013; May and Harrison 2013; Lane et al. 2016; Carr and Maggini 2017; Bonnet 2019).

Most of the animals and plants synthesize ascorbate from glucose (Drouin et al. 2011). In reptiles, amphibians, and primitive fish, the synthesis of ascorbate takes place in the kidney; while the site of synthesis in mammals is in the liver (Ching et al. 2015). However, guinea pigs, other primates, humans, and a few species of fruit-eating bats cannot synthesize ascorbate because of the dysfunction of the gene encoding L-gulonolactone oxidase (GLO) (Nishikimi et al. 1994). GLO is the enzyme required for the last step in ascorbate synthesis (Cui et al. 2011; Drouin et al. 2011). Therefore, dietary intake is vitally important. It can be acquired through the consumption of many plants, particularly in vegetables such as broccoli and

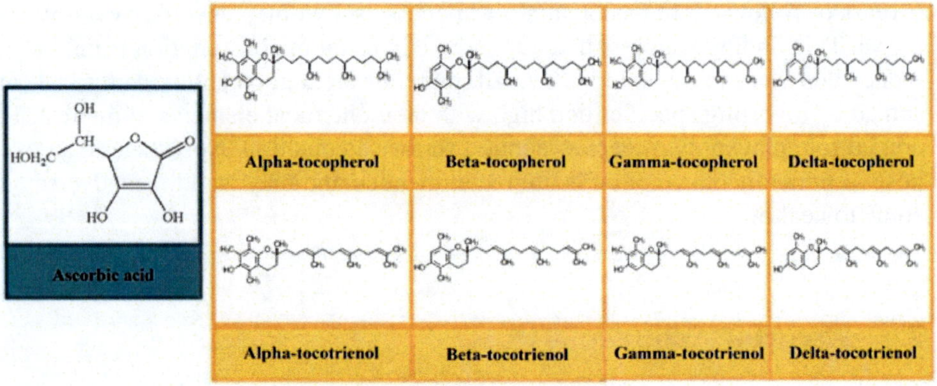

Fig. 7.5 Molecular structures of ascorbic acid and vitamin E congeners, namely tocopherols (α-tocopherol, β-tocopherol, γ-tocopherol, and δ-tocopherol) and tocotrienols (α-tocotrienol, β-tocotrienol, γ-tocotrienol, and δ-tocotrienol)

tomatoes, and citrus fruits including orange and lemon (Aditi and Graham 2012). A typical human diet contains both DHA and ascorbate, in which the absorption occurs in the enterocytes of the small intestine (Vanderslice and Higgs 1991). In general, the DHA is absorbed through Na^+-independent facilitative glucose transporters (GLUTs) followed by intracellular reduction (Navale and Paranjape 2016); whereas the ascorbate is accumulated in cells via Na^+-dependent vitamin C transporters (SVCTs) (Wohlrab et al. 2017). At physiological circumstances, the plasma DHA levels are very low (≤2 μM); whereas the plasma glucose concentrations are relatively higher (2–5 mM). Therefore, high intracellular ascorbate levels are mainly due to the uptake of ascorbate by SVCT2 (SLC23A2) and SVCT1 (SLC23A1) (May 2011).

One of the prominent functions of ascorbic acid in animal and plant metabolism is to mediate several enzymatic reactions (Smirnoff 2018). In general, these enzymes are di- or monooxygenases, which contain copper or iron at the active site and require ascorbic acid for maximal activity. Ascorbic acid plays an important role in maintaining the transition metal ion centers of these enzymes in a reduced form (Du et al. 2012). In animals, the Fe-dioxygenases involved in the collagen biosynthesis and require ascorbic acid for maximal activity (Davey et al. 2000). Ascorbic acid is needed for the mediation of adrenal hormone and folate synthesis, wound healing, carnitine biosynthesis, and osteoblasts and fibroblasts activity (Davey et al. 2000). The previous study further revealed that ascorbic acid can suppress the catechol-*O*-methyl transferase, and thereby promoting the bioavailability of adrenaline (Kern and Bernards 1997). This finding suggests that ascorbic acid could play a crucial role in neuroendocrine control. In support of this, the central nervous system and adrenal medulla, the adrenaline synthesis areas, showed a high concentration of ascorbic acid (May 2012).

Ascorbic acid possesses an ability to protect both membrane components of cells and cytosolic from oxidant damage (Ali et al. 2020). In particular, ascorbate serves as a predominant antioxidant to scavenge the free radicals that are produced as by-products from cellular metabolism (Akram et al. 2017). Indeed, one of the crucial features of ascorbate is the synergistic action with vitamin E (Vineetha et al. 2020). Ascorbic acid may play an indirect role against α-tocopheroxyl radical in cellular membranes (Traber and Stevens 2011). In this regard, ascorbate is critical in modulating vitamin E to suppress lipid oxidation. It has been demonstrated that increased plasma ascorbate levels through supplementation reduced rates of disappearance of vitamin E in smokers (Bruno et al. 2006), suggesting that ascorbate could maintain vitamin E, more likely via "recycling". Ascorbic acid is a powerful antioxidant that exerts the ability to donate a hydrogen atom and produces a relatively stable ascorbyl-free radical (Nweze et al. 2019). Ascorbic acid is thought to be a critical water-soluble antioxidant in decreasing oxidative stress and neutralizing ROS (Peng et al. 2019).

Ascorbic acid has been suggested to reduce oxidative damage as well as decrease the risk of age-related diseases (Halabi et al. 2018; Delrobaei et al. 2019; Shivavedi et al. 2019). High concentrations of intravenous ascorbic acid (10 g/day) have been used as a complementary agent in cancer patients since 1970s (Cameron and Campbell 1974; Cameron and Pauling 1976; Padayatty et al. 2006; Daniel 2011). The study showed that the rate of survival in patients with advanced cancer was improved following supplementation of a high dose of ascorbic acid (Cameron and Campbell 1974; Cameron and Pauling 1976). However, the methodology of several aspects, for instance, data analysis and collection were criticized in these studies, and thereby limited the use of ascorbic acid in cancer patients (van Gorkom et al. 2019). Subsequently, oral supplementation of ascorbic acid was evaluated in cancer patients (Creagan et al. 1979). Intriguingly, based on the evidence from a systematic review, intravenous administration appears to have a better effect compared to oral administration (van Gorkom et al. 2019).

The therapeutic potential of ascorbic acid has been consistently reported in *in vitro* (Pires et al. 2018), *in vivo* (Ribeiro et al. 2009), and clinical studies (Li et al. 2019b). The previous study has demonstrated that ascorbic acid can suppress the viability of colon cancer (C2BBe1) cells, with half-maximal inhibitory concentrations (IC_{50}) of 0.95 mM after 24 h incubation (Pires et al. 2016). In line with this, the proliferation of breast cancer (MCF-7) cell lines was significantly reduced following administration of a high concentration of ascorbic acid (10 mM or 20 mM for 2 h) (Lee et al. 2019). Ascorbic acid has also been evaluated as a component of combination therapy with chemosensitization agents or primary treatment (Hoffer et al. 2015; Baillie et al. 2018). It has been shown that combining high dose of ascorbic acid (5, 10, and 20 mM for 2 h) with tamoxifen (1, 2, and 4 μM for 24 h) or fulvestrant (25 and 50 μM for 24 h) further reduced the viability of MCF-7 cell lines, suggesting that combination of high concentration of ascorbic acid and chemotherapy may have therapeutic potential against breast cancer (Lee et al. 2019). In an animal study, feeding Balb/c nu/nu mice with ascorbic acid (150 mg/kg for 14 days) significantly suppressed the colon tumor growth compared to the control group (Pires et al. 2018) (Table 7.10). In human study, a relatively low plasma ascorbic acid level was found in patients with endometrial cancer (Kuiper et al. 2010). From the study reviewed, high-grade tumors had low amounts of ascorbate

Table 7.10 *In vitro*, *in vivo*, and clinical studies conducted in ascorbic acid supplements and their effects on cancer

Cell lines/animal models/ subjects	Treatment	Findings	References
Advanced cancer patients	10 g/day	↑ The rate of survival	Cameron and Campbell (1974); Cameron and Pauling (1976)
Colon cancer cells (C2BBe1)	0.95 mM for 24 h	↓ Cell proliferation	Pires et al. (2016)
Different stages of cancers (breast, nasopharynx carcinoma, lung, ovarian, and liver)	25–100 g/day	Improved the quality of life and reduce the risk of metastasis	Raymond et al. (2016)
Stage IV poorly-differentiated pancreatic ductal adenocarcinoma male patient	75–125 g/infusion for 2–3 times/week	Ameliorates the progression of the disease and improved the survival	Drisko et al. (2018)
Balb/c nu/nu mice	150 mg/kg for 14 days	↓ Colon tumor growth	Pires et al. (2018)
Breast cancer cells (MCF-7)	10 mM or 20 mM for 2 h	↓ Cell viability	Lee et al. (2019)

concentrations (Kuiper et al. 2010). Importantly, reduced plasma ascorbic acid level was linked to a high concentration of the inflammatory marker including CRP (Halabi et al. 2018). Ascorbic acid deficiency is commonly observed in 30% of patients with advanced cancer (Mayland et al. 2005). In stage IV poorly-differentiated pancreatic ductal adenocarcinoma male patient aged 68 years old, high concentration of ascorbic acid (75–125 g/infusion for 2–3 times/week) ameliorates the progression of the disease and improved the survival of the patients (about 4 years after diagnosis) (Drisko et al. 2018). A study reported by Raymond et al. (2016) evaluated intravenous ascorbic acid (25–100 g/day) in relation to several types of cancers of different stages including breast, nasopharynx carcinoma, lung, ovarian, and liver. The study showed that intravenous ascorbic acid therapy improved the quality of life and thereby may prolong their life expectancy (Raymond et al. 2016). The study further revealed that intravenous ascorbic acid may reduce the risk of metastasis and stabilize cancer (Raymond et al. 2016). In the context of dietary ascorbic acid from foods, the data from systematic review and meta-analysis involving 17 studies suggested that dietary intake of citrus fruits, one of the excellent sources of ascorbic acid, reduced the oral cavity and pharyngeal cancer risk by 50% (Cirmi et al. 2018) (Table 7.11). In support of this, 10 case-control studies involving 18,207 participants have shown that intake of ascorbic acid derived from a variety of vegetables and fruits such as pears/apples, green salad, tomatoes, kiwi, citrus fruits, potatoes, cabbage, mandarin orange, Japanese persimmon, and spinach (mean intake of ascorbic acid = 133.40 mg/day) was negatively associated with head and neck cancer (Favero et al. 1997; Imaeda et al. 1999; Edefonti et al. 2015)

Table 7.11 Effects of dietary consumption of foods rich in ascorbic acid on cancers in human studies

Types of cancers	Dietary food intake	Findings	References
Pancreatic cancer	Ascorbic acid from diet (22.9–226 mg/day)	↓ Risk of pancreatic cancer	Hua et al. (2016)
Pancreatic cancer	Ascorbic acid from diet (51.3–678.55 mg/day)	No effect	Hua et al. (2016)
Oral cavity and pharyngeal cancer	Citrus fruits	↓ Oral cavity and pharyngeal cancer risk by 50%	Cirmi et al. (2018)
Head and neck cancer	Pears/apples, green salad, tomatoes, kiwi, citrus fruits, potatoes, cabbage, mandarin orange, Japanese persimmon, and spinach (mean intake of vitamin C = 133.40 mg/day)	↓ Risk of head and neck cancer	Favero et al. (1997); Imaeda et al. (1999); Edefonti et al. (2015)

(Table 7.11). Likewise, the meta-analysis of case-control studies conducted between 1988 and 2013 involving 10,115 controls and 3818 pancreatic cancer had revealed that ascorbic acid from the diet (22.9–226 mg/day) was inversely associated with the risk of pancreatic cancer (Hua et al. 2016) (Table 7.11). However, Hua et al. (2016) did not identify an association of ascorbic acid obtained from the diet (51.3–678.55 mg/day) and pancreatic cancer in a follow-up study conducted from 2002 to 2013 involving a total of 278,000 subjects and 1140 cancer participants (Table 7.11). Compared to those who consume ascorbic acid from supplements (60–450 mg/day), individuals who consume ascorbic acid-rich diet have relatively strong protection against pancreatic cancer (Hua et al. 2016). The preventive effect of ascorbic acid on cancers could be attributed to its antioxidant and anti-inflammatory actions. Yet, the anticancer ability of ascorbic acid is modulated via a few mechanisms, for instance, inducing apoptosis, mediating growth factor signaling, and regulating cell cycle and phase II detoxifying enzymes (Kim et al. 2015; Piersma et al. 2017; Gao et al. 2019; Zhou et al. 2020a).

Intakes of vegetables and fruits, especially rich in ascorbic acid, are thought to have a protective effect on cardiovascular health (Tan et al. 2018) (Table 7.12). Data from a meta-analysis of prospective studies have shown that a diet high in ascorbic acid (45–375.8 mg/day) is negatively associated with the risk of stroke (Chen et al. 2013). From the study reviewed, it showed that high circulating ascorbic acid concentration (3.6–73.36 μmol/L) may reduce the risk of stroke (Chen et al. 2013). Ascorbic acid prevents stroke via a few mechanisms including protect the membrane from peroxidation, suppress the growth of smooth muscle, and decrease the oxidation of LDL activity (Siow et al. 1998). Importantly, systemic inflammation is implicated in the pathogenesis of stroke (Dziedzic 2015). Ascorbic acid has been suggested to have anti-inflammatory activity (Diomede et al. 2020). In this regard, ascorbic acid may delay the development of stroke through pressure-lowering

Table 7.12 Clinical studies conducted in ascorbic acid supplements and foods rich in ascorbic acid and their effects on cardiovascular health

Cardiovascular health	Source of ascorbic acid	Findings	References
Coronary heart disease	Ascorbic acid-rich vegetables and fruits (>30 mg ascorbic acid/serving)	Protect against coronary heart disease	Joshipura et al. (2001)
Healthy young adults	Ascorbic acid (≥150 mg/day) from fruits and vegetables intakes (including tangerine/orange, kiwi, avocado, papaya, mango, melon, watermelon, nectarine/apricot/peach, grape-fruit, strawberry, pear/apple, banana, pineapple, fig, guava, plum, cherry, cucumber/zucchini/aubergines, peppers, green beans, pumpkin/carrot, tomatoes, cabbage, broccoli/cauliflower, chicory/lettuce, dark-green leaves/spinach, onion, 'gazpacho', beetroot, and asparagus)	↑ Glutathione peroxidase activity and total antioxidant capacity ↓ Plasma ox-LDL levels	Hermsdorff et al. (2012)
Adults	Ascorbic acid (500–4000 mg/day and 500–2000 mg/day)	↓ Blood pressure and improve endothelial function	Juraschek et al. (2012); Ashor et al. (2014)
Stroke	Diet high in ascorbic acid (45–375.8 mg/day)	↓ Risk of stroke	Chen et al. (2013)
Stroke	Circulating ascorbic acid concentration (3.6–73.36 μmol/L)	↓ Risk of stroke	Chen et al. (2013)
Hypertensive and/or diabetic obese adults	Ascorbic acid supplements (500–1000 mg/day)	No significant effects on several cardiovascular endpoints	Sesso et al. (2008); Ellulu et al. (2015); Buijsse et al. (2015)
Adults aged 18 years or older	Increased 50 mg/day of dietary ascorbic acid intake	↓ 8% of CVD mortality risk	Jayedi et al. (2019)
Adults aged 18 years or older	Increased 20 μmol/L in circulating ascorbic acid levels	↓ 13% risk of CVD mortality	Jayedi et al. (2019)

CVD cardiovascular disease, *ox-LDL* oxidized-low density lipoprotein

activity (Uesugi et al. 2017). Based on the clinical evidence, ascorbic acid supplementation is not likely to prevent stroke (Myung et al. 2013). Increasing consumption of ascorbic acid-rich foods and adhering to healthy lifestyles and dietary habits may markedly decrease the burden of stroke and other CVD (Chen et al. 2013). The previous study stated that circulating ascorbic acid concentrations may act as a good predictor of diet status and stroke risk (Chen et al. 2013). Similarly, the prospective cohort study involving 42,148 men aged 40 to 75 years old and 84,251 women aged 34 to 59 years old and followed up for 8 and 14 years, respectively, on ascorbic acid-rich vegetables and fruits (>30 mg ascorbic acid/serving) in relation to the risk of coronary heart disease (Joshipura et al. 2001). The study showed that consumption of ascorbic acid-rich vegetables and fruits may protect against the development of coronary heart disease (Joshipura et al. 2001). Intriguingly, increased 1 serving/day of vegetables and fruits reduced the risk of coronary heart disease by 4%, suggesting that ascorbic acid-rich vegetables and fruits and green leafy vegetables played a pivotal cardioprotective role in total vegetables and fruits (Joshipura et al. 2001). In support of this, the cross-sectional study has demonstrated that ascorbic acid (≥150 mg/day) from fruits and vegetables intake (including tangerine/orange, kiwi, avocado, papaya, mango, melon, watermelon, nectarine/apricot/peach, grapefruit, strawberry, pear/apple, banana, pineapple, fig, guava, plum, cherry, cucumber/zucchini/aubergines, peppers, green beans, pumpkin/carrot, tomatoes, cabbage, broccoli/cauliflower, chicory/lettuce, dark-green leaves/spinach, onion, 'gazpacho', beetroot, and asparagus) significantly increased GPx activity and total antioxidant capacity and reduced plasma ox-LDL levels in healthy young adults, suggesting that ascorbic acid has the potential to regulate antioxidant-mediated mechanism and combat oxidative stress (Hermsdorff et al. 2012). Hypertension and CVD are often linked to low GPx levels (Jin et al. 2011), indicating that GPx can prevent free radical production in the redox homeostatic balance of cellular metabolism. In a further study focused on the risk of total cardiovascular mortality outcomes, a dose-response meta-analysis of 15 prospective observational studies showed that a relatively low risk (8%) for CVD mortality for those with increased 50 mg/day of dietary ascorbic acid intake (Jayedi et al. 2019). From the study reviewed, it showed that increased 20 μmol/L in circulating ascorbic acid levels is linked to the 13% lower risk of CVD mortality (Jayedi et al. 2019). Importantly, a dose-response meta-analysis suggested that a possible threshold of ascorbic acid intake of about 200 mg/day (Jayedi et al. 2019). Some research has also emerged to suggest that the optimal dietary intake of ascorbic acid is about 200 mg/day; the concentration in which the blood cells are completely saturated and the plasma is relatively saturated with ascorbic acid (Frei et al. 2012). Mangels et al. (1993) exploring the bioavailability of synthetic ascorbic acid and food rich in ascorbic acid. Interestingly, the bioavailability of ascorbic acid from cooked broccoli, orange juice, and oranges is similar to ascorbic acid supplements (Mangels et al. 1993). The data from clinical trials (Sesso et al. 2008; Ellulu et al. 2015) and observational studies (Buijsse et al. 2015) have shown that supplementation of ascorbic acid (500–1000 mg/day) did not lead to any significant effects on several cardiovascular endpoints. By contrast, studies from meta-analyses revealed that a high concentration of ascorbic acid (500–4000 mg/day and 500–2000 mg/day) can reduce blood pressure (Juraschek et al. 2012) and improve endothelial function (Ashor et al. 2014), respectively.

Research evidence has demonstrated the potential protective function of ascorbic acid in diabetes (Mason et al. 2019) (Table 7.13). Ascorbic acid deficiency ($<$0.3 mg/dL) is linked to type 2 diabetes mellitus (Praveen et al. 2020). It has been shown that individuals with ascorbic acid deficiency have high serum MDA levels, suggesting that increased oxidative stress (Praveen et al. 2020). In a further study focused on diabetes outcomes, Praveen et al. (2020) showed that high serum ascorbic acid levels are negatively linked to glycated hemoglobin (HbA1c), suggesting that ascorbic acid may be beneficial in managing glycemic control and prevent micro- and macrovascular complications. These complications contribute to the mortality and

Table 7.13 Clinical studies conducted in ascorbic acid supplements and foods rich in ascorbic acid and their effects on diabetes mellitus

Sources	Dosages	Findings	References
Ascorbic acid and vitamin E	Ascorbic acid (1250 mg) and vitamin E (680 IU) for 4 weeks	Inhibited albuminuria	Gaede et al. (2001)
Sour orange juice	240 mL/day for 4 weeks	↓ Fasting blood glucose levels	Ravanshad et al. (2006)
Fruits	86.2–165.5 mg/day	↓ Risk of diabetic retinopathy	Tanaka et al. (2013)
Watermelon, bananas, berries (grapes, blueberries, and strawberries), drupes (Japanese apricots, cherries, and peaches), citrus fruits (lemon, grapefruit, and orange), and pome fruits (Japanese persimmons, Japanese pears, and apples), and other melons	22.6–253 g/day	↓ Incidence of diabetic retinopathy	Tanaka et al. (2013)

morbidity of diabetes mellitus patients, in which all these factors are linked to oxidative stress (Ullah et al. 2016). Importantly, patients with diabetic nephropathy show reduced levels of ascorbic acid (Varma et al. 2014; Chou and Tseng 2017). The exclusion of ascorbic acid from the tubular epithelial cells via the competition of dehydroascorbate and glucose, and thereby deprive the cells from antioxidant capacity and ultimately results in the accumulation of ROS in diabetes (Chen et al. 2005). The deficiency of ascorbic acid stimulates the transforming growth factor beta (TGF-β) signaling pathway, and thereby aggravates the diabetic mesangial cellular expansion and deposits of extracellular matrix (Ji et al. 2017). Dietary ascorbic acid has demonstrated a positive impact on diabetes (Mason et al. 2019). The consumption of sour orange juice (240 mL/day for 4 weeks) significantly increased the ascorbic acid intake by 24% (Ravanshad et al. 2006). The study further revealed that the fasting blood glucose levels were significantly reduced after intakes of sour orange juice, suggesting that short-term dietary consumption of sour orange juice may reduce the fasting blood glucose levels (Ravanshad et al. 2006). Dietary ascorbic acid not only decreases blood glucose levels but also declines the risk of diabetic retinopathy (May 2016). A follow-up study for 8 years including 978 patients with type 2 diabetes aged 40–70 years old had revealed that ascorbic acid in fruits (86.2–165.5 mg/day) decreased the risk of diabetic retinopathy (Tanaka et al. 2013). Tanaka et al. (2013) further demonstrated that increased consumption of fruits (22.6–253 g/day) such as watermelon, bananas, berries (grapes, blueberries, and strawberries), drupes (Japanese apricots, cherries, and peaches), citrus fruits (lemon, grapefruit, and orange), and pome fruits (Japanese persimmons, Japanese pears, and apples), and other melons is negatively linked to the incidence of diabetic retinopathy. Such protection has been accredited to antioxidant vitamins including ascorbic acid. High fruits and vegetables consumption are known to trigger the elevation of plasma ascorbic acid levels (Djuric et al. 2006). Nonetheless, Lee et al. (2010) did not identify an association between ascorbic acid and diabetic retinopathy. In support of this, a systematic review involving 4094 participants found that no association between dietary intake of ascorbic acid and diabetic retinopathy (Lee et al. 2010). While, a systematic review including 3 randomized controlled trials, 15 cross-sectional, 4 case-control, and 9 prospective studies performed by Wong et al. (2018a) suggest that the relationships of ascorbic acid intakes and diabetic retinopathy are remained unclear, suggesting that further studies are required to elucidate these relationships. In addition to the effects reported on diabetic retinopathy, supplementation of ascorbic acid has also shown a beneficial effect on diabetic nephropathy (Hirsch et al. 1998). Diabetic nephropathy is a microvascular complication of diabetes mellitus that contributes to end-stage renal disease (Singh et al. 2011; Rondeva and Wolf 2014). For instance, ascorbic acid (1250 mg) and vitamin E (680 IU) for 4 weeks were significantly inhibited albuminuria in patients with type 2 diabetes with macro- or microalbuminuria (Gaede et al. 2001).

Oxidative stress is implicated in the etiology of neurodegenerative disease such as Parkinson's disease (Butterfield and Halliwell 2019). The plasma ascorbate levels were linked to cognition in the general population. Compared to cognitively intact individuals, the cognitively impaired group had relatively low amounts of ascorbate concentrations (Travica et al. 2017). In general, brain ascorbate levels are maintained at higher levels compared to other organs (Harrison and May 2009). Research evidence indicates that patients with Parkinson's disease showed high oxidative stress markers and low plasma ascorbate levels compared to healthy individuals (Sudha et al. 2003; Medeiros et al. 2016). A study including 215 participants aged 50–90 years old had revealed a significant role of ascorbate levels in Parkinson's disease. It has been shown that ascorbate levels (>23 μmol/L) were positively related to the cognitive function in Parkinson's disease (Spencer et al. 2020). This favorable effect is more likely due to the other non-antioxidant acts as ascorbate, which may contribute to the cognitive function (Spencer et al. 2020). Ascorbic acid has been shown to prevent Alzheimer's disease-induced by oxidative stress (Monacelli et al. 2017) (Table 7.14). The implication of ascorbic acid in relation to the pathophysiology of dementia and Alzheimer's disease has been widely demonstrated in both *in vivo* and *in vivo* studies. Ascorbic acid delays disease progression through several mechanisms including reduce neuronal cell death, decrease oxidative stress, and

Table 7.14 Clinical studies conducted in foods rich in ascorbic acid and their effects on neurological disorders

Sources	Dosages	Findings	References
Ascorbic acid from cabbage, broccoli, sprouts, kiwi, and citrus fruits	121.6 mg/day for 6 years	↓ Risk of Alzheimer's disease	Engelhart et al. (2002)
Ascorbic acid-rich fruits and vegetables	90.3–238.4 g/day for 13 years	Improved verbal memory performance	Péneau et al. (2011)
Ascorbic acid derived from oranges and orange juice	95–213 g/day for a mean of 17 years	Not associated with the cognitive decline	Devore et al. (2013)

inhibit proinflammatory cytokines (Zhang et al. 2018c). A follow-up study for 6 years including 5395 participants had revealed a significant role of ascorbic acid in Alzheimer's disease (Engelhart et al. 2002). It has been shown that for individuals aged ≥55 years old, dietary intake of ascorbic acid (121.6 mg/day) from cabbage, broccoli, sprouts, kiwi, and citrus fruits may reduce the risk of Alzheimer's disease (Engelhart et al. 2002). In support of this, a prospective study analyzed of 2533 participants aged 45–60 years has demonstrated that dietary intake of ascorbic acid-rich fruits and vegetables (90.3–238.4 g/day) for 13 years shows a better verbal memory performance (Péneau et al. 2011). This finding indicates that verbal memory performance could be improved by increasing fruits and vegetables consumption (Péneau et al. 2011). Intriguingly, supplementation of ascorbic acid (124–761 g/day) for a mean of 17 years reduced cognitive decline in the elderly (Devore et al. 2013). Although research evidence indicates that ascorbic acid shows a better cognitive function, not all experimental studies demonstrated such an association. Devore et al. (2013) did not identify an association between dietary consumption of ascorbic acid and cognitive decline. The study involving 16,010 participants aged ≥70 years old from 1995 to 2001 revealed that dietary consumption of ascorbic acid (95–213 g/day) derived from oranges and orange juice for a mean of 17 years was not linked to the cognitive decline (Devore et al. 2013). Together, the research evidence indicates that ascorbic acid has significant beneficial outcomes in diseases suffered by the elderly. Ascorbic acid should be included in the diet of the elderly through the consumption of foods rich in ascorbic acid. Indeed, the potential implications of ascorbic acid on age-related diseases are nonetheless worthy of further elucidation to explore the pathophysiology of age-related diseases.

7.7 Vitamin E

Vitamin E is a lipid-soluble and plant-derived compound comprised of eight structurally related lipophilic chromanol congeners with a side chain located at the C2 position (Niki and Abe 2019). Vitamin E usually found in fat-rich food, for instance, seeds and edible oils, or present in fortified food (Rizvi et al. 2014), including α-, β-, γ-, and δ-tocopherols and four corresponding tocotrienols (Shahidi and de Camargo 2016) (Fig. 7.5). The four tocopherols possess a saturated phytyl side chain; whereas four tocotrienols exert an unsaturated isoprenyl side chain with three double bonds at C11′, C7′, and C3′ (Shahidi and de Camargo 2016). The double bonds of tocotrienols' side chains at both C7′ and C3′ containing *trans*-configuration. Both α-, β-, γ-, and δ-tocotrienols and tocopherols are classified based on the position and the number of methyl substitution in the chromanol ring (Ahsan et al. 2015). The δ-forms of tocotrienols and tocopherols possess one methyl group and γ- and β-forms exert two methyl groups; whereas the α-forms have three methyl groups at C8, C7, and C5-position in their chromanol ring (Niki and Abe 2019). Of all isoforms of vitamin E, γ-tocopherol is the predominant form of vitamin E found in the diet; whereas α-tocopherol is the most common form of vitamin E in the mammalian tissues (Jiang et al. 2001; Jiang 2014). Vitamin E has been recognized as one of the crucial free radical scavenging antioxidants in humans to protect the biological molecules from detrimental oxidative modifications (Niki and Abe 2019).

Vitamin E exerts a unique anti-atherosclerotic property (Elbeltagy et al. 2019). Much information indicates that vitamin E can modulate the activity of several protein kinases such as protein kinase C and protect against platelet aggregation, lipid deposition in the aorta, and inflammation (Khadangi and Azzi 2019). A systematic review and dose-response meta-analysis of prospective observational studies comprising of 7777 cases and 228,531 participants showed an inverse association of CVD mortality risk for 10 μmol/L increment in circulating α-tocopherol concentration (Jayedi et al. 2019). However, in a further study focused on the risk of CVD mortality outcomes, Jayedi et al. (2019) found that every 5 mg/day increment in dietary vitamin E intake was not linked to the risk of CVD mortality. These findings suggested that the circulating α-tocopherol may have a better predictive value in relation to the risk of total CVD mortality compared to the dietary vitamin E intakes (Jayedi et al. 2019). A significant decrease in ATP-binding membrane cassette transporter A1 (*ABCA1*) DNA methylation levels among Japanese women was found to be related to the high dietary vitamin E intake (6.88–9.43 mg/day) (Fujii et al. 2019). Research evidence indicates that the antioxidant property of vitamin E is involved in the mediation of transcriptional activity via DNA methyltransferase-dependent pathway (Remely et al. 2017). In this regard, it is plausible that the negative association of *ABCA1* DNA methylation and dietary intake of vitamin E modulates DNA methyltransferase activity (Fujii et al. 2019). ABCA1 is a predominant transporter that modulates the formation of HDL particles via interaction with apolipoprotein A1 (apoA1) (Fujii et al. 2019). The previous study stated that high levels of ABCA1 DNA methylation are negatively linked to HDL-cholesterol levels and subsequently increased the risk of coronary heart disease in humans (Guay et al. 2012, 2014). In addition, dietary vitamin E from vegetable oils, nuts, cereal grains, and seeds (median intake = 12.9 IU/day during 4 years of follow-up) has also been reported to decrease the risk of coronary heart disease among men who did not consume vitamin supplements (Rimm et al. 1993). Table 7.15 summarizes the *in vitro*, animal, and clinical studies of vitamin E on age-related diseases.

Research evidence has demonstrated that a vitamin E-rich diet has the potential to protect against neurodegenerative disease (Frank et al. 2012; Boccardi et al. 2016). In a human study, Engelhart et al. (2002) followed 5395 participants aged 55 years and above for 6 years had found that high consumption of vitamin E-rich foods (13.8 mg/day) such as egg yolk, milk, nuts, and grains, reduced the risk of Alzheimer's disease. Inflammation and oxidative stress are implicated in age-related cognitive impairment (Costa d'Avila et al. 2018). A single-blinded and randomized-controlled trial among healthy and middle-aged/old adults showed that intakes of 3 oz/day (84 g) almond for 6 months significantly increased serum α-tocopherol level by 8% compared to the baseline (Rakic et al. 2021). Although the study reported by Rakic et al. (2021) did not find an association of serum α-tocopherol concentration with the improved cognition performance, the pivotal role played by α-tocopherol on brain function should not be neglected. Alpha-tocopherol acts as a

Table 7.15 *In vitro*, animal, and clinical studies of vitamin E on age-related diseases

Age-related diseases	Study conditions	Source of vitamin E	Durations	Outcomes	References
CVD	Systematic review and dose-response meta-analysis of prospective observational studies comprising of 7777 cases and 228,531 adults aged 18 years and above	Dietary vitamin E	Every 5 mg/day increment	Not associated with the CVD mortality risk	Jayedi et al. (2019)
CVD	Systematic review and dose-response meta-analysis of prospective observational studies comprising of 6542 cases and 34,285 participants	Circulating α-tocopherol	Every 10 μmol/L increment	↓ CVD mortality	Jayedi et al. (2019)
CVD	Cross-sectional study involving 225 participants (117 women and 108 men) aged 39 years old and above	Dietary vitamin E	6.88–9.43 mg/day	↓ ABCA1 DNA methylation levels	Fujii et al. (2019)
Coronary disease	39,910 U.S. male health professionals aged 40 to 75 years old and follow-up for 4 years	Dietary vitamin E from vegetable oils, nuts, cereal grains, and seeds	Median intake = 12.9 IU/day	↓ Coronary disease risk in men	Rimm et al. (1993)
Alzheimer's disease	Prospective cohort study including 5395 participants aged 55 years and above and follow-up for an average of 6 years	Vitamin E-rich foods such as egg yolk, milk, nuts, and grains	13.8 mg/day	↓ Risk of Alzheimer's disease	Engelhart et al. (2002)
Age-related cognitive impairment	Single-blinded, controlled, randomized-trial including 60 men and postmenopausal women aged 50–75 years	Almond	3 oz./day or 84 g for 6 months	↑ Serum α-tocopherol level by 8%	Rakic et al. (2021)
Neurodegenerative disease	Systematic review of 10 animal studies	Tocotrienol-rich fraction from palm oil	100–200 mg/kg per day for 10 weeks-8 months orally	Improved cognitive function	Ismail et al. (2020)
Osteoporosis	MC3T3-E1 cells	Annatto-derived tocotrienol	0.001–1 μg/mL	↓ RhoA activation and downregulating HMGR transcriptional activity	Hasan et al. (2020)

Osteoporosis	Male Sprague-Dawley rats	Annatto-derived tocotrienol	60 and 100 mg/kg	Enhances bone calcium content and femoral biomechanical strength	Mohamad et al. (2018)
Osteoporosis	Male Sprague-Dawley rats	Palm tocotrienol	60 mg/kg/day for 2 months	Preserved bone strength and structure and maintained serum resorption levels	Elvy Suhana et al. (2018)
Osteoporosis	Male Wistar rats	Palm vitamin E	60 or 100 mg/kg for 12 weeks	Improved trabecular thickness, bone volume, and osteoblast surface	Wong et al. (2018b)
Prostate cancer	PC-3 cell line	Annatto tocotrienol	20 μg/mL for 24 h	↓ Proliferation of PC-3	Sugahara et al. (2015)
Lung cancer	Dose-response meta-analysis involving 9 prospective cohort studies including 431,359 controls and 4164 cases	Dietary vitamin E mainly from soybean oil, soybeans, eggs, and leafy greens	Every 2 mg/day increment	↓ 5% of lung cancer risk	Zhu et al. (2017)

ABCA1 ATP-binding membrane cassette transporter A1, *CVD* cardiovascular disease, *HMGR* HMG-CoA reductase, *MC3T3-E1* murine pre-osteoblastic cells, *PC-3* prostate cancer cell lines

protector of the integrity of polyunsaturated fatty acid as well as modulating the cell membrane function and morphology, suggesting a critical in maintaining adequate α-tocopherol concentration in a brain (Traber and Atkinson 2007). In support of this, a systematic review revealed that feeding healthy animals with a tocotrienol-rich fraction from palm oil (100–200 mg/kg per day for 10 weeks–8 months orally) had better cognitive function (Ismail et al. 2020). The study further demonstrated that tocotrienol-rich fraction (25, 50, or 100 mg/kg for 10 weeks) possesses anti-inflammatory and antioxidant activities, improved cognition performance by ameliorating the oxidative stress and decreasing neuroinflammation in diabetes-induced rats (Ismail et al. 2020). This finding is further supported by Kuhad et al. (2009) and Tiwari et al. (2009) who found that α-tocotrienol and tocotrienol-rich fraction had similar antioxidant activity and improved cognitive function in diabetes-induced rats, implied that α-tocotrienol is the isomer that may responsible for the antioxidant activities and better cognitive function of the tocotrienol-rich fraction. The previous study also found that a mutual underlying pathophysiological event between neurological disorders and diabetes mellitus (Nasrolahi et al. 2019) is likely due to the homeostatic imbalance of insulin-glucagon in diabetes mellitus, and thereby triggers amyloid deposition in the brain and pancreatic islets of Langerhans (Morsi et al. 2019). Feeding diabetic rats with tocotrienol-rich fraction enhanced the cholinergic function by decreasing the cerebrocortical concentration of acetylcholinesterase (Kuhad et al. 2009). Acetylcholinesterase is the enzyme that catalyzes the breakdown of acetylcholine neurotransmitter, and thereby leads to the development of neurodegeneration and dementia (Bohnen et al. 2018).

In addition to the age-related diseases mentioned above, the deficiency of vitamin E is linked to the diseases closely linked to aging, for instance, osteoporosis (Mata-Granados et al. 2013). *In vitro* study showed that annatto-derived tocotrienol (0.001–1 μg/mL) inhibits the mevalonate pathway by suppressing the RhoA activation and downregulating 3-hydroxy-3-methyl-glutarylcoenzyme-A (HMG-CoA) reductase (HMGR) gene expression in murine pre-osteoblastic (MC3T3-E1) cells, suggesting that annatto-derived tocotrienol may modulate the osteoblast mineralization (Hasan et al. 2020). The inhibition of the mevalonate pathway by annatto-derived tocotrienol indicates that the annatto may serve as a potent bone anabolic agent. Annatto-derived tocotrienol is found in the seeds of achiote plant native to tropical America, containing 10% γ- and 90% δ-tocotrienol (Raddatz-Mota et al. 2017). Consistent with the study reported by Hasan et al. (2020), Mohamad et al. (2018) also found that supplementation of annatto-derived tocotrienol (60 and 100 mg/kg) significantly decreased the degeneration of cortical and trabecular bone thickness in buserelin-induced rats. From the study reviewed, it showed that annatto-derived tocotrienol (60 and 100 mg/kg for 3 months) enhances the bone calcium content and femoral biomechanical strength, suggests that it may serve as a potential anti-osteoporotic agent for men who received androgen deprivation therapy (Mohamad et al. 2018). Further, a beneficial role of palm tocotrienols has also been shown on osteoporosis. An animal study showed that palm tocotrienol (60 mg/kg/day for 2 months) preserved bone strength and structure and maintained serum resorption levels in rats, implied that palm tocotrienol has potential against glucocorticoid-induced osteoporosis by modulating osteoclast and osteoblast at the transcriptional level (Elvy Suhana et al. 2018). Likewise, palm vitamin E (60 or 100 mg/kg for 12 weeks) also increased trabecular thickness, bone volume, and osteoblast surface in high-carbohydrate high-fat-treated rats (Wong et al. 2018b).

Dietary intake of vitamin E not only can modulate CVD and neurological disorders, it also inhibits several cancers. Annatto tocotrienol (20 μg/mL for 24 h) was shown to inhibit the proliferation of PC-3 prostate cancer cell lines via the suppression of STAT3 and Src (Sugahara et al. 2015). In support of this, a study by Tan and Norhaizan (2021) has revealed that prostate cancer and oxidative stress could be prevented by dietary components rich in antioxidants. While for lung cancer, a dose-response meta-analysis has demonstrated the risk of lung cancer was decreased by 5% for every 2 mg/day increment of dietary vitamin E intake (Zhu et al. 2017). Vitamin E prevents cancer via induction of apoptosis, inhibition of invasiveness, and mediate immune response (Jiang 2017). These data suggest that dietary vitamin E as natural cancer protector. Taken together, the evidence demonstrated that regular intakes of food rich in tocopherols and tocotrienols may potentially decrease the risk of age-related diseases. However, further studies are warranted to explore the overall long-term effects in comparative randomized clinical trials.

7.8 Ubiquinone

Ubiquinone or known as coenzyme Q10 is an endogenously synthesized and unique fat-soluble vitamin-like antioxidant that is produced in the body (Laredj et al. 2014). Ubiquinone was first isolated from beef heart mitochondria in 1957. The ubiquinone is found in organs with high rates of metabolism including kidney, liver, brain, and heart (Parrado-Fernández et al. 2011). Of all the food sources, meat and fish are the richest sources of dietary ubiquinone (Pravst et al. 2010). Ubiquinone is comprised of a polyisoprenoid side chain and benzoquinone ring containing 6 to 10 subunits with species-specific length and confers stability to the molecule inside the phospholipid bilayer (Hernández-Camacho et al. 2018) (Fig. 7.6). Ubiquinone is a ubiquitous component in the mitochondrial ETC placed in the inner mitochondrial membrane, in which it transfers electrons to complex III from complexes II and I to provide energy to the intermembrane space for proton translocation (López-Lluch et al. 2010). Ubiquinone is a potent antioxidant in the mitochondria that involve in energy synthesis via ETC (Saini 2011). It can neutralize ROS and protect plasma lipoproteins and cell membranes (Noh et al. 2013). It has been suggested that ubiquinone exerts the potential to protect against CVD via reduction of lipid peroxidation of LDL-cholesterol (Dludla et al. 2020). However, the concentration of ubiquinone is decreased with advancing age and thereby increased the risk to develop the symptoms associated with aging (Niklowitz et al. 2016). In this regard, decreasing ubiquinone concentration during aging could be one of the major contributors to the development of age-related diseases (de Barcelos and Haas 2019). It has been proposed that NAD(P)H:quinone oxidoreductase 1 (NQO1) may serve as

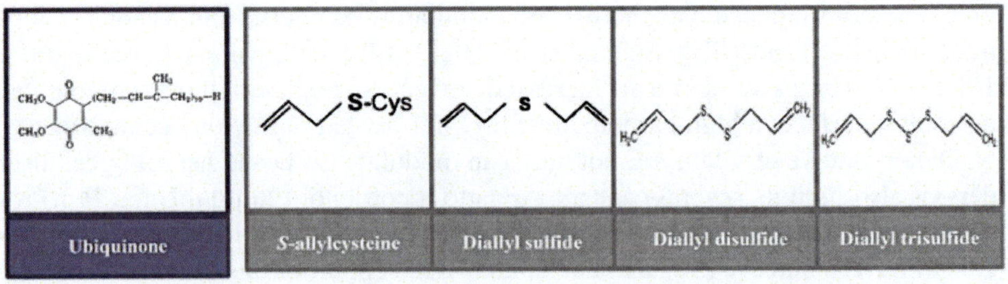

Fig. 7.6 Molecular structures of ubiquinone and organosulfur compounds (*S*-allylcysteine, diallyl sulfide, diallyl disulfide, and diallyl trisulfide)

a redox-sensitive switch to mediate the cell response in relation to the alteration of the redox status (Ross and Siegel 2017).

Emerging research evidence indicates that ubiquinone is beneficial toward neuroinflammation (Nyariki et al. 2019; Yang et al. 2020c) (Table 7.16). A study by Yamagishi et al. (2014a) followed nearly 6000 participants aged 40–69 years for 5 years revealed that serum ubiquinone levels were inversely associated with the risk of disabling dementia in the Japanese population, implied that higher serum ubiquinone concentration may have a potential effect in preventing dementia. The preventive ability of ubiquinone against dementia is more likely due to the antioxidant effect. Yet, ubiquinone prevents dementia progression via a few mechanisms, for instance, improved behavioral performance and decreased deposition of amyloid-β and brain oxidative stress (Dumont et al. 2011; Yang et al. 2016). Indeed, the rats with damaged cerebral cortices and hippocampi showed a detrimental impact on the markers of oxidative damage (Onodera et al. 2003). Matthews et al. (1998) showed that dietary intake of ubiquinone of 200 mg/kg/day for 2 months elevated the ubiquinone concentration in brain mitochondria of rats. Supplementation of 0–100 μM ubiquinone for 48 h protects the detrimental effects of Aβ in a dose-dependent manner, primarily by suppressing oxidative stress through stimulation of the PI3K/Akt pathway in the rat cortical neurons (Choi et al. 2012). An animal study also found that ubiquinone can reduce the total oxidant levels and serum MDA levels as well as increased total antioxidant capacity levels (Komaki et al. 2019). This study suggests that ubiquinone may offer neuroprotection against Aβ-induced neurotoxicity on hippocampal synaptic plasticity through antioxidant activity (Komaki et al. 2019). In human study, ubiquinone concentrations were relatively low in both Parkinson and Alzheimer's disease patients (Mischley et al. 2012; Manzar et al. 2020). In line with this, Yamagishi et al. (2014b) also found that individuals with low ubiquinone levels in the serum have a relatively high risk of dementia. Increased oxidative stress and mitochondrial impairment are thought to be involved in the pathogenesis of neurological disorder (Geldenhuys et al. 2017).

It has been suggested that the synthesis of ubiquinone is changed throughout the entire life of organisms, and is significantly decreased in the initial stages of aging (Alho and Lonnrot 2000; Crane 2001). In mice, reduced mitochondrial ubiquinone concentration of skeletal muscle and liver homogenates (Onur et al. 2014) was

Table 7.16 Animal and clinical studies of ubiquinone on age-related diseases

Age-related diseases	Study conditions	Source of ubiquinone	Durations	Outcomes	References
Neurodegenerative disease	Male Sprague-Dawley rats	Dietary ubiquinone	200 mg/kg for 2 months	↑ Ubiquinone concentration in brain mitochondria	Matthews et al. (1998)
Disabling dementia	Case-control study involving about 6000 Japanese aged 40–69 years and follow-up for 5 years	Serum ubiquinone	–	↓ Risk of disabling dementia	Yamagishi et al. (2014a)
Dementia	Japanese population	Serum ubiquinone	–	Negative associated with the risk of dementia	Yamagishi et al. (2014b)
Neurodegenerative disease	Rat cortical neurons	Ubiquinone supplement	0–100 μM ubiquinone for 48 h	Protects the detrimental effects of Aβ in a dose-dependent manner	Choi et al. (2012)
Oral cancer	Cross-sectional study involving patients more than 20 years old and under 80 years old with oral cancer (mean age = 54 years old)	Serum/plasma ubiquinone	Median ubiquinone = 280 nmol/L	Negative associated with oral cancer	Chan et al. (2020)
Hepatocellular carcinoma	A total of 71 primary hepatocellular carcinoma patients (mean age = 59 ± 11 years old)	Plasma ubiquinone	After surgery: 280 nmol/L	Negative associated with the hepatocellular carcinoma, especially in patients after surgery	Liu et al. (2017b)
Lung cancer	Prospective case-control study involving 596 participants, including 395 control subjects and 201 incident lung cancer cases	Plasma ubiquinone	–	Inverse association between ubiquinone and risk of lung cancer among current smokers	Shidal et al. (2021)

(continued)

Age-related diseases	Study conditions	Source of ubiquinone	Durations	Outcomes	References
Coronary artery disease	Systematic review and meta-analysis of 13 placebo-controlled and double-blind randomized controlled trials (364 individuals from intervention group and 349 participants in control group)	Ubiquinone supplements	60–300 mg/day of ubiquinone for 4 to 48 weeks	Reduced diene and MDA concentration, and increased catalase and SOD levels	Jorat et al. (2019)
Coronary artery disease	A randomized, placebo-controlled trial including 51 patients who were identified with cardiac catheterization with at least 50% stenosis of one major coronary artery and having statins for at least 1 month (24 subjects in placebo and 27 participants from ubiquinone group)	Ubiquinone supplement	300 mg/day for 12 weeks	Decreased TNF-α and increased GPx, catalase, and SOD activities	Lee et al. (2013)
CVD	Systematic review and meta-analysis of randomized controlled trials including 12 studies with a total of 650 diabetic patients	Ubiquinone supplement	20–200 mg/day between 8 weeks and 6 months	Reduced LDL-C and total cholesterol level in diabetic patients	Dludla et al. (2020)
Heart failure	Systematic review of 11 randomized controlled trials including 1573 participants	Ubiquinone supplement	Varied from 2 to 10 mg/kg/day in 2 or 3 divided concentration for 6 months to 400 mg/day for 3 months	Reduce the risk of mortality from all causes and hospitalization related to heart failure	Al Saadi et al. (2021)
Chronic heart failure	Meta-analysis involving 2149 patients with chronic heart failure	Ubiquinone supplement	100–200 mg/day for 3–12 months	Enhanced exercise capacity and reduced mortality	Lei and Liu (2017)

CVD cardiovascular disease, *GPx* glutathione peroxidase, *LDL-C* low-density lipoprotein cholesterol, *MDA* malondialdehyde, *SOD* superoxide dismutase, *TNF-α* tumor necrosis factor-α observed during aging, but not in the heart, brain, and kidney (Lass et al. 1999). However, Onur et al. (2014) found that *C. elegans*, in which ubiquinone concentration in the elderly is increased with age. Importantly, decreased circulating ubiquinone levels were found in the elderly, in which the redox status of ubiquinone significantly shifted to the oxidized form (Niklowitz et al. 2016). A study by Nagase et al. (2018) evaluated ubiquinone deficiency and increased oxidative stress among Japanese centenarians. The data showed that the percentages of the oxidized form of ubiquinone were significantly increased in centenarians compared to individuals aged 76 years, implied that an elevation of oxidative stress in centenarians (Nagase et al. 2018). The study further revealed that total serum ubiquinone level in centenarians was significantly decreased compared to individuals aged 76 years old (Nagase et al. 2018). Reduced ubiquinone levels in old age were related to the loss of antioxidant capacity in cell organelles, human tissues, or serum/plasma levels (Niklowitz et al. 2016).

Research evidence has indicated that ubiquinone deficiency is associated with cancer (Trueba et al. 2004; Cooney et al. 2011). A cross-sectional study has shown that patients with oral cancer had low ubiquinone levels, and higher ubiquinone status was significantly related to the reduced risk of metabolic syndrome, lipid profiles, and waist circumference (Chan et al. 2020). The study also found that the median ubiquinone concentration in patients with oral cancer was 280 nmol/L, which is lower than that of the healthy adults (500–1700 nmol/L) (Molyneux et al. 2008). In line with this, patients with hepatocellular carcinoma had significantly lower ubiquinone levels (280 nmol/L) after surgery (Liu et al. 2017b). The ubiquinone deficiency observed in patients with oral cancer is likely due to the high oxidative stress (Korde et al. 2011). Interestingly, the previous study stated that ubiquinone concentration was negatively linked to the risk of metabolic syndrome and increased HDL-C levels (Chan et al. 2020). These findings imply that intake of ubiquinone not only decreases oxidative stress, they also modulate lipid metabolism (Chan et al. 2020), suggesting the enormous functional potential of ubiquinone. A prospective case-control study involving 596 participants, including 395 control subjects and 201 incident lung cancer cases demonstrated an inverse association between ubiquinone and risk of lung cancer, particularly in current smokers (Shidal et al. 2021), implied that elevation of ubiquinone may provide an antioxidant reservoir that is capable in scavenging ROS produced by cigarette smoke in lung tissues (Shidal et al. 2021). This finding suggests that ubiquinone is implicated in the pathogenesis of lung cancer and may be related to the disease progression (Shidal et al. 2021).

Besides its effects on neurological disorders and cancer, the previous study has also demonstrated the role of ubiquinone on the incidence of CVD (Table 7.16). Oxidative stress via increased ROS production can affect the availability of endothelial NO, and thereby compromise vascular function (Lubos et al. 2008). The increased arterial atherosclerotic buildup is linked to endothelial dysfunction and thereby elevated the risk of heart failure (Boudina and Abel 2010; Akoumi et al. 2017; Borghetti et al. 2018; American Heart Association 2019). Data from systematic review and meta-analysis of randomized controlled trials including 12 studies with a total of 650 diabetic patients showed that the LDL-C and TC levels were significantly decreased in diabetic patients supplemented with ubiquinone (20–200 mg/day between 8 weeks and 6 months) compared to those who consume placebo (Dludla et al. 2020), suggesting that ubiquinone may potentially reduce the risk of CVD in diabetic patients by alleviating the oxidative stress and enhancing endothelial function. In particular, heart muscle is relatively prone to ubiquinone deficiency and utilizes more energy compared to other tissues (Hargreaves et al. 2020). The weakening of the heart muscle may result in the swelling of the liver, lower feet and legs, the lining of the intestine, and lung. Heart failure is characterized by a loss of contractile function due to the depletion of energy in mitochondria associated with ubiquinone deficiency (Sharma et al. 2016). The systematic review of 11 randomized controlled trials including 1573 participants evaluated the efficacy and safety of ubiquinone in heart failure (Al Saadi et al. 2021). The study showed that ubiquinone (varied from 2 to 10 mg/kg/day in 2 or 3 divided concentration for 6 months to 400 mg/day for 3 months) may reduce the risk of mortality from all causes and hospitalization related to heart failure (Al Saadi et al. 2021). Emerging research indicates that heart failure patients suffer a deficiency of ubiquinone levels coupled with an elevation of ROS within the heart (Al Saadi et al. 2021). Removal of ROS by ubiquinone can facilitate the mediation of redox status in the heart (Ulla et al. 2017). Ubiquinone may also stabilize myocardial calcium-dependent ion channels and thereby prevent the depletion of metabolites essential for the synthesis of ATP. Likewise, a meta-analysis involving 2149 patients with chronic heart failure demonstrated that intake of 100–200 mg/day ubiquinone for 3–12 months significantly enhanced exercise capacity and reduced mortality compared to those who consume placebo (Lei and Liu 2017). Importantly, Di Lorenzo et al. (2020) showed that ubiquinone may confer prognostic advantage and did not lead to any adverse outcomes in heart failure patients. These observations imply that ubiquinone could be beneficial to patients with heart failure despite may not serve as a primary treatment, suggesting that ubiquinone may be considered as an adjunct to conventional treatment.

In a further study focused on inflammation outcomes, Jorat et al. (2019) evaluated ubiquinone in relation to inflammatory markers and proinflammatory cytokine in coronary artery disease patients. A recent systematic review and meta-analysis of placebo-controlled and double-blind randomized controlled trials involving 364 individuals from the intervention group and 349 participants in the control group demonstrated that supplementation 60–300 mg/day of ubiquinone for 4 to 48 weeks reduced diene and MDA concentration, and increased catalase and SOD levels in coronary artery disease patients (Jorat et al. 2019). From the study reviewed, Jorat et al. (2019) found that supplementation of ubiquinone did not show any improvement in GPx, IL-6, TNF-α, and CRP concentration among coronary artery disease patients. However, a randomized, placebo-controlled trial conducted by Lee et al. (2013) demonstrated that administration of 300 mg/day of ubiquinone supplement for 12 weeks reduced TNF-α concentration and increased GPx, catalase, and SOD levels in patients who had coronary artery disease during statins therapy. The findings on these CVD-related outcomes are inconclusive due to some limitations of clinical trials that have accessed these crucial CVD-related outcomes, including (1) short duration of treatment; (2) different concentration of ubiquinone used; (3) the characteristics of study population; (4) small sample size; and (5) the differences in study design. Taken together, aging is linked to increased oxidative stress and reduced ubiquinone levels in mitochondria. Ubiquinone not only modulates inflammatory mediators, it has also shown its efficacy in heart failure. Despite many studies have demonstrated the promising finding of ubiquinone intake on several age-related diseases and oxidative stress markers, the efficacy of ubiquinone on clinical trials remains obscure. Further, research evidence available on the reduced risk of age-related diseases via ubiquinone-rich containing food is limited. Therefore, the clinical benefits of food rich in ubiquinone in relation to age-related diseases required further elucidation in long-term randomized clinical trials.

Mitochondria-targeted antioxidants have been identified to play a crucial role in ameliorating the mitochondrial oxidative damage caused by ROS production that is linked to various diseases (Nazarov et al. 2017). The mitochondria-targeted antioxidants provide greater protection towards oxidative damage is likely due to their abilities to cross the mitochondria phospholipid bilayer and thereby eliminating ROS (Jiang et al. 2020). In particular, mitochondria-targeted antioxidants contain a broad spectrum of compounds that contain an antioxidant group associated with the mitochondria-targeted moiety including triphenylphosphonium (TPP) cation (Nazarov et al. 2017). Emerging evidence indicates that mitochondria-targeted antioxidants are widely used to evaluate the impact of mitochondria on different pathological processes involve oxidative stress (Murphy and Smith 2007). It has been shown that mitochondria-targeted lipophilic ubiquinone (MitoQ) had greater protection towards oxidative damage in the mitochondria compared to untargeted antioxidants (Hao et al. 2018). In general, a broad spectrum of antioxidants could be targeted to mitochondria through TPP-conjugated ubiquinone. Substantial studies revealed that ubiquinol moiety of MitoQ can protect against peroxynitrite and react with superoxide (Maiti et al. 2018). A human study has demonstrated that MitoQ can protect against mitochondrial ROS produced in vascular tissues (Chen et al. 2020). Rossman et al. (2018) evaluated the efficacy and safety of MitoQ in healthy older adults aged 60–79 years with impaired endothelial function. The data showed that plasma MitoQ was increased after consuming 20 mg/day MitoQ for 6 weeks and was well-tolerated compared to the placebo group (Rossman et al. 2018) (Table 7.17). Notably, intakes of MitoQ significantly increased brachial artery flow-mediated dilation and decreased plasma ox-LDL, a biomarker of oxidative stress, compared to placebo. This improvement is likely due to the alleviation of mitochondrial ROS-related inhibition of endothelial function (Rossman et al. 2018). Based on the evidence, MitoQ may hold promise in treating age-related vascular dysfunction. Compared to those who consume ubiquinone, elderly aged >55 years who consume MitoQ at 200 mg/day for 14 days significantly increased the plasma ubiquinone levels in older men (Zhang et al. 2018d) (Table 7.17). The human study further demonstrated that intake of 200 mg/day of MitoQ for 4 weeks showed a better absorption in adults compared to ubiquinone (Langsjoen and Langsjoen 2014) (Table 7.17). Indeed, the data from two acute pharmacokinetics studies indicate that MitoQ shows better bioavailability compared to ubiquinone (Miles et al.

Table 7.17 Effects of MitoQ treatment on health

Age-related diseases	Study conditions/ experimental model	Source of MitoQ	Durations	Outcomes	References
CVD	Randomized, placebo-controlled, double-blinded, cross-over design study involving 20 healthy older adults (60–79 years old)	MitoQ supplement	20 mg/day for 6 weeks	Increased branchial artery flow-mediated dilation and decreased plasma oxidized LDL	Rossman et al. (2018)
Alzheimer's disease	Transgenic *Caenorhabditis elegans* model of Alzheimer's disease	MitoQ supplement	1 and 5 μM	Delays Aβ-induced paralysis and ameliorates depletion of the mitochondrial cardiolipin levels	Ng et al. (2014)
–	Healthy subjects	MitoQ supplement	200 mg/day for 4 weeks	Shows a better absorption compared to ubiquinone	Langsjoen and Langsjoen (2014)
–	Elderly aged >55 years	MitoQ supplement	200 mg/day for 14 days	Increased the plasma ubiquinone levels	Zhang et al. (2018d)

CVD cardiovascular disease, *LDL* low-density lipoprotein

2004; Evans et al. 2009). Besides its effects on CVD, previous studies have also indicated the role of MitoQ in the mediation of neuroinflammation. In the context of Alzheimer's disease, MitoQ (1 and 5 μM) delays Aβ-induced paralysis in transgenic *Caenorhabditis elegans* model of Alzheimer's disease as well as prolongs the lifespan of transgenic Aβ-overexpressing nematodes, suggesting that MitoQ not only promote healthspan but also increase lifespan (Ng et al. 2014) (Table 7.17). Importantly, loss of cardiolipin may contribute to mitochondrial dysfunction in aging and neurodegenerative disease (Falabella et al. 2021). Another study further demonstrated that treatment of MitoQ (1 and 5 μM) ameliorates depletion of the mitochondrial cardiolipin levels, indicates that MitoQ may protect the mitochondrial matrix (Ng et al. 2014). Collectively, the findings showed that MitoQ may decrease the mitochondrial dysfunction related to age-related diseases.

7.9 Organosulfur Compounds

Organosulfur compounds (Fig. 7.6) are predominantly found in vegetables species belonging to *Brassica* (Cruciferous) and *Allium* genus, for instance, cauliflower, cabbage, broccoli, garlic, and onion (Ruhee et al. 2020). The vegetables of the *Allium* genus possess a typical pungent aroma and flavor, due to the presence of volatile or oil-soluble sulfur compounds (Itakura et al. 2001). While the less odorous organosulfur compounds such as S-allyl mercapto cysteine (SAMC) and S-allyl cysteine (SAC) found in aqueous extract of garlic are water-soluble in nature and usually present in vegetables of the *Allium* genus (Wang et al. 2016b). Organosulfur compounds are liberated when the garlic is crushed or cut into pieces (Garrow et al. 2001). In onion, several intermediary sulfur compounds, for instance, tri-, di-, and monosulfides, thiosulfonates, and thiosulfinates are identified, which contributes to the strong flavor (Liguori et al. 2017). Dietary organic sulfur compounds from *Allium* species such as propyl disulfide, dimethyl disulfide, and diallyl disulfide have vasodilating, anti-inflammatory in addition to prevent several chronic diseases (Chu et al. 2017) (Table 7.18). Indeed, Fe-S clusters play a crucial role in the origin of life including RNA, acetyl-CoA, and DNA (Fuss et al. 2015). Glucosinolates are one of the naturally occurring compounds found in the *Brassica* genus, which are known as the precursors of isothiocyanates (Ruhee et al. 2020). In general, the hydrolysis of glucosinolates into thiocyanates and isothiocyanates is caused by insects and food preparation (Cabello-Hurtado et al. 2012).

Emerging research evidence indicates that organosulfur compounds exert many health benefits including anticancer properties (Alkreathy 2020). The preclinical data has demonstrated that Alliaceous vegetables containing organosulfur compounds can protect against carcinogenesis via several mechanisms such as induction of xenobiotic-metabolizing enzymes, scavenging free radicals, inhibiting the metabolism of nitrosamines and hydrocarbons, modulating cell cycle arrest, alleviating the detoxification of carcinogens, decreasing bioactivation of carcinogens, induction of apoptosis, and mediation of redox modification (Somade et al. 2018). The previous studies suggest that high broccoli sprout, garlic, and onion consumption provides greater protection in cancer risk (Zhang et al. 2020a), including breast (Desai et al. 2020), prostate (Beaver et al. 2018), and colorectal (Zhou et al. 2020b). A meta-analysis involving 12,558 cases included a study from inception to May 2019 demonstrated that intake of garlic was negatively linked to colorectal cancer (Zhou et al. 2020b). Of all phytochemicals from the *Allium* genus, organosulfur compounds were shown a greater protective effect on metabolic disorders (Ahmad et al. 2019). Much information indicates that most of the organosulfur compounds from garlic possess anticancer activity. S-allylmercaptocysteine and S-allyl-L-cysteine are abundant organosulfur compounds identified in aged garlic extract (Yan et al. 2013). In neuroblastoma cancer cells, aged garlic extract (2 mg/mL) and its derived compounds, S-allyl-L-cysteine (50 mM for 48 h) triggered apoptosis by inducing mitochondrial permeability transition (Kanamori et al. 2020). The protective effect of S-allyl-L-cysteine against neuroblastoma cancer cells could be

Table 7.18 Effects of organosulfur compounds on age-related diseases in *in vitro*, *in vivo*, and human studies

Age-related diseases	Experimental model	Source of organosulfur compounds	Durations	Outcomes	References
Hypertension	SVEC4-10	Dimethyl disulfide and diallyl disulfide from *Allium* species	500 μM for 4 h	Positively triggered PGI_2 production	Chu et al. (2017)
Breast cancer	Transgenic mouse models including SV40 and Her2/neu mice	AIN-93G diet containing 26% of sulforaphane-based broccoli sprout	Began from conception and continued throughout pregnancy until the weaning period	Exhibited a maximal preventive against breast cancer development compared to postnatal early life treatment	Li et al. (2018b)
Prostate cancer	Phase II study involving 20 patients with recurrent prostate cancer	Sulforaphane-rich broccoli sprout extracts	200 μmoles/day for 20 weeks	Did not lead to ≥50% PSA declines	Alumkal et al. (2015)
Prostate cancer	TRAMP mice	AIN-93G diet containing 15% broccoli sprout powder	9–25 weeks	Decreased the prostate cancer incidence	Beaver et al. (2018)
Prostate cancer	Randomized, double-blinded 3-arm parallel intervention study involving 49 localized prostate cancer patients on active surveillance	Glucoraphanin-rich broccoli soup	300 mL/week for 12 months	Inversely related to the risk of prostate cancer progression	Traka et al. (2019)
Neuroblastoma cancer	SJ-N-KP and IMR5 cell lines	Aged garlic extract and its derived compounds, S-allyl-L-cysteine	Aged garlic extract (2 mg/mL) and its derived compounds, S-allyl-L-cysteine (50 mM for 48 h)	Triggered apoptosis by inducing mitochondrial permeability transition	Kanamori et al. (2020)
CVD	A systematic review of 10 randomized clinical trials between 2004 and 2018	Garlic or aged garlic extract	400–2400 mg/day varied from 15 days to 3 months	Improved vascular function	Emamat et al. (2020)

CVD	36 Male Wistar albino rats, the metabolic syndrome rats were fed with high-fat diet for 4 weeks followed by intraperitoneal single streptozotocin injection in a dose of 25 mg/kg	Diallyl trisulfide derived from garlic	40 mg/kg every second day for 3 weeks	Increased HDL-C and reduced LDL-C and TG levels	Jeremic et al. (2020)
Type 2 diabetes mellitus	Adults from Tehran Lipid and Glucose Study (2006–2008 to 2009–2011)	*Allium* vegetables (raw onion and garlic)	≥142 g/week	Reduced incidence of hyperinsulinemia and insulin resistance	Mirmiran et al. (2019)
Type 2 diabetes mellitus	8-week-old female BALB/c mice	Diallyl disulfide	30 mg/kg body weight/day for 23 days	Decreased urinary 8-OHDG and proteinuria levels, alleviated doxorubicin-induced renal injury, and increased catalase activities in renal cortical tissues	Lin et al. (2019)
Neurodegenerative disorders	Male ICR mice	Sulforaphane-enriched broccoli sprouts	200 mg/kg for 2 weeks, 7 times/week	Suppress neuronal apoptosis and modulate scopolamine-induced memory impairment	Subedi et al. (2019)
Neuroinflammation	Streptozotocin-induced diabetic rats	Garlic	0.5 g/kg of for 10 days	Increased GPx and SOD activities in brain	Rahmani et al. (2020)

AIN-93G American Institute of Nutrition, *CVD* cardiovascular disease, *GPx* glutathione peroxidase, *HDL-C* high-density lipoprotein cholesterol, *LDL-C* low-density lipoprotein cholesterol, *PGI_2* prostacyclin, *PSA* prostate-specific antigen, *SOD* superoxide dismutase, *SVEC4-10* endothelial cells, *TG* triglycerides, *TRAMP* transgenic adenocarcinoma of the mouse prostate, *8-OHDG* 8-hydroxy-2′-deoxyguanosine attributed to its antioxidant properties, and thereby protecting the cells from protein oxidation and lipid peroxidation (Kanamori et al. 2020). The data from animal study also found that feeding mice with a diet containing 15% broccoli sprout or 400 mg sulforaphane/kg diet for 9–25 weeks decreased the prostate cancer incidence and HDAC3 protein expression in the anterior lobes and ventral of the prostate epithelial cells (Beaver et al. 2018). HDAC3 is predominantly expressed in patients with prostate cancer and an increase of class-I HDAC is thought to be associated with the initial stage of prostate carcinogenesis (Weichert et al. 2008). Sulforaphane, a natural dietary isothiocyanate abundantly found in cruciferous vegetables (Dacosta and Bao 2017). A study has shown that sulforaphane at 10 or 20 μM can effectively decrease NO, iNOS, and proinflammatory cytokines expression via stimulation of Nrf2/HO-1 signaling pathways in LPS-induced RAW 264.7 cells (Ruhee et al. 2019). Compared to postnatal early-life treatment, administration a diet containing 26% of sulforaphane-based broccoli sprout during maternal/prenatal (began from conception and continued throughout pregnancy until the weaning period) exhibited a maximal preventive against breast cancer development, suggesting that feeding mice with sulforaphane-based broccoli sprout diet during prenatal/maternal may affect the gene expression through transplacental epigenetic modulation and thereby delay breast tumor development in later life (Li et al. 2018b). The preventive ability of sulforaphane-based broccoli sprout in prenatal/maternal transgenic mice is mediated by multiple cellular pathways including modulation of epigenetic pathways, apoptosis, cell cycle, DNA repair, and regulation of transcription activity (Li et al. 2018b). While for prostate cancer, Traka et al. (2019) found that dietary intake of 300 mL/week of glucoraphanin-rich broccoli soup for 12 months was inversely linked to the risk of prostate cancer progression by modulating the transcriptional activity in the prostate of men on active surveillance. Intriguingly, Alumkal et al. (2015) did not identify a beneficial role of sulforaphane-rich broccoli sprout extracts against prostate cancer. The men with high levels of PSA in serum were associated with a higher incidence of prostate cancer (Catalona et al. 1991). Alumkal et al. (2015) found that the consumption of 200 μmoles/day sulforaphane-rich broccoli sprout extracts for 20 weeks did not lead to ≥50% PSA declines in men with recurrent prostate cancer.

In a study focused on inflammation outcomes, Chu et al. (2017) explored the anti-inflammatory effects of propyl disulfide, dimethyl disulfide, and diallyl disulfide from *Allium* species in LPS-treated RAW 264.7 cells. The data showed that propyl disulfide, dimethyl disulfide, and diallyl disulfide (50–500 μM for 6 h) from *Allium* species significantly suppressed the prostaglandin E_2 (PGE_2) production in LPS-induced RAW 264.7 cells. Dietary propyl disulfide, dimethyl disulfide, and diallyl disulfide (500 μM for 6 h) was found to inhibit COX-2 and iNOS expression in activated RAW 264.7 cells, implied that these dietary organosulfur compounds can prevent the LPS-induced inflammatory response in RAW 264.7 cells (Chu et al. 2017).

Besides its effects on cancer and inflammation, a recent systematic review suggests that consumption of 400–2400 mg/day garlic or aged garlic extract between 15 days and 3 months has the potential to improve vascular function, especially for those individuals with cardiovascular risk factor (Emamat et al. 2020). Indeed, a wide range of organosulfur compounds was found in different garlic products with possibly mediate different physiological actions. Raw garlic is rich in γ-glutamylcysteines, which thereby convert to alliin (Amagase et al. 2001). When garlic is crushed, cut, or ingested, allinase enzyme is stimulated and thereby converts alliin to allicin. Subsequently, allicin is decomposed to other organosulfur compounds including ajoene (Amagase et al. 2001). Diallyl trisulfide, a secondary metabolite of allicin, is a volatile organosulfur compound produced by crushed garlic (Abe et al. 2020). A recent finding has suggested that diallyl trisulfide can effectively suppress platelet aggregation by regulating the reaction of sulfhydryl groups (Hosono et al. 2020). Jeremic et al. (2020) further demonstrated that feeding metabolic syndrome rats with 40 mg/kg of diallyl trisulfide derived from garlic every second day for 3 weeks increased HDL-C and reduced LDL-C and TG levels. The protective effect of diallyl trisulfide could be attributed to the suppression of endogenous cholesterol synthesis (Aouadi et al. 2000). The previous study has revealed that intake of garlic is associated with the inhibition of hepatic fatty acid synthesis primarily by regulating the critical enzymes and thereby decreasing plasma TG levels and lipid accumulation in the liver (Thomson et al. 2006). Jeremic et al. (2020) further demonstrated that diallyl trisulfide protects the heart against *ex-vivo* treated I/R injury in metabolic syndrome rats by alleviating the inflammation, apoptosis, and oxidative stress in the heart. This finding suggests that diallyl trisulfide has the potential to prevent cardiovascular complications and metabolic disorders (Jeremic et al. 2020). Prostacyclin (PGI_2) is a vasoactive substance derived from the endothelium that plays a crucial role in alleviating hypertension (Mitchell et al. 2014). An *in vitro* study has revealed that dietary dimethyl disulfide and diallyl disulfide (500 μM for 4 h) was positively triggered PGI_2 production in an endothelial cell line (SVEC4-10) (Chu et al. 2017). The data also demonstrated the upregulation of COX-2 expression in SVEC4-10 cells after administration of 500 μM of dimethyl disulfide or diallyl disulfide from *Allium* species for 4 h, suggesting that these organosulfur compounds can stimulate the COX-2 levels in SVEC4-10 cells (Chu et al. 2017). Indeed, COX-2 plays a critical role in the biosynthesis of PGI_2 from arachidonic acid (Nørregaard et al. 2015). In this regard, the upregulation of PGI_2 expression after treatment of dimethyl disulfide or diallyl disulfide from *Allium* species in SVEC4-10 cells could be facilitated by the increase of COX-2 levels (Chu et al. 2017), suggesting that dimethyl disulfide or diallyl disulfide may potentially act as a vasodilative mediator.

Organosulfur compounds not only reduce cancer and CVD, they also suppress hyperglycemia and insulin resistance (Melino et al. 2019). Dysfunctions of β-cell mass and progressive decrease in β-cell function are linked to the development of diabetes (Russo et al. 2014; Miranda-Osorio et al. 2016). Consumption of organosulfur compounds-rich diet was negatively linked to diabetes mellitus (Ülger and Çakiroglu 2020). A longitudinal follow-up study by Mirmiran et al. (2019) evaluated *Allium* vegetables (raw onion and garlic) in relation to hyperinsulinemia, β-cell dysfunction, and insulin resistance among adults. The study showed that high consumption of raw onion and garlic (≥142 g/week) significantly reduced the risk of hyperinsulinemia and insulin resistance compared to low intake of raw onion and garlic (˂8.0 g/week). This finding suggests that a high intake of *Allium* vegetables may improve insulin homeostasis. However, intake of *Allium* vegetable (raw onion and garlic) was not associated with a better β-cell function (Mirmiran et al. 2019). The previous study stated that diabetic rats had relatively high amounts of DNA damage and low total antioxidant capacity and GSH levels compared to non-diabetic rats (Suresh et al. 2017). Feeding doxorubicin-induced nephrotoxicity mice with diallyl disulfide, one of the crucial components in garlic, at a dose of 30 mg/kg body weight/day for 23 days, decreased urinary 8-hydroxy-2′-deoxyguanosine (8-OHDG) and proteinuria levels, alleviated doxorubicin-induced renal injury, and increased catalase activities in renal cortical tissues, suggesting that garlic could prevent oxidative stress (Lin et al. 2019).

The previous studies demonstrated the beneficial effects of garlic and its organosulfur compounds in neurodegenerative diseases (Sripanidkulchai 2019; Bigham et al. 2021). An animal study showed that administration of 0.5 g/kg of garlic for 10 days increased GPx and SOD activities in the brain of streptozotocin-induced diabetic rats (Rahmani et al. 2020). The study further revealed that garlic (0.5 g/kg for 10 days) can improve depressive- and anxiety-like behaviors in diabetic rats possibly by ameliorating brain oxidative stress (Rahmani et al. 2020), demonstrating the enormous functional potential of organosulfur compounds. Moreover, research evidence indicates that sulforaphane-enriched broccoli sprouts (200 mg/kg for 2 weeks, 7 times/week) are of benefit in the suppression of neuronal apoptosis and modulation of scopolamine-induced memory impairment in mice, particularly via the mediation of caspase-3 and Nrf2 signaling pathways (Subedi et al. 2019). These findings suggest that sulforaphane-enriched broccoli sprouts may protect against neuroinflammation. Taken together, regular consumption of food rich in organosulfur compounds can decrease age-related diseases in addition to offer healthy aging in terms of nutrition.

7.10 Ergothioneine

Ergothioneine is a naturally occurring amino acids that are produced in certain fungi, cyanobacteria, and actinobacteria (Fahey 2001; Hseu et al. 2020). It is a thiourea derivative of histidine with a sulfur atom on its imidazole ring (Hseu et al. 2020) (Fig. 7.7). Ergothioneine serves as a powerful scavenger of hydroxyl radicals (•OH) (Mishra et al. 2018). Due to its excellent antioxidant activity, ergothioneine plays a prominent role in preventing various chronic diseases (Han et al. 2021). Ergothioneine is ubiquitously present in plants, animals, and bacteria and is thought to protect against ultraviolet radiation (Hseu et al. 2020). It also accumulates in various parts of the human body, for instance, skin, eyes, seminal fluids, kidney, erythrocytes, liver, and bone marrow (Markova et al. 2009; Cheah and Halliwell 2012). Oyster mushroom and bolete are the richest sources of dietary ergothioneine (Ey et al. 2007; Nakamichi et al. 2016). In particular, mushroom is capable in the

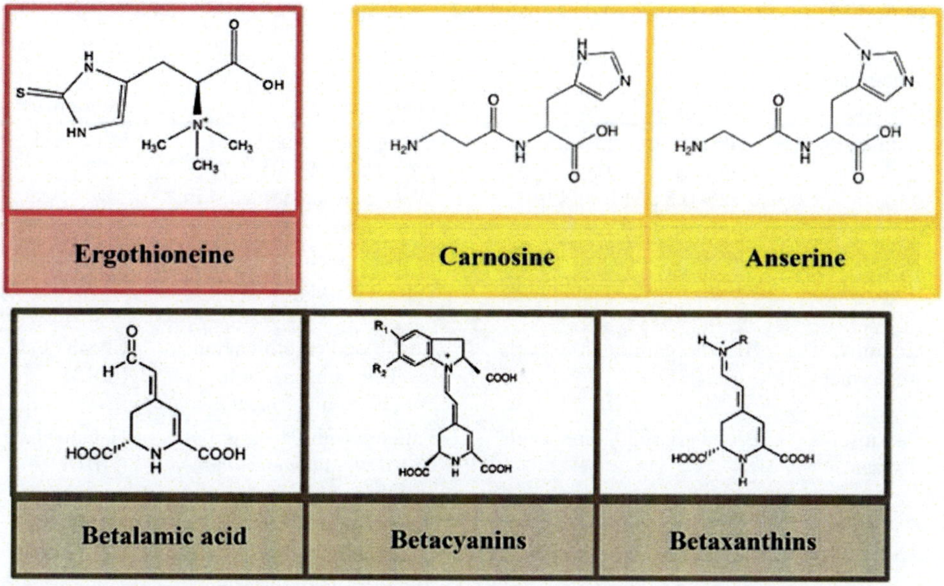

Fig. 7.7 Molecular structures of ergothioneine, betalamic acid, betacyanins, betaxanthins, carnosine, and anserine

biosynthesis of ergothioneine along with other fungi (Cheah and Halliwell 2012; Weigand-Heller et al. 2012; Kalaras et al. 2017). Ergothioneine can also be obtained from the diet such as kidney beans, oat bran, black beans, kidney, and liver (Hseu et al. 2020). Despite ergothioneine is found in many cells, plasma membranes seem to be intrinsically impermeable to ergothioneine, and thus the mode of action for accumulation and uptake in cells must exist (Cheah and Halliwell 2012). The previous study stated that ergothioneine is preferentially produced in secretions, cells, and organs predisposed to high amounts of inflammation and oxidative stress such as kidney and liver (Cheah and Halliwell 2012).

It has been suggested that ergothioneine was effective in protecting the cells from photoaging (Bazela et al. 2014). Data from *in vitro* study showed that ergothioneine (0.125–0.5 μM for 24 h) upregulated type I procollagen levels in UVA-exposed human skin fibroblast (HSF) cells in a concentration-dependent manner, suggesting that ergothioneine may protect the HSF cells from UVA radiation-induced skin aging and collagen degradation (Hseu et al. 2020). Table 7.19 summarizes the effects of ergothioneine on age-related diseases in *in vitro* and human studies.

Many studies revealed that mushroom exerts beneficial effects to the brain (Borodina et al. 2020; Zhang et al. 2017b; Feng et al. 2019b) as well as reduced the incidence of age-related diseases such as diabetes (Calvo et al. 2016; Okada et al. 2019; Hodge and Calvo 2019) and prostate cancer (Zhang et al. 2020b). Compared to those who rarely consume mushrooms, elderly who consume mushroom more than 2 portions/week have a significantly lower incidence of mild cognitive impairment (Feng et al. 2019b). This neuroprotective effect could be mainly ascribed to

Table 7.19 *In vitro* and human studies conducted in ergothioneine and their effects on age-related diseases

Age-related diseases	Experimental model	Outcomes	References
Skin aging	HSF cells treated with ergothioneine at a dose of 0.125–0.5 μM for 24 h	Upregulated type I procollagen levels in UVA-exposed human skin fibroblast	Hseu et al. (2020)
Mild cognitive impairment	Elderly who consume mushroom more than 2 portions/week	Reduced incidence of mild cognitive impairment	Feng et al. (2019b)
Cognitive impairment	Blood ergothioneine levels	Low blood ergothioneine levels in subjects with cognitive impairment	Cheah et al. (2016)
Parkinson's disease	Blood ergothioneine levels	Patients with Parkinson's disease had lower blood ergothioneine levels	Hatano et al. (2016)
CVD	3236 participants free from stroke, coronary artery disease, and type 2 diabetes mellitus aged 57.4 years and follow-up for 21.4 years	Ergothioneine metabolic level decreased the risk of overall mortality, cardiovascular mortality, and coronary disease	Smith et al. (2020)
Frailty	Elderly	Low in ergothioneine metabolic levels	Kameda et al. (2020)

CVD cardiovascular disease, *HSF* human skin fibroblast, *UVA* Ultraviolet A

their ergothioneine composition (Borodina et al. 2020). Research evidence has also revealed a potential protective function of ergothioneine in neurological disorders (Cheah et al. 2019). Ergothioneine supplementation was found to reduce neuronal injury (Yang et al. 2012) and protect rat phaeochromocytoma cells against Aβ-induced apoptotic death (Jang et al. 2004). In human study, individuals with cognitive impairment had lower blood ergothioneine levels compared to individuals with no cognitive impairment (Cheah et al. 2016). Likewise, the previous study also revealed that Parkinson's disease patients had relatively lower blood ergothioneine levels compared to control (Hatano et al. 2016). These findings indicate that ergothioneine transporter is present in the brain, in which the ergothioneine can pass through the blood-brain barrier (Tang et al. 2018), suggesting that low ergothioneine levels may predispose individuals to neurodegenerative disease.

Notably, intake of ergothioneine was associated with longevity (Beelman et al. 2019). A recent study suggested that ergothioneine can serve as a "longevity vitamin" based on the Triage Theory. This theory speculates that the human body utilizes specific micronutrients as it was in a triage phenomenon where the priority is given to its role on survival and reproduction (Ames 2018). When the ergothioneine supply is low, the functions that support long-term health can be compromised, and thereby may result in a decrease in life expectancy (Beelman et al. 2020). Similarly, Paul and Snyder (2010) revealed that ergothioneine is a crucial physiological cytoprotectant and could be served as a vitamin.

In addition to the effects observed above, a follow-up study for 21.4 years including 3236 participants free from stroke, coronary artery disease, and type 2 diabetes mellitus aged 57.4 years had revealed that ergothioneine metabolic level was associated with the lower risk of overall mortality, cardiovascular mortality, and coronary disease (Smith et al. 2020). This finding indicates that high ergothioneine levels may have a reactionary manner to protect from oxidative stress, and is thought to be a crucial factor in the pathogenesis of CVD (Higashi et al. 2009). The study further revealed a correlation between ergothioneine and health-conscious food patterns, particularly vegetable intakes (Smith et al. 2020). A study by Playdon et al. (2017) further supported the association of ergothioneine and higher consumption of seafood and vegetables with a low intake of added sugar and solid fats. Indeed, an interaction between metabolome and healthy dietary intake is particularly interesting because the circulating of specific metabolites have been demonstrated to predict future risk of premature mortality (Cheng et al. 2015), coronary artery disease (Ottosson et al. 2018), and type 2 diabetes mellitus (Wang et al. 2011). In another study, Kameda et al. (2020) evaluated 131 blood metabolites on frailty among the elderly. The data showed that ergothioneine metabolic levels were relatively low in frail individuals compared to non-frail (Kameda et al. 2020). Collectively, more studies are imminently required to explore means to increase ergothioneine in the diet. Given a strong relationship between mushroom intake and blood ergothioneine level (Pallister et al. 2016), the beneficial roles of mushroom are likely due to ergothioneine. Nevertheless, further studies are warranted to support this speculation.

7.11 Betalains

Betalain is a water-soluble pigment consist of nitrogen that is characterized by two structural units, namely red-violet betacyanins and yellow-orange betaxanthins (Fu et al. 2020). Betalains have uniform structural characteristics derived from betalamic acid (Fig. 7.7), with radical R_2 or R_1, in which the substituents can be a radical or hydrogen (Fu et al. 2020). In general, red-violet betacyanins such as neobetanin, isobetanin, prebetanin, and betanin, absorb wavelengths at 538 nanometers; whereas yellow-orange betaxanthins absorb wavelengths at 480 nm (Fu et al. 2020). Betacyanins and betaxanthins (Fig. 7.7) are considered as immonium conjugates of cyclo-3,4-dihydroxyphenylalanine (cyclo-DOPA) with betalamic acid and amino compounds, for instance, amino acids, amines, or their derivatives. Betalains are sensitive toward enzymes, light, pH, heat, and oxygen, which can affect their stability (Ravichandran et al. 2013). Betalain pigments are widely found in red beet and are thought to exert potent immune and intestinal regulatory effect, anti-inflammatory, antibacterial, hepatoprotective, anticarcinogenic, and antioxidant activities (Sanchez-Gonzalez et al. 2013; Tesoriere et al. 2013; Liñero et al. 2017; Bao et al. 2019; Rahimi et al. 2019a). Further, betalains can also protect cells from DNA damage and peroxidation (Winkler et al. 2005; Esatbeyoglu et al. 2014). Due to these health-promoting effects, betalains can be utilized as cosmetics, food additives, colorants, and drugs in the form of beet powder and beet juice (Janiszewska 2014).

Betalains exert potent antioxidant ability in scavenging free radicals (Mikolajczyk-Bator and Czapski 2017). The potent antioxidant activity of betalains could be due to their molecular structures, in which it exerts its ability to donate hydrogen to reactive species. Betacyanins show a strong antioxidant activity compared to betaxanthins (Stintzing and Carle 2004). Betacyanins and betaxanthins derived from red beetroot demonstrated an antiradical activity by scavenging 2,2′-azino-bis(3-ethylbenzthiazo-line-6-sulfonic acid) (ABTS) radical (Escribano et al. 1998). Indeed, betalains possess unsaturated double bonds and aromatic rings and are potential to quench free radicals (Fu et al. 2020). Due to its antioxidant properties, betalains can effectively decrease the pathogenesis of several diseases (Table 7.20) and prevent LDL oxidation by modulating the redox system, and subsequently creating a good cellular redox environment in modulating the oxidative stress in organisms (Zielinska-Przyjemska et al. 2009; Esatbeyoglu et al. 2014).

Betalains are not only possessed antioxidant properties, they are also exerted anti-inflammatory activity by suppressing the oxidants, hypochlorous acid scavenging, and COX generated by neutrophils during an inflammatory response (Reddy et al. 2005; Allegra et al. 2005). Administration of 50 mg/kg of beetroot betanin for 30 min intraperitoneally prior to ischemia of the superior mesenteric artery for 1 h, followed by the reperfusion for 1, 4, and 24 h in rats model revealed that betanin protects the lung parenchyma and jejunal mucosa and decreases the inflammatory cell density after intestinal ischemic-reperfusion injury (Toth et al. 2019) (Table 7.20). In human study, Tesoriere et al. (2005) evaluated the betalain pigments in red blood cells of healthy individuals after consuming a diet rich in cactus pear fruit. Betanin was shown in red blood cells isolated at 3 h after consuming a cactus pear fruit meal (Tesoriere et al. 2005). The study further revealed that red blood cells isolated at 3 h after intake of cactus pear fruit significantly delay the onset of an *ex vivo* cumene hydroperoxide-induced hemolysis, suggesting that betalains extracted from cactus pear fruits may combat oxidative stress (Tesoriere et al. 2005) (Table 7.20). Oxidative stress can also stimulate inflammatory response via the NF-κB signaling pathway (Mo et al. 2020). The previous study has demonstrated that administration of 100 mg/kg of betalain-rich beetroot subcutaneously or orally reduced carrageenan-induced paw edema, increased IL-10 levels, and decreased IL-1β levels, TNF-α, and carrageenan-induced superoxide anion, indicating that betalain-rich beetroot possesses anti-inflammatory potential on carrageenan-induced peritonitis and paw edema (Martinez et al. 2015).

Emerging studies have demonstrated that betalains exerted beneficial physiological activities in relation to atherosclerosis and hypertension (Fu et al. 2020). An animal study showed that feeding rats with 150 and 300 mg/kg betalain-rich beetroot juice for 28 days restored cardiac endogenous antioxidants and decreased nitrosative and oxidative stress in isoproterenol-induced myocardial injury rats (Raish et al. 2019). Betalain-rich beetroot juice has also been reported to ameliorate the edema, myonecrosis, and histological damage, upregulate Bcl-2 protein expression, and decrease MMP-9 protein levels (Raish et al. 2019). In this regard, this finding

Table 7.20 Effects of betalains in *in vitro*, *in vivo*, and human studies

Diseases	Experimental model	Source of betalains	Durations	Outcomes	References
Intestinal inflammation	Male specific pathogen-free Charles River Wistar rats	Beetroot betanin	50 mg/kg for 30 min	Decreased inflammatory cell density after intestinal ischemic-reperfusion injury	Toth et al. (2019)
Inflammation	Healthy volunteers	Cactus pear fruit meal	–	Delay the onset of an *ex vivo* cumene hydroperoxide-induced hemolysis	Tesoriere et al. (2005)
Myocardial ischemia	Rats	Betalain-rich beetroot juice	150 and 300 mg/kg for 28 days	Restored cardiac endogenous antioxidants and decreased nitrosative and oxidative stress	Raish et al. (2019)
Osteoarthritis	Osteoarthritis participants	Betalain-rich red beet extract	35–100 mg twice/day for 10 days	Improved discomfort associated with osteoarthritis	Pietrzkowski et al. (2010)
Joint mobility	40 people aged 35–75 years with moderate chronic muscle/joint related pain	Betalain-rich Nopal cactus fruit juice	3 oz/day for 8 weeks	Improved joint range of motion and mobility in lumbar, thoracic, and cervical regions	Jensen (2020)
CVD	Systematic review of 7 human and 9 animal studies	Betalain-rich cacti (cactus pear and dragon fruit)	Animal = varied from 1 to 500 mg/kg body weight between 3 and 60 days; human = varied from 25 mg to 3000 mg between 14 and 56 days	Improved endothelial function, reduced vascular stiffness, and modulated heart rate	Cheok et al. (2020)
Dyslipidemia	Male Wistar rats aged approximately 4 weeks	3% beetroot crisps rich in betalains	4 weeks	Reduced hepatic total cholesterol and serum glucose levels	Wroblewska et al. (2011)
Liver cancer	Rats	Red beetroot juice	8 mL/kg body weight/day for 28 days	Decreased the NDEA-induced DNA damage and liver injury	Krajka-Kuźniak et al. (2012)
Liver cancer	THLE-2 cells	Betalain	2, 10, and 20 μM	Induced Nrf2 activation and downstream phase II detoxification gene expression	Krajka-Kuźniak et al. (2013)

CVD cardiovascular disease, *NDEA* N-nitrosodiethylamine, *Nrf2* nuclear factor erythroid 2-related factor 2, *THLE-2* human liver cell line implies that betalain-rich beetroot juice alleviates cardiac structural damages and dysfunction by reducing apoptosis, inflammation, and oxidative stress in cardiac tissues (Raish et al. 2019). Consistent with the study reported by Raish et al. (2019), Wroblewska et al. (2011) also found that hepatic total cholesterol and serum glucose concentration were reduced after 4 weeks of 3% beetroot crisps rich in betalains administration in the hyperlipidemia mouse model. A recent systematic review of 7 human and 9 animal studies suggested that dietary intake of betalain-rich cacti (cactus pear and dragon fruit) (animal = varied from 1 to 500 mg/kg body weight between 3 and 60 days; human = varied from 25 mg to 3000 mg between 14 and 56 days) improved endothelial function, reduced vascular stiffness, and modulated heart rate in both populations at risk of CVD and healthy individuals (Cheok et al. 2020), suggesting that betalain-rich cacti may improve vascular health.

In addition, betalains have the potential to protect against osteoarthritis. Betalains from plant extract exert anti-inflammatory activity by inactivating COX and lipoxygenase (Vidal et al. 2014; Rahimi et al. 2019b). In human study, consumption of betalain-rich red beet extract at a dose of 35–100 mg twice/day for 10 days improved discomfort associated with osteoarthritis (Pietrzkowski et al. 2010). It has been suggested that mobility and joint function are negatively affected by inflammation (Chow and Chin 2020). Low-grade inflammation has been implicated in the development of chronic pain in muscles and joints (Christensen et al. 2021). Data from a double-blind, placebo-controlled study involving 40 people aged 35–75 years with moderate chronic muscle/joint-related pain revealed that consumption of 3 oz/day of betalain-rich Nopal cactus fruit juice for 8 weeks improved joint range of motion and mobility in lumbar, thoracic, and cervical regions (Jensen 2020). Notably, patients consuming betalain-rich Nopal cactus fruit juice relied less on pain medication to complete daily activities and improved some daily activities including lying, sitting, and walking, compared to those who consume placebo (Jensen 2020). The protective effects of betalain-rich Nopal cactus fruit juice toward physical and joint mobility functioning are likely due to the anti-inflammatory and antioxidant properties of betalain-rich Nopal cactus fruit juice. However, Jensen (2020) found that betalain-rich Nopal cactus fruit juice failed to show any benefit in improving the mobility of shoulders, knees, and hips in patients with chronic muscle/joint-related pain.

Notably, research evidence indicates that dietary intake of betalain-rich containing food may decrease the risk of cancer. Betalains in red beetroot are of benefit in the activation of Nrf2 transcriptional factor and phase II enzymes to trigger the antioxidant defense activity of endogenous cells (Lechner and Stoner 2019). An animal study demonstrated that feeding rats with 8 mL/kg body weight/day of red beetroot juice for 28 days decreased the N-nitrosodiethylamine (NDEA)-induced DNA damage and liver injury (Krajka-Kuźniak et al. 2012). Moreover, betalain (2, 10, and 20 μM) was found to trigger Nrf2 activation and downstream phase II detoxification gene expression in human liver (THLE-2) cells (Krajka-Kuźniak et al. 2013), indicates that betalain may exert anticarcinogenic and hepatoprotective activity. In colon cancer, Farabegoli et al. (2017) revealed that red beetroot betalains induced cytotoxicity of colon cancer (CaCo-2) cells by increased Bcl2-like protein 4, cleaved caspase-3 and poly ADP-ribosyl polymerase 1 levels, and dampened IL-8 and COX-2 mRNA expression in LPS-induced Caco-2 cells. Taken together, the effects mentioned above indicate that betalains from beetroot have the potential to protect against metabolic ailments and oxidative stress. Based on the evidence, the reported findings are mainly focused on the bioactivities of red beetroot extracts. Most of the nutritional values of red beetroot are also contributed by micronutrients and bioactive compounds such as betaine, ferulic acid, lutein, and betalains (Lechner and Stoner 2019). Thus, further studies are warranted to evaluate the bioactivities of individual betalains in daily betalain-rich diets.

7.12 Carnosine and Anserine

Carnosine was first identified as a dipeptide, β-alanyl-L-histidine in 1918 (Barger and Tutin 1918; Baumann and Ingvaldsen 1918). Carnosine consists of three ionizable groups, namely imidazole ring in histidine, amino group of the β-alanine residue, and carboxylic group (Tanokura et al. 1976) (Fig. 7.7). Under physiological pH, carnosine exists mainly as zwitterionic and exerts a net positive charge (Wu 2020). Dietary carnosine is absorbed through enterocytes of the small intestine across the apical membrane mediated by peptide transporter-1 (PepT1) (Wu 2020). In the enterocytes, a small level of carnosine is hydrolyzed into histidine and β-alanine by carnosinase-2 (Sadikali et al. 1975). Carnosine is actively hydrolyzed into the histidine and β-alanine by carnosinase-1, an enzyme that is released and synthesized from the liver, subsequently taken up by extra-intestinal cells through specific transporters. Indeed, dietary consumption of carnosine can increase its concentration in heart, brain, and skeletal muscle (Boldyrev et al. 2013). Intriguingly, the previous study found that intake of 4 g synthetic carnosine does not influence the plasma concentration in the human body, possibly due to the activity of plasma carnosinase-1 activity, in which it is rapidly hydrolyzed into histidine and β-alanine (Gardner et al. 1991). In general, carnosine is synthesized from histidine and β-alanine which catalyzed by ATP-dependent carnosine synthetase (Wu 2013); both of the amino acids are reused for the synthesis of carnosine by the olfactory bulb of the brain, heart, and skeletal muscle (Harding and O'Fallon 1979; Drozak et al. 2010; Boldyrev et al. 2013). Beta-alanine can be obtained from the diet or via endogenous syntheses from the catabolism of polyamines, pyrimidines, coenzyme A, malonic acid semialdehyde, and aspartate (Wu 2013); whereas histidine can also be found in the diet and the degradation of myosin, actin, hemoglobin, and other proteins in the body (Wu 2020). Compared to those who rarely consumption of meat, adults who consume 150 g chicken broth or beef had a relatively higher urinary carnosine concentration within 7 h after consumption was 15- and 13-fold, respectively (Yeum et al. 2010).

Carnosine contains a functional imidazole ring, which can donate hydrogen to free radicals and convert in the form of non-radical substances (Kohen et al. 1988). There are several physiological roles of carnosine that have been reported, specifically in organs and tissues, including protects against protein oxidation, lipid peroxidation, and the formation of lipoxidation end products and advanced protein glycation, scavenging peroxyl radicals and ROS, activation of muscle ATPase to provide energy, mediation of homeostasis and metal-ion chelation, and acting as an intracellular pH buffer (Boldyrev et al. 2013; Barca et al. 2019; Nelson et al. 2019).

Following the discovery of carnosine in beef, the researchers attempt to evaluate this compound in other animal species. In 1929, Ackermann et al. (1929) and Tolkatschevskaya (1929) identified a carnosine-like compound, namely methyl carnosine or β-alanyl-1-methyl-L-histidine, in goose skeletal muscle. Subsequently, this compound was named anserine after the taxonomic name for the goose. Similar to carnosine, this peptide is characterized by three ionizable groups, namely imidazole ring in histidine, amino group of the β-alanine residue, and carboxylic group (Bertinaria et al. 2011) (Fig. 7.7). Under physiological pH, anserine exists mainly as zwitterionic and has a net positive charge (Wu 2020). Most of the anserine in the human diet comes from beef (Wu et al. 2016) and fish such as trout, tuna, and salmon (Boldyrev et al. 2013), but it is absent from human tissues, for instance, brain, heart, and skeletal muscle (Mannion et al. 1992). Notably, anserine has a physiological function similar to carnosine (Kohen et al. 1988; Boldyrev et al. 2013; Everaert et al. 2019). Nonetheless, some of the biochemical properties of anserine are different from carnosine. For instance, anserine did not chelate copper and may not modulate the availability of nitric oxide in cells (Boldyrev et al. 2013).

Anserine and carnosine are potent antioxidants (Wu et al. 2003) and are widely found in beef skeletal muscle. In particular, anserine and carnosine are absent from all plant sources such as white rice, wheat flour, sweet potatoes, soybeans, peanuts, and corn grains (Hou et al. 2019). In this regard, vegetarians were shown to have a greater likelihood of deficiencies of anserine and carnosine, especially to those physically active individuals (Rogerson 2017). Substantial studies indicate that anserine and carnosine play a pivotal role in protecting mammalian cells from injury and oxidative stress (Abplanalp et al. 2019; Alkhatib et al. 2020).

Research evidence demonstrated that carnosine's precursors, L-histidine and β-alanine, can easily be taken up into the brain from circulation via amino acid transporters in the blood-brain barrier (Hawkins et al. 2018). Besides carnosine's precursors, carnosine can also cross the blood-brain barrier (Jin et al. 2005). Brain carnosine is vitally important for the proper modulation of brain function and may serve as a neuroprotective molecule, neuromodulator, and endogenous antioxidant (Berezhnoy et al. 2019). Carnosine is believed to have a neuroprotective effect by improving the acetylcholinesterase activity (Ma et al. 2015), strengthening the antioxidant system (Maiese et al. 2007), and decreasing oxidative stress (Colín-Barenque et al. 2018). Data from systematic review and meta-analysis including 454 animals revealed that carnosine exerts neuroprotective efficacy in decreasing infarct size (Davis et al. 2016). The data showed that administration of 1000 mg/kg of carnosine led to a 38.1% reduction in infarct size, suggesting that carnosine may protect against brain ischemia (Davis et al. 2016) (Table 7.21). One potential biological mechanism linking neuroinflammation and carnosine has been described by Wang et al. (2018). Wang et al. (2018) demonstrated that dietary intake of 600 mg/kg beef decoction rich in carnosine for 7 days reduced neurological deficits,

Table 7.21 Effects of carnosine, anserine, and its precursors in *in vivo* and human studies

Age-related diseases	Experimental model	Carnosine/Anserine and its precursors	Concentration/Durations	Outcomes	References
Alzheimer's disease	Aged AβPPswe/PSEN1dE9 Alzheimer's-model mice	Anserine	10 mg for 8 weeks	Inhibited glial inflammatory response, modulated pericyte coverage on endothelial cells in the brain, and improved memory deficits	Kaneko et al. (2017)
Ischemic stroke	Systematic review and meta-analysis including 454 animals	Carnosine	1000 mg/kg	Reduced 38.1% infarct size	Davis et al. (2016)
Neurodegenerative disorder	12 middle-aged subjects (8 men and 4 women)	β-alanine	2.4 g/day for 28 days	Reduced exercise-induced declines in executive functions	Furst et al. (2018)
Neurodegenerative disorder	7 healthy vegetarians (3 women and 4 men) and 19 trained male cyclists	β-alanine	6.4 g/day for 28 days	No effect on brain carnosine/homocarnosine signal in vegetarians and omnivores and the cognitive function after or before exercise in trained cyclists	Solis et al. (2015)
Ischemic stroke	Sprague-Dawley rats	Beef decoction	600 mg/kg for 7 days	Reduced neurological deficits	Wang et al. (2018)
Atherosclerosis	Apolipoprotein E-null mice	Carnosine	60 mg/kg in drinking water for 6 weeks	Reduced atherosclerotic lesion formation in aortic valves	Barski et al. (2013)
Type 2 diabetes mellitus	Randomized, double-blinded, placebo-controlled trial of 100 obese women with metabolic syndrome aged 33–51 years	Histidine	4 g/day for 12 weeks	Reduced insulin resistance	Feng et al. (2013)
Type 2 diabetes mellitus	Male C57BL/KsJm/Leptdb (db/db) mice	Anserine	3 doses of 100 mg/kg every other day intravenously	Decreased proteinuria and blood glucose levels and improved vascular permeability	Peters et al. (2018)

decreased the proinflammatory cytokines including interferon-γ, TNF-α, and IL-6 expression, and increased anti-inflammatory cytokine such as IL-4 levels in rats. In human study, an increase in exercise capacity and reduced exercise-induced declines in executive functions was found in middle-aged individuals supplemented with 2.4 g/day of β-alanine for 28 days (Furst et al. 2018). As mentioned above, β-alanine is one of the carnosine's precursors. A study by Hoffman et al. (2015) further supported the role of β-alanine in the modulation of the thalamus, hypothalamus, amygdala, hippocampus, and cerebral cortex in the rodent model. Improved intracranial carnosine was correlated with better executive functioning (Furst et al. 2018). Nonetheless, Solis et al. (2015) did not identify the association between β-alanine and neurodegenerative disorders. The data showed that intake of 6.4 g/day β-alanine for 28 days did not affect the brain carnosine/homocarnosine signal in vegetarians and omnivores as well as the cognitive function after or before exercise in trained cyclists (Solis et al. 2015). In addition, the neuroprotective effect was also shown in anserine (Schön et al. 2019; Masuoka et al. 2021). The data from animal study revealed that feeding AβPPswe/PSEN1dE9 mice with 10 mg of anserine for 8 weeks inhibited the glial inflammatory response, modulated pericyte coverage on endothelial cells in the brain, and improved memory deficits in aged mice (Kaneko et al. 2017) (Table 7.21). This finding implies that anserine may improve spatial memory and alleviate neurovascular dysfunction in aged animals.

A study by Yeum et al. (2010) evaluated the systemic delivery of anserine, carnosine, and histidine dipeptides in humans after intake of food rich in histidine dipeptides. A significant increase of plasma anserine concentration among healthy women was found to be associated with 150 g chicken or chicken broth from 150 g chicken (Yeum et al. 2010). However, the plasma carnosine levels were not detected after intake of chicken broth from 150 g chicken, 150 g chicken, 150 g beef, or 450 mg pure carnosine (Yeum et al. 2010). The undetectable carnosine in the blood is likely due to the rapid uptake of dipeptides into the tissues or hydrolysis by carnosinase (Yeum et al. 2010). The study further revealed that intake of 150 g beef increased the urinary concentration of carnosine and anserine (Yeum et al. 2010). Similarly, the urinary levels of anserine and carnosine were also increased after consuming chicken broth from 150 g chicken. This finding indicates that dietary histidine-dipeptides are rapidly hydrolyzed by carnosinase in plasma and excreted through urine may serve as a reactive carbonyl species sequestering agent. In another study, Park et al. (2005) found that the serum carnosine levels were increased after 15 min from 0 to 46.0 mM/L of plasma in participants who received beef patties, and reaching the highest after 3.5 h beef consumption. The ability of carnosine to absorb intactly suggests that carnosine could be a crucial dietary intervention to suppress advanced glycation end products (AGEs) formation (Freund et al. 2018). A recent study by Cripps et al. (2017) revealed that carnosine is effective in scavenging the glucolipotoxic free radicals, for instance, RONS. It has been suggested that carnosine can regulate glucose homeostasis by modulating skeletal muscle glucose uptake and increasing insulin secretion (Cripps et al. 2017). Feng et al. (2013) exploring the impact of serum histidine on oxidative stress and insulin resistance in randomized, double-blinded, placebo-controlled trials. The study showed that obese women with metabolic syndrome who received 4 g/day of histidine for 12 weeks decreased GPx, SOD, IL-6, TNF-α, fat mass, waist circumference, BMI, and HOMA-IR and increased adiponectin and serum histidine, implied that histidine may improve insulin resistance by suppressing proinflammatory cytokine, possibly via NF-κB signaling pathway in adipocytes (Feng et al. 2013) (Table 7.21). Moreover, intravenous injections of male C57BL/KsJm/Leptdb (db/db) mice with anserine (3 doses of 100 mg/kg every other day) decreased proteinuria and blood glucose levels and improved vascular permeability (Peters et al. 2018). Heat shock protein (Hsp70) deficiency is often associated with stress in the target organs of diabetic complications (Bellini et al. 2017). The inability of the diabetic glomeruli to modulate stress response may increase the susceptibility of the diabetic lesion (Barutta et al. 2008). Intriguingly, the *in vitro* study demonstrated that anserine (1 mM) is only effective to stimulate the Hsp70-expression in H_2O_2-stressed human tubular cells (HK-2), but it is not observed in carnosine (Peters et al. 2018).

Besides its effects mentioned above, a beneficial effect of carnosine has also been described on the incidence of CVD. Carnosine inhibits Cu^{2+}-induced LDL oxidation (Decker et al. 2001). Consistent with the study reported by Decker et al. (2001), Barski et al. (2013) also found that carnosine (60 mg/kg in drinking water for 6 weeks) reduced atherosclerotic lesion formation in aortic valves and alleviated the accumulation of protein-4-hydroxynonenal adducts, protein-4-hydroxyhexenal, and protein-acrolein in apolipoprotein E-null mice, possibly by modulating aldehyde removal from atherosclerotic lesions. Further, a systematic review and meta-analysis of randomized controlled trials included 21 studies from inception to January 25, 2019, demonstrated that a relatively low level of TG for those with a greater intake of carnosine supplements compared to the control (Menon et al. 2021). However, no effects were observed in the total cholesterol to HDL-C ratio, HDL-C, LDL-C, heart rate, and blood pressure after consuming carnosine (Menon et al. 2021). Based on the evidence, intakes of carnosine and anserine might be a promising strategy to improve age-related diseases conditions and ameliorate oxidative stress. However, there was limited evidence on the implication of the dietary consumption of carnosine and anserine-rich containing food on age-related diseases. Therefore, the beneficial effects of carnosine and anserine-rich food on pathological changes of mitochondrial oxidative stress related to chronic disease conditions are worth further elucidation in randomized clinical trials.

References

Abe I, Seki T, Umehara K et al (2000) Green tea polyphenols: novel and potent inhibitors of squalene epoxidase. Biochem Biophys Res Commun 268:767–771

Abe K, Hori Y, Myoda T (2020) Volatile compounds of fresh and processed garlic. Exp Ther Med 19:1585–1593

Abokyi S, To C-H, Lam TT et al (2020) Central role of oxidative stress in age-related macular degeneration: evidence from a review of the molecular mechanisms and animal models. Oxidative Med Cell Longev 2020, Article ID 7901270, 19 pages

Abplanalp W, Haberzettl P, Bhatnagar A et al (2019) Carnosine supple-mentation mitigates the deleterious effects of particulate matter exposure in mice. J Am Heart Assoc 8:e013041

Ackermann D, Timpe O, Poller K (1929) Uber das Anserin, einen neun Bestandteil der Vogelmuskulatur. Z Physiol Chem 183:1–10

Adham AN, Abdelfatah S, Naqishbandi AM et al (2021) Cytotoxicity of apigenin toward multiple myeloma cell lines and suppression of iNOS and COX-2 expression in STAT1-transfected HEK293 cells. Phytomedicine 80:153371

Aditi A, Graham DY (2012) Vitamin C, gastritis, and gastric disease: a historical review and update. Dig Dis Sci 57:2504–2515

Age-Related Eye Disease Study 2 Research G (2013) Lutein+zeaxanthin and omega-3 fatty acids for age-related macular degeneration: the age-related eye disease study 2 (AREDS2) randomized clinical trial. JAMA 309:2005–2015

Ahmad P, Alvi SS, Khan MS (2019) Functioning of organosulfur compounds from garlic (*Allium sativum* Linn) in targeting risk factor-mediated atherosclerosis: a cross talk between alternative and modern medicine. In: Akhtar M, Swamy M, Sinniah U (eds) Natural bioactive compounds. Springer, Singapore, pp 561–585

Ahrazem O, Gómez-Gómez L, Rodrigo MJ et al (2016) Carotenoid cleavage oxygenases from microbes and photosynthetic organisms: features and functions. Int J Mol Sci 17:1781

Ahsan H, Ahad A, Siddiqui WA (2015) A review of characterization of tocotrienols from plant oils and foods. J Chem Biol 8:45–59

Aizawa K, Liu C, Tang S et al (2016) Tobacco carcinogen induces both lung cancer and non-alcoholic steatohepatitis and hepatocellular carcinomas in ferrets which can be attenuated by lycopene supplementation. Int J Cancer 139:1171–1181

Akhlaghi M, Nasab MG, Riasatian M et al (2020) Soy isoflavones prevent bone resorption and loss, a systematic review and meta-analysis of randomized controlled trials. Crit Rev Food Sci Nutr 60:2327–2341

Akkarachiyasit S, Charoenlertkul P, Yibchok-Anun S et al (2010) Inhibitory activities of cyanidin and its glycosides and synergistic effect with acarbose against intestinal alpha-glucosidase and pancreatic alpha-amylase. Int J Mol Sci 11:3387–3396

Akoumi A, Haffar T, Mousterji M et al (2017) Palmitate mediated diacylglycerol accumulation causes endoplasmic reticulum stress, Plin2 degradation, and cell death in H9C2 cardiomyoblasts. Exp Cell Res 354:85–94

Akram S, Rehman F (2018) Hardness in drinking-water, its sources, its effects on humans and its household treatment. J Chem Appl 4:4

Akram NA, Shafiq F, Ashraf M (2017) Ascorbic acid-A potential oxidant scavenger and its role in plant development and abiotic stress tolerance. Front Plant Sci 8:613

Al Saadi T, Assaf Y, Farwati M et al (2021) Coenzyme Q10 for heart failure. Cochrane Database Syst Rev 2:CD008684

Alho H, Lonnrot K (2000) Coenzyme Q supplementation and longevity. In: Kagan VE, Quinn PJ (eds) Coenzyme Q: molecular mechanisms in health and disease. CRC Press, Boca Raton, pp 371–380

Ali SS, Ahsan H, Zia MK et al (2020) Understanding oxidants and antioxidants: classical team with new players. J Food Biochem 44:e13145

Alisio A, Cañas F, de Bronia DH et al (1997) Effect of vitamin D deficiency on lipid composition and calcium transport in basolateral membrane vesicles from chick intestine. Biochem Mol Biol Int 42:339–347

Alkhatib A, Feng W-H, Huang Y-J et al (2020) Anserine reverses exercise-induced oxidative stress and preserves cellular homeostasis in healthy men. Nutrients 12:1146

Alkreathy HM (2020) Potential anticancer effects of aged garlic extract and its water-soluble organosulfur compounds. J Pharm Res Int 32:108–121

Allegra M, Furtmüller PG, Jantschko W et al (2005) Mechanism of interaction of betanin and indicaxanthin with human myeloperoxidase and hypochlorous acid. Biochem Biophys Res Commun 332:837–844

Alumkal JJ, Slottke R, Schwartzman J et al (2015) A phase II study of sulforaphane-rich broccoli sprout extracts in men with recurrent prostate cancer. Investig New Drugs 33:480–489

Alvi SA, Ansaria IA, Ahmad MK et al (2017) Lycopene amends LPS induced oxidative stress and hypertriglyceridemia via modulating PCSK-9 expression and Apo-CIII mediated lipoprotein lipase activity. Biomed Pharmacother 96:1082–1093

Amagase H, Petesch BL, Matsuura H et al (2001) Intake of garlic and its bioactive components. J Nutr 131:955s–962s

American Heart Association (2019) Cardiovascular disease and diabetes. https://www.heart.org/en/health-topics/diabetes/why-diabetes-matters/cardiovascular-disease-diabetes. Accessed 9 Oct 2019

Ames BN (2018) Prolonging healthy aging: longevity vitamins and proteins. Proc Natl Acad Sci U S A 115:10836–10844

Aneja R, Hake PW, Burroughs TJ et al (2004) Epigallocatechin, a green tea polyphenol, attenuates myocardial ischemia reperfusion injury in rats. Mol Med 10:55–62

Anita T, Suwandono A, Ariyanti I et al (2017) Effect of consuming tomato (*Lycopersium comune*) juice in lowering blood pressure in pregnant mothers with hypertension. Belitung Nurs J 3:707–711

Aouadi R, Aouidet A, Elkadhi A et al (2000) Effect of fresh garlic (*allium sativum*) on lipid metabolism in male rats. Nutr Res 20:273–280

Ardura-Fabregat A, Boddeke EWGM, Boza-Serrano A et al (2017) Targeting neuroinflammation to treat Alzheimer's disease. CNS Drugs 31:1057–1082

Arslan S, Kadayifcilar S, Samur G (2019) The potential role of dietary antioxidant capacity in preventing age-related macular degeneration. J Am Coll Nutr 38:424–432

Ashor AW, Lara J, Mathers JC et al (2014) Effect of vitamin C on endothelial function in health and disease: a systematic review and meta-analysis of randomised controlled trials. Atherosclerosis 235:9–20

Aune D, Keum N, Giovannucci E et al (2018) Dietary intake and blood concentrations of antioxidants and the risk of cardiovascular disease, total cancer, and all-cause mortality: a systematic review and dose-response meta-analysis of prospective studies. J Clin Nutr 108:1069–1091

Ayton S, Lei P, Appukuttan AT et al (2020) Brain zinc deficiency exacerbates cognitive decline in the R6/1 model of Huntington's disease. Neurotherapeutics 17:243–251

Baillie N, Carr A, Peng S (2018) The use of intravenous vitamin C as a supportive therapy for a patient with glioblastoma multiforme. Antioxidants 7:115

Bajic VP, Van Neste C, Obradovic M et al (2019) Glutathione "redox homeostasis" and its relation to cardiovascular disease. Oxidative Med Cell Longev 2019, Article ID 5028181, 14 pages

Balk EM, Adam GP, Langberg VN et al (2017) Global dietary calcium intake among adults: a systematic review. Osteoporos Int 28:3315–3324

Ballatori N, Krance SM, Notenboom S et al (2009) Glutathione dysregulation and the etiology and progression of human diseases. Biol Chem 390:191–214

Bao XW, Han XX, Du GM et al (2019) Antioxidant activities and immunomodulatory effects in mice of betalain *in vivo*. Food Sci Chin 40:196–201

Barca A, Ippati S, Urso E et al (2019) Carnosine modulates the Sp1-Slc31a1/Ctr1 copper-sensing system and influences copper homeostasis in murine CNS-derived cells. Am J Physiol Cell Physiol 316:C235–C245

Barger G, Tutin F (1918) Carnosine, constitution and synthesis. Biochem J 12:402–407

Barman S, Srinivasan K (2017) Attenuation of oxidative stress and cardioprotective effects of zinc supplementation in experimental diabetic rats. Br J Nutr 117:335–350

Barp J, Araujo AS, Fernandes TR et al (2002) Myocardial antioxidant and oxidative stress changes due to sex hormones. Braz J Med Biol Res 35:1075–1081

Barski OA, Xie Z, Baba SP et al (2013) Dietary carnosine prevents early atherosclerotic lesion formation in apolipoprotein E-null mice. Arterioscler Thromb Vasc Biol 33:1162–1170

Barsony J, Sugimura Y, Verbalis JG (2011) Osteoclast response to low extracellular sodium and the mechanism of hyponatremia-induced bone loss. J Biol Chem 286:10864–10875

Bartolini D, Dallaglio K, Torquato P et al (2018) Nrf2-p62 autophagy pathway and its response to oxidative stress in hepatocellular carcinoma. Transl Res 193:54–71

Barutta F, Pinach S, Giunti S et al (2008) Heat shock protein expression in diabetic nephropathy. Am J Physiol Renal Physiol 295:F1817–F1824

Batra P, Sharma AK (2013) Anti-cancer potential of flavonoids: recent trends and future perspectives. 3 Biotech 3:439–459

Baumann L, Ingvaldsen T (1918) Concerning histidine and carnosine. The synthesis of carnosine. J Biol Chem 35:263–276

Bazela K, Solyga-Zurek A, Debowska R et al (2014) L-Ergothioneine protects skin cells against UV-induced damage—a preliminary study. Cosmetics 1:51–60

Beaver LM, Lȯhr C, Clarke JD et al (2018) Broccoli sprouts delay prostate cancer formation and decrease prostate cancer severity with a concurrent decrease in HDAC3 protein expression in transgenic adenocarcinoma of the mouse prostate (TRAMP) mice. Curr Dev Nutr 2:nzy002

Beelman RB, Kalaras MD, John P et al (2019) Micronutrients and bioactive compounds in mushrooms: a recipe for healthy aging? Nutr Today 54:16–22

Beelman RB, Kalaras MD, Phillips AT et al (2020) Is ergothioneine a 'longevity vitamin' limited in the American diet? J Nutr Sci 9:e52

Bella DL, Hahn C, Stipanuk MH (1999) Effects of nonsulfur and sulfur amino acids on the regulation of hepatic enzymes of cysteine metabolism. Am J Phys 277:E144–E153

Bellini S, Barutta F, Mastrocola R et al (2017) Heat shock proteins in vascular diabetic complications: review and future perspective. Int J Mol Sci 18:2709

Belovic MM, Girones-Vilaplana A, Moreno DA et al (2016) Tomato (*Solanum lycopersicum* L.) processing main product (juice) and by-product (pomace) bioactivity potential measured as antioxidant activity and angiotensin converting enzyme inhibition. J Food Process Pres 40:1229–1237

Berezhnoy DS, Stvolinsky SL, Lopachev AV et al (2019) Carnosine as an effective neuroprotector in brain pathology and potential neuromodulator in normal conditions. Amino Acids 51:139–150

Bertinaria M, Rolando B, Giorgis M et al (2011) Synthesis, physicochemical characterization and biological activities of new carno-sine derivatives stable in human serum as potential neuroprotective agents. J Med Chem 54:611–621

Bhatia N, Singh B, Koul A (2018) Lycopene treatment stalls the onset of experimentally induced hepatocellular carcinoma: a radioisotopic, physiological and biochemical analysis. J Hepatoma Res 4:1–16

Bhattacharjee A, Chakraborty K, Shukla A (2017) Cellular copper homeostasis: current concepts on its interplay with glutathione homeostasis and its implication in physiology and human diseases. Metallomics 9:1376–1388

Bigham M, Mohammadipour A, Hosseini M et al (2021) Neuroprotective effects of garlic extract on dopaminergic neurons of substantia nigra in a rat model of Parkinson's disease: motor and non-motor outcomes. Metab Brain Dis 36:927–937

Bjelakovic G, Nikolova D, Gluud LL et al (2012) Antioxidant supplements for prevention of mortality in healthy participants and patients with various diseases. Cochrane Database Syst Rev 3:CD007176

Bjelakovic G, Nikolova D, Gluud C (2013) Meta-regression analyses, meta-analyses, and trial sequential analyses of the effects of supplementation with beta-carotene, vitamin A, and vitamin E singly or in different combinations on all-cause mortality: do we have evidence for lack of harm? PLoS One 8:e74558

Boccardi V, Baroni M, Mangialasche F et al (2016) Vitamin E family: role in the pathogenesis and treatment of Alzheimer's disease. Alzheimers Dement Transl Res Clin Interv 2:182–191

Bøhn SK, Myhrstad MC, Thoresen M et al (2010) Blood cell gene expression associated with cellular stress defense is modulated by antioxidant-rich food in a randomised controlled clinical trial of male smokers. BMC Med 8:54

Bohnen NI, Grothe MJ, Ray NJ et al (2018) Recent advances in cholinergic imaging and cognitive decline—revisiting the cholinergic hypothesis of dementia. Curr Geriatr Rep 7:1–11

Boldyrev AA, Aldini G, Derave W (2013) Physiology and pathophysiology of carnosine. Physiol Rev 93:1803–1845
Bondonno NP, Bondonno CP, Blekkenhorst LC et al (2018) Flavonoid-rich apple improves endothelial function in individuals at risk for cardiovascular disease: a randomized controlled clinical trial. Mol Nutr Food Res 62:1700674
Bonnet U (2019) The sour side of vitamin C might mediate neuroprotective, anticonvulsive and antidepressant-like effects. Med Hypotheses 131:109320
Borghetti G, von Lewinski D, Eaton DM et al (2018) Diabetic cardiomyopathy: current and future therapies. Beyond glycemic control. Front Physiol 9:1514
Borodina I, Kenny LC, McCarthy CM et al (2020) The biology of ergothioneine, an antioxidant nutraceutical. Nutr Res Rev 33:190–217
Borowiec K, Michalak A (2021) Flavonoids from edible fruits as therapeutic agents in neuroinflammation – a comprehensive review and update. Crit Rev Food Sci Nutr 1–19
Boudina S, Abel ED (2010) Diabetic cardiomyopathy, causes and effects. Rev Endocr Metab Disord 11:31–39
Brasier AR (2010) The nuclear factor-κB–interleukin-6 signalling pathway mediating vascular inflammation. Cardiovasc Res 86:211–218
Breithaupt DE, Alpmann A, Carrière F (2007) Xanthophyll esters are hydrolysed in the presence of recombinant human pancreatic lipase. Food Chem 103:651–656
Bristow SM, Horne AM, Gamble GD et al (2019) Dietary calcium intake and bone loss over 6 years in osteopenic postmenopausal women. J Clin Endocrinol Metab 104:3576–3584
Britton G, Liaaen-Jensen S, Pfander H (1998) Carotenoids, volume 3: Biosynthesis and metabolism. Birkhäuser, Basel
Britton G, Liaaen-Jensen S, Pfander H (2008) Carotenoids, volume 4: Natural functions. Birkhäuser, Basel
Brownlee M (2001) Biochemistry and molecular cell biology of diabetic complications. Nature 414:813–820
Bruno RS, Leonard SW, Atkinson J et al (2006) Faster plasma vitamin E disappearance in smokers is normalized by vitamin C supplementation. Free Radic Biol Med 40:689–697
Buijsse B, Jacobs DR, Steffen LM et al (2015) Plasma ascorbic acid, a Priori Diet Quality Score, and incident hypertension: a prospective cohort study. PLoS One 10:e0144920
Bule M, Abdurahman A, Nikfar S et al (2019) Antidiabetic effect of quercetin: a systematic review and meta-analysis of animal studies. Food Chem Toxicol 125:494–502
Bungau S, Abdel-Daim MM, Tit DM et al (2019) Health benefits of polyphenols and carotenoids in age-related eye diseases. Oxidative Med Cell Longev 2019, Article ID 9783429, 22 pages
Butterfield DA, Halliwell B (2019) Oxidative stress, dysfunctional glucose metabolism and Alzheimer disease. Nat Rev Neurosci 20:148–160
Cabello-Hurtado F, Gicquel M, Esnault M-A (2012) Evaluation of the antioxidant potential of cauliflower (*Brassica oleracea*) from a glucosinolate content perspective. Food Chem 132:1003–1009
Calabrese G, Morgan B, Riemer J (2017) Mitochondrial glutathione: regulation and functions. Antioxid Redox Signal 27:1162–1177
Calvo MS, Mehrotra A, Beelman RB et al (2016) A retrospective study in adults with metabolic syndrome: diabetic risk factor response to daily consumption of *Agaricus bisporus* (White button mushrooms). Plant Foods Hum Nutr 71:245–251
Cameron E, Campbell A (1974) The orthomolecular treatment of cancer. II. Clinical trial of high-dose ascorbic acid supplements in advanced human cancer. Chem Biol Interact 9:285–315
Cameron E, Pauling L (1976) Supplemental ascorbate in the supportive treatment of cancer: prolongation of survival times in terminal human cancer. Proc Natl Acad Sci U S A 73:3685–3689
Cammerer MA, Gonçalves SC, de Araujo GN et al (2018) The effects of a flavonoid-rich diet on oxidative stress, inflammation, and lipid profile after elective percutaneous coronary intervention: a randomized clinical trial. Prev Nutr Food Sci 23:108–114
Cao Z, Wang P, Gao X et al (2019) Lycopene attenuates aluminum-induced hippocampal lesions by inhibiting oxidative stress-mediated inflammation and apoptosis in the rat. J Inorg Biochem 193:143–151

Carr AC, Maggini S (2017) Vitamin C and immune function. Nutrients 9:1211
Carretero J, Obrador E, Anasagasti MJ et al (1999) Growth-associated changes in glutathione content correlate with liver metastatic activity of B16 melanoma cells. Clin Exp Metastasis 17:567–574
Carretero J, Obrador E, Esteve JM et al (2001) Tumoricidal activity of endothelial cells: inhibition of endothelial nitric oxide production abrogates tumor cytotoxicity induced by hepatic sinusoidal endothelium in response to B16 melanoma adhesion *in vitro*. J Biol Chem 276:25775–25782
Carrillo-López A, Yahia EM (2014) Changes in color-related compounds in tomato fruit exocarp and mesocarp during ripening using HPLC-APcI$^+$-mass spectrometry. J Food Sci Technol 51:2720–2726
Catalona WJ, Smith DS, Ratliff TL et al (1991) Measurement of prostate-specific antigen in serum as a screening test for prostate cancer. N Engl J Med 324:1156–1161
Cebova M, Pechanova O (2020) Protective effects of polyphenols against ischemia/reperfusion injury. Molecules 25:3469
Cha JH, Kim WK, Ha AW et al (2017) Anti-inflammatory effect of lycopene in SW480 human colorectal cancer cells. Nutr Res Pract 11:90–96
Chabosseau P, Rutter GA (2016) Zinc and diabetes. Arch Biochem Biophys 611:79–85
Chan M-Y, Lee B-J, Chang P-S et al (2020) The risks of ubiquinone and β-carotene deficiency and metabolic disorders in patients with oral cancer. BMC Cancer 20:310
Chasapis CT, Ntoupa P-SA, Spiliopoulou CA et al (2020) Recent aspects of the effects of zinc on human health. Arch Toxicol 94:1443–1460
Cheah IK, Halliwell B (2012) Ergothioneine; antioxidant potential, physiological function and role in disease. Biochim Biophys Acta Mol basis Dis 1822:784–793
Cheah IK, Feng L, Tang RMY et al (2016) Ergothioneine levels in an elderly population decrease with age and incidence of cognitive decline; a risk factor for neurodegeneration? Biochem Biophys Res Commun 478:162–167
Cheah IK, Ng L-T, Ng L-F et al (2019) Inhibition of amyloid-induced toxicity by ergothioneine in a transgenic *Caenorhabditis elegans* model. FEBS Lett 593:2139–2150
Chen L, Jia RH, Qiu CJ et al (2005) Hyperglycemia inhibits the uptake of dehydroascorbate in tubular epithelial cell. Am J Nephrol 25:459–465
Chen G-C, Lu D-B, Pang Z et al (2013) Vitamin C intake, circulating vitamin C and risk of stroke: a meta-analysis of prospective studies. J Am Heart Assoc 2:e000329
Chen W, Guo C, Jia Z et al (2020) Inhibition of mitochondrial ROS by MitoQ alleviates white matter injury and improves outcomes after intracerebral haemorrhage in mice. Oxidative Med Cell Longev 2020, Article ID 8285065, 12 pages
Chen X, Huang C, Sun H et al (2021) Puerarin suppresses inflammation and ECM degradation through Nrf2/HO-1 axis in chondrocytes and alleviates pain symptom in osteoarthritic mice. Food Funct 12:2075–2089
Cheng S, Larson MG, McCabe EL et al (2015) Distinct metabolomic signatures are associated with longevity in humans. Nat Commun 6:6791
Cheng HM, Koutsidis G, Lodge JK et al (2017) Tomato and lycopene supplementation and cardiovascular risk factors: a systematic review and meta-analysis. Atherosclerosis 257:100–108
Cheok A, George TW, Rodriguez-Mateos A et al (2020) The effects of betalain-rich cacti (dragon fruit and cactus pear) on endothelial and vascular function: a systematic review of animal and human studies. Food Funct 11:6807–6817
Chiavarini M, Naldini G, Fabiani R (2020) The role of diet in osteoporotic fracture healing: a systematic review. Curr Osteoporos Rep 18:138–147
Chin H-K, Horng C-T, Liu Y-S et al (2018) Kaempferol inhibits angiogenic ability by targeting VEGF receptor-2 and downregulating the PI3K/AKT, MEK and ERK pathways in VEGF-stimulated human umbilical vein endothelial cells. Oncol Rep 39:2351–2357
Ching B, Chew SF, Ip YK (2015) Ascorbate synthesis in fishes: a review. IUBMB Life 67:69–76
Chitchumroonchokchai C, Bomser JA, Glamm JE et al (2004) Xanthophylls and α-tocopherol decrease UVB-induced lipid peroxidation and stress signaling in human lens epithelial cells. J Nutr 134:3225–3232

Choi H, Park H-H, Koh S-H et al (2012) Coenzyme Q10 protects against amyloid beta-induced neuronal cell death by inhibiting oxidative stress and activating the PI3K pathway. Neurotoxicology 33:85–90

Chou ST, Tseng ST (2017) Oxidative stress markers in type 2 diabetes patients with diabetic nephropathy. Clin Exp Nephrol 21:283–292

Chow YY, Chin K-Y (2020) The role of inflammation in the pathogenesis of osteoarthritis. Mediat Inflamm 2020, Article ID 8293921, 19 pages

Christensen K, Gleason CE, Mares JA (2020) Dietary carotenoids and cognitive function among US adults, NHANES 2011–2014. Nutr Neurosci 23:554–562

Christensen JO, Nilsen KB, Hopstock LA et al (2021) Shift work, low-grade inflammation, and chronic pain: a 7-year prospective study. Int Arch Occup Environ Health 94:1013–1022

Chu A, Foster M, Samman S (2016) Zinc status and risk of cardiovascular diseases and type 2 diabetes mellitus—a systematic review of prospective cohort studies. Nutrients 8:707

Chu C-C, Wu W-S, Shieh J-P et al (2017) The anti-inflammatory and vasodilating effects of three selected dietary organic sulfur compounds from *Allium* Species. J Funct Biomater 8:5

Chung RWS, Leandersson P, Lundberg AK et al (2017) Lutein exerts anti-inflammatory effects in patients with coronary artery disease. Atherosclerosis 262:87–93

Circu ML, Aw TY (2012) Glutathione and modulation of cell apoptosis. Biochim Biophys Acta, Mol Cell Res 1823:1767–1777

Cirmi S, Ferlazzo N, Lombardo GE et al (2016) Neurodegenerative diseases: might citrus flavonoids play a protective role? Molecules 21:1312

Cirmi S, Navarra M, Woodside JV et al (2018) Citrus fruits intake and oral cancer risk: a systematic review and meta-analysis. Pharmacol Res 133:187–194

Cirmi S, Maugeri A, Lombardo GE et al (2021) A flavonoid-rich extract of mandarin juice counteracts 6-OHDA-induced oxidative stress in SH-SY5Y cells and modulates Parkinson-related genes. Antioxidants 10:539

Colín-Barenque L, Bizarro-Nevares P, Villalva AG et al (2018) Neuroprotective effect of carnosine in the olfactory bulb after vanadium inhalation in a mouse model. Int J Exp Pathol 99:180–188

Collins JK, Arjmandi BH, Claypool PL et al (2004) Lycopene from two food sources does not affect antioxidant or cholesterol status of middle-aged adults. Nutr J 3:15

Cook NR, Albert CM, Gaziano JM et al (2007) A randomized factorial trial of vitamins C and E and beta carotene in the secondary prevention of cardiovascular events in women: results from the Women's Antioxidant Cardiovascular Study. Arch Intern Med 167:1610–1618

Cooney RV, Dai Q, Gao Y-T et al (2011) Low plasma coenzyme Q(10) levels and breast cancer risk in Chinese women. Cancer Epidemiol Biomark Prev 20:1124–1130

Cos P, Ying L, Calomme M et al (1998) Structure–activity relationship and classification of flavonoids as inhibitors of xanthine oxidase and superoxide scavengers. J Nat Prod 61:71–76

Costa d'Avila J, Siqueira LD, Mazeraud A et al (2018) Age-related cognitive impairment is associated with long-term neuroinflammation and oxidative stress in a mouse model of episodic systemic inflammation. J Neuroinflammation 15:28

Costa-Rodrigues J, Pinho O, Monterio PRR (2018) Can lycopene be considered an effective protection against cardiovascular disease? Food Chem 245:1148–1153

Counotte G, Holzhauer M, Carp-van Dijken S et al (2019) Levels of trace elements and potential toxic elements in bovine livers: a trend analysis from 2007 to 2018. PLoS One 14:e0214584

Crane FL (2001) Biochemical functions of coenzyme Q10. J Am Coll Nutr 20:591–598

Creagan ET, Moertel CG, O'Fallon JR et al (1979) Failure of high-dose vitamin c (ascorbic acid) therapy to benefit patients with advanced cancer. A controlled trial. N Engl J Med 301:687–690

Cripps MJ, Hanna K, Lavilla C et al (2017) Carnosine scavenging of glucolipotoxic free radicals enhances insulin secretion and glucose uptake. Sci Rep 7:13313

Cristóvão JS, Santos R, Gomes CM (2016) Metals and neuronal metal binding proteins implicated in Alzheimer's disease. Oxidative Med Cell Longev 2016, Article ID 9812178, 13 pages

Crowe-White KM, Voruganti VS, Talevi V et al (2020) Variation of serum lycopene in response to 100% watermelon juice: an exploratory analysis of genetic variants in a randomized controlled crossover study. Curr Dev Nutr 4:nzaa102

Cruz KJC, de Oliveira ARS, do Nascimento Marreiro D (2015) Antioxidant role of zinc in diabetes mellitus. World J Diabetes 6:333–337

Cruz KJC, Morais JBS, de Oliveira ARS et al (2017) The effect of zinc supplementation on insulin resistance in obese subjects: a systematic review. Biol Trace Elem Res 176:239–243

Cui J, Yuan X, Wang L et al (2011) Recent loss of vitamin C biosynthesis ability in bats. PLoS One 6:e27114

Cui L, Xu F, Wang M et al (2019) Dietary natural astaxanthin at an early stage inhibits *N*-nitrosomethylbenzylamine–induced esophageal cancer oxidative stress and inflammation via downregulation of NFκB and COX2 in F344 rats. Oncol Targets Ther 12:5087–5096

Cutler GJ, Nettleton JA, Ross JA et al (2008) Dietary flavonoid intake and risk of cancer in postmenopausal women: the Iowa Women's Health Study. Int J Cancer 123:664–671

Czech A, Zarycka E, Yanovych D et al (2020) Mineral content of the pulp and peel of various citrus fruit cultivars. Biol Trace Elem Res 193:555–563

da Costa Pereira Soares N, de Barros EM, Machado CL et al (2019) Comparative analysis of lycopene content from different tomato-based food products on the cellular activity of prostate cancer cell lines. Foods 8:201

Dacosta C, Bao Y (2017) The role of MicroRNAs in the chemopreventive activity of sulforaphane from cruciferous vegetables. Nutrients 9:902

Dalgaard F, Bondonno NP, Murray K et al (2019) Associations between habitual flavonoid intake and hospital admissions for atherosclerotic cardiovascular disease: a prospective cohort study. Lancet Planet Health 3:e450–e459

Daniel G (2011) Ascorbate exerts anti-proliferative effects through cell cycle inhibition and sensitizes tumor cells towards cytostatic drugs. Cancer Chemother Pharmacol 67:1157–1166

Datla KP, Christidou M, Widmer WW et al (2001) Tissue distribution and neuroprotective effects of citrus flavonoid tangeretin in a rat model of Parkinson's disease. Neuroreport 12:3871–3875

Davey MW, Van Montagu M, Inzé D et al (2000) Plant L-ascorbic acid: chemistry, function, metabolism, bioavailability and effects of processing. J Sci Food Agric 80:825–860

Davis CK, Laud PJ, Bahor Z et al (2016) Systematic review and stratified meta-analysis of the efficacy of carnosinein animal models of ischemic stroke. J Cereb Blood Flow Metab 36:1686–1694

de Barboza GD, Guizzardi S, de Talamoni NT (2015) Molecular aspects of intestinal calcium absorption. World J Gastroenterol 21:7142–7154

de Barcelos IP, Haas RH (2019) CoQ10 and aging. Biol 8:28

De Benedictis CA, Vilella A, Grabrucker AM (2019) The role of trace metals in Alzheimer's disease. In: Wisniewski T (ed) Alzheimer's disease. Codon Publications, Brisbane, p 31895516

de Castro Peixoto J, Neves BJ, Vasconcelos FG et al (2019) Flavonoids from Brazilian Cerrado: biosynthesis, chemical and biological profile. Molecules 24:2891

Decker EA, Ivanov V, Zhu B-Z et al (2001) Inhibition of low-density lipoprotein oxidation by carnosine and histidine. J Agric Food Chem 49:511–516

Deding U, Baatrup G, Christensen LP et al (2020) Carrot intake and risk of colorectal cancer: a prospective cohort study of 57,053 Danes. Nutrients 12:332

Delrobaei F, Fatemi I, Shamsizadeh A et al (2019) Ascorbic acid attenuates cognitive impairment and brain oxidative stress in ovariectomized mice. Pharmacol Rep 71:133–138

Desai G, Schelske-Santos M, Nazario CM et al (2020) Onion and garlic intake and breast cancer, a case-control study in Puerto Rico. Nutr Cancer 72:791–800

Desari S, Ganjayi MS, Yellanurkonda P et al (2018) Role of glutathione S-transferases in detoxification of a polycyclic aromatic hydrocarbon, methylcholanthrene. Chemico Biol Interact 294:81–90

Desideri E, Ciccarone F, Ciriolo MR (2019) Targeting glutathione metabolism: partner in crime in anticancer therapy. Nutrients 11:1926

Devi S, Kumar V, Singh SK et al (2021) Flavonoids: potential candidates for the treatment of neurodegenerative disorders. Biomedicines 9:99

Devore EE, Kang JH, Stampfer MJ et al (2013) The association of antioxidants and cognition in the Nurses' Health study. Am J Epidemiol 177:33–41
Di Lorenzo A, Iannuzzo G, Parlato A et al (2020) Clinical evidence for Q10 coenzyme supplementation in heart failure: from energetics to functional improvement. J Clin Med 9:1266
Diomede F, Marconi GD, Guarnieri S et al (2020) A novel role of ascorbic acid in anti-inflammatory pathway and ROS generation in HEMA treated dental pulp stem cells. Materials 13:130
Djuric Z, Ren J, Mekhovich O et al (2006) Effects of high fruit-vegetable and/or low-fat intervention on plasma micronutrient levels. J Am Coll Nutr 25:178–187
Dludla PV, Nyambuya TM, Orlando P et al (2020) The impact of coenzyme Q10 on metabolic and cardiovascular disease profiles in diabetic patients: a systematic review and meta-analysis of randomized controlled trials. Endocrinol Diabetes Metab 3:e00118
do Nascimento Marreiro D, Cruz KJC, Morais JBS et al (2017) Zinc and oxidative stress: current mechanisms. Antioxidants 6:24
Dou W, Zhang J, Li H et al (2014) Plant flavonol isorhamnetin attenuates chemically induced inflammatory bowel disease via a PXR-dependent pathway. J Nutr Biochem 25:923–933
Dowell FJ, Hamilton CA, McMurray J et al (1993) Effects of a xanthine oxidase/hypoxanthine free radical and reactive oxygen species generating system on endothelial function in New Zealand white rabbit aortic rings. J Cardiovasc Pharmacol 22:792–797
Drisko JA, Serrano OK, Spruce LR et al (2018) Treatment of pancreatic cancer with intravenous vitamin C: a case report. Anti-Cancer Drugs 29:373–379
Drouin G, Godin J-R, Pagé B (2011) The genetics of vitamin C loss in vertebrates. Curr Genomics 12:371–378
Drozak J, Veiga-da-Cunha M, Vertommen D et al (2010) Molecular identification of carnosine synthase as ATP-grasp domain-containing protein 1 (ATPGD1). J Biol Chem 285:9346–9356
Du J, Cullen JJ, Buettner GR (2012) Ascorbic acid: chemistry, biology and the treatment of cancer. Biochim Biophys Acta 1826:443–457
Dükel M, Tavsan Z, Kayali HA (2021) Flavonoids regulate cell death-related cellular signaling via ROS in human colon cancer cells. Process Biochem 101:11–25
Dumont M, Kipiani K, Yu F et al (2011) Coenzyme Q10 decreases amyloid pathology and improves behavior in a transgenic mouse model of Alzheimer's disease. J Alzheimers Dis 27:211–223
Dziedzic T (2015) Systemic inflammation as a therapeutic target in acute ischemic stroke. Expert Rev Neurother 15:523–531
Edefonti V, Hashibe M, Parpinel M et al (2015) Natural vitamin C intake and the risk of head and neck cancer: a pooled analysis in the International Head and Neck Cancer Epidemiology Consortium. Int J Cancer 137:448–462
Eisenhauer B, Natoli S, Liew G et al (2017) Lutein and zeaxanthin—food sources, bioavailability and dietary variety in age-related macular degeneration protection. Nutrients 9:120
El-Ashmony SMA, Morsi HK, Abdelhafez AM (2012) Effect of zinc supplementation on glycemic control, lipid profile, and renal functions in patients with type II diabetes: a single blinded, randomized, placebo-controlled, trial. J Biol Agric Health 2:33
Elbeltagy MAF, Elkholy WB, Salman AS (2019) Effect of atherosclerosis and the protective effect of the antioxidant vitamin E on the rabbit cerebellum. Microscopy 68:369–378
Elhady SS, Eltamany EE, Shaaban AE et al (2020) Jaceidin flavonoid isolated from *Chiliadenus montanus* attenuates tumor progression in mice via VEGF inhibition: *in vivo* and *in silico* studies. Plan Theory 9:1031
Ellulu MS, Rahmat A, Ismail P et al (2015) Effect of vitamin C on inflammation and metabolic markers in hypertensive and/or diabetic obese adults: a randomized controlled trial. Drug Des Devel Ther 9:3405–3412
Ellwanger JH, Molz P, Dallemole DR et al (2015) Selenium reduces bradykinesia and DNA damage in a rat model of Parkinson's disease. Nutrition 31:359–365
Elsaed WM, Mohamed HA (2017) Dietary zinc modifies diabetic-induced renal pathology in rats. Ren Fail 39:246–257

Elvy Suhana MR, Fairus A, Norazlina M et al (2018) Protective effects of palm tocotrienol against glucocorticoid induced osteoporosis via regulation of gene expressions. Med Heal 13:175–197
Emamat H, Tangestani H, Totmaj AS et al (2020) The effect of garlic on vascular function: a systematic review of randomized clinical trials. Clin Nutr 39:3563–3570
Engelhart MJ, Geerlings MI, Ruitenberg A et al (2002) Dietary intake of antioxidants and risk of Alzheimer disease. JAMA 287:3223–3229
Ershow AG, Skeaff SA, Merkel JM et al (2018) Development of databases on iodine in foods and dietary supplements. Nutrients 10:100
Esatbeyoglu T, Wagner AE, Motafakkerazad R et al (2014) Free radical scavenging and antioxidant activity of betalain: electron spin resonance spectroscopy studies and studies in cultured cells. Food Chem Toxicol 73:119–126
Escribano J, Pedreno MA, Garcia-Carmona F et al (1998) Characterization of the antiradical activity of batalains from *Beta vulgaris L.* roots. Phytochem Anal 9:124–127
Espinosa-Díez C, Miguel V, Vallejo S et al (2018) Role of glutathione biosynthesis in endothelial dysfunction and fibrosis. Redox Biol 14:88–99
Estrela JM, Mena S, Obrador E et al (2017) Polyphenolic phytochemicals in cancer prevention and therapy: bioavailability versus bioefficacy. J Med Chem 60:9413–9436
Evans M, Baisley J, Barss S et al (2009) A randomized, double-blind trial on the bioavailability of two CoQ10 formulations. J Funct Foods 1:65–73
Everaert I, Baron G, Barbaresi S et al (2019) Development and validation of a sensitive LC–MS/MS assay for the quantification of anserine in human plasma and urine and its application to pharmacokinetic study. Amino Acids 51:103–114
Ey J, Schomig E, Taubert D (2007) Dietary sources and antioxidant effects of ergothioneine. J Agric Food Chem 55:6466–6474
Fahey RC (2001) Novel thiols of prokaryotes. Annu Rev Microbiol 55:333–356
Falabella M, Vernon HJ, Hanna MG et al (2021) Cardiolipin, mitochondria, and neurological disease. Trends Endocrinol Metab 32:P224–P237
Fang J-Y, Lin C-H, Huang T-H et al (2019) *In vivo* rodent models of type 2 diabetes and their usefulness for evaluating flavonoid bioactivity. Nutrients 11:530
Farabegoli F, Scarpa ES, Frati A et al (2017) Betalains increase vitexin-2-O-xyloside cytotoxicity in CaCo-2 cancer cells. Food Chem 218:356–364
Farhood HB, Aljabery RN, Majid A (2019) The study of oxidant-antioxidant status in type 2 diabetes mellitus. J Phys Conf Ser 1294:052037
Fathy SM, Drees EA (2016) Protective effects of Egyptian cloudy apple juice and apple peel extract on lipid peroxidation, antioxidant enzymes and inflammatory status in diabetic rat pancreas. BMC Complement Altern Med 16:8
Favero A, Salvini S, Russo A et al (1997) Sources of macro- and micronutrients in Italian women: results from a food frequency questionnaire for cancer studies. Eur J Cancer Prev 6:277–287
Favero G, Paganelli C, Buffoli B et al (2014) Endothelium and its alterations in cardiovascular diseases: life style intervention. Biomed Res Int 2014, Article ID 801896, 28 pages
Feng RN, Niu YC, Sun XW et al (2013) Histidine supplementation improves insulin resistance through suppressed inflammation in obese women with the metabolic syndrome: a randomised controlled trial. Diabetologia 56:985–994
Feng L, Nie K, Jiang H et al (2019a) Effects of lutein supplementation in age-related macular degeneration. PLoS One 14:e0227048
Feng L, Cheah IK-M, Ng MM-X et al (2019b) The association between mushroom consumption and mild cognitive impairment: a community-based cross-sectional study in Singapore. J Alzheimers Dis 68:197–203
Fernández-López JA, Fernández-Lledó V, Angosto JM (2020) New insights into red plant pigments: more than just natural colorants. RSC Adv 10:24669–24682
Fiedor J, Burda K (2014) Potential role of carotenoids as antioxidants in human health and disease. Nutrients 6:466–488
Figueroa A, Wong A, Jaime SJ et al (2017) Influence of L-citrulline and watermelon supplementation on vascular function and exercise performance. Curr Opin Clin Nutr Metab Care 20:92–98

Finley JW, Gao S (2017) A perspective on *crocus sativus* L. (saffron) constituent crocin: a potent water-soluble antioxidant and potential therapy for Alzheimer's disease. J Agric Food Chem 65:1005–1020

Fischer R, Maier O (2015) Interrelation of oxidative stress and inflammation in neurodegenerative disease: role of TNF. Oxidative Med Cell Longev 2015, Article ID 610813, 18 pages

Flanagan E, Müller M, Hornberger M et al (2018) Impact of flavonoids on cellular and molecular mechanisms underlying age-related cognitive decline and neurodegeneration. Curr Nutr Rep 7:49–57

Folch J, Ettcheto M, Petrov D et al (2018) Review of the advances in treatment for Alzheimer disease: strategies for combating β-amyloid protein. Neurologia 33:47–58

Forster MJ, Dubey A, Dawson KM et al (1996) Age-related losses of cognitive function and motor skills in mice are associated with oxidative protein damage in the brain. Proc Natl Acad Sci U S A 93:4765–4769

Frank J, Chin XWD, Schrader C et al (2012) Do tocotrienols have potential as neuroprotective dietary factors? Ageing Res Rev 11:163–180

Fraser PD, Enfissi EMA, Halket JM et al (2007) Manipulation of phytoene levels in tomato fruit: effects on isoprenoids, plastids, and intermediary metabolism. Plant Cell 19:3194–3211

Frederickson CJ (1989) Neurobiology of zinc and zinc-containing neurons. Int Rev Neurobiol 31:145–238

Frei B, Birlouez-Aragon I, Lykkesfeldt J (2012) Authors' perspective: what is the optimum intake of vitamin C in humans? Crit Rev Food Sci Nutr 52:815–829

Freund MA, Chen B, Decker EA (2018) The inhibition of advanced glycation end products by carnosine and other natural dipeptides to reduce diabetic and age-related complications. Compr Rev Food Sci Food Saf 17:1367–1378

Fu Y, Shi J, Xie S-Y et al (2020) Red beetroot betalains: perspectives on extraction, processing, and potential health benefits. J Agric Food Chem 68:11595–11611

Fujii R, Yamada H, Munetsuna E et al (2019) Associations between dietary vitamin intake, ABCA1 gene promoter DNA methylation, and lipid profiles in a Japanese population. Am J Clin Nutr 110:1213–1219

Furst T, Massaro A, Miller C et al (2018) β-Alanine supplementation increased physical performance and improved executive function following endurance exercise in middle aged individuals. J Int Soc Sports Nutr 15:32

Fuss JO, Tsai C-L, Ishida JP et al (2015) Emerging critical roles of Fe–S clusters in DNA replication and repair. Biochim Biophys Acta, Mol Cell Res 1853:1253–1271

Gaede P, Poulsen HE, Parving HH et al (2001) Double-blind, randomized study of the effect of combined treatment with vitamin C and E on albuminuria in type 2 diabetic patients. Diabet Med 18:756–760

Galarraga V, Boffetta P (2016) Coffee drinking and risk of lung cancer—a meta-analysis. Cancer Epidemiol Biomark Prev 25:951–957

Gamcsik MP, Kasibhatla MS, Teeter SD et al (2012) Glutathione levels in human tumors. Biomarkers 17:671–691

Ganai AA, Farooqi H (2015) Bioactivity of genistein: a review of *in vitro* and *in vivo* studies. Biomed Pharmacother 76:30–38

Ganger R, Garla R, Mohanty BP et al (2016) Protective effects of zinc against acute arsenic toxicity by regulating antioxidant defense system and cumulative metallothionein expression. Biol Trace Elem Res 169:218–229

Gao S, Qin T, Liu Z et al (2011) Lutein and zeaxanthin supplementation reduces H_2O_2-induced oxidative damage in human lens epithelial cells. Mol Vis 17:3180–3190

Gao X, Wei K, Hu B et al (2019) Ascorbic acid induced HepG2 cells' apoptosis via intracellular reductive stress. Theranostics 9:4233–4240

Gardeazabal I, Romanos-Nanclares A, Martinez-Gonzalez MA et al (2019) Total polyphenol intake and breast cancer risk in the Seguimiento Universidad de Navarra (SUN) cohort. Br J Nutr 122:542–551

Gardner ML, Illingworth KM, Kelleher J et al (1991) Intestinal absorption of the intact peptide carnosine in man, and comparison with intestinal permeability to lactulose. J Physiol 439:411–422
Garrow JS, James WPT, Ralph A (2001) Phytoprotectants. In: Human nutrition and dietetics. Churchill Livingstone, Edinburgh, pp 417–425
Gaudiano G, Koch TH, LoBello M et al (2000) Lack of glutathione conjugation to adriamycin in human breast cancer MCF-7/DOX cells. Inhibition of glutathione S-transferase p1-1 by glutathione conjugates from anthracyclines. Biochem Pharmacol 60:1915–1923
Geldenhuys WJ, Benkovic SA, Lin L et al (2017) MitoNEET (CISD1) knockout mice show signs of striatal mitochondrial dysfunction and a Parkinson's disease phenotype. ACS Chem Neurosci 8:2759–2765
Gharibzahedi SMT, Jafari SM (2017) The importance of minerals in human nutrition: bioavailability, food fortification, processing effects and nanoencapsulation. Trends Food Sci Technol 62:119–132
Glutsch V, Hamm H, Goebeler M (2019) Zinc and skin: an update. JDDG 17:589–596
Godwin AK, Meister A, O'Dwyer PJ et al (1992) High resistance to cisplatin in human ovarian cancer cell lines is associated with marked increase of glutathione synthesis. Proc Natl Acad Sci U S A 89:3070–3074
Golden MHN, Waterlow JC (1977) Total protein synthesis in elderly people: a comparison of results with [^{15}N]glycine and [^{14}C]leucine. Clin Sci Mol Med 53:277–288
Gong Y, Yang Y (2020) Activation of Nrf2/AREs-mediated antioxidant signalling, and suppression of profibrotic TGF-β1/Smad3 pathway: a promising therapeutic strategy for hepatic fibrosis—a review. Life Sci 256:117909
Gong X, Smith JR, Swanson HM et al (2018) Carotenoid lutein selectively inhibits breast cancer cell growth and potentiates the effect of chemotherapeutic agents through ROS-mediated mechanisms. Molecules 23:905
Goñi I, Hernández-Galiot A (2019) Intake of nutrient and non-nutrient dietary antioxidants. Contribution of macromolecular antioxidant polyphenols in an elderly Mediterranean population. Nutrients:11–2165
Grammenandi K, Kyriazi M, Katsarou-Katsari A et al (2016) Low-molecular-weight hydrophilic and lipophilic antioxidants in nonmelanoma skin carcinomas and adjacent normal-looking skin. Skin Pharmacol Physiol 29:324–331
Grassi D, Mulder TP, Draijer R et al (2009) Black tea consumption dose-dependently improves flow-mediated dilation in healthy males. J Hypertens 27:774–781
Grimble RF, Jackson AA, Persaud C et al (1992) Cysteine and glycine supplementation modulate the metabolic response to tumor necrosis factor alpha in rats fed a low protein diet. J Nutr 122:2066–2073
Grotto D, Carneiro MFH, de Castro MM et al (2018) Long-term excessive selenium supplementation induces hypertension in rats. Biol Trace Elem Res 182:70–77
Grune T, Schröder P, Biesalski HK (2005) Low molecular weight antioxidants. In: Grune T (ed) Reactions, processes. The handbook of environmental chemistry. Springer, Berlin/Heidelberg, pp 77–90
Grune T, Lietz G, Palou A et al (2010) β-carotene is an important vitamin A source for humans. J Nutr 140:2268S–2285S
Gu Y, Han J, Jiang C et al (2020) Biomarkers, oxidative stress and autophagy in skin aging. Ageing Res Rev 59:101036
Guay SP, Brisson D, Munger J et al (2012) *ABCA1* gene promoter DNA methylation is associated with HDL particle profile and coronary artery disease in familial hypercholesterolemia. Epigenetics 7:464–472
Guay SP, Légaré C, Houde AA et al (2014) Acetylsalicylic acid, aging and coronary artery disease are associated with ABCA1 DNA methylation in men. Clin Epigenetics 6:14
Guest J, Grant R (2016) Carotenoids and neurobiological health. In: The benefits of natural products for neurodegenerative diseases, vol 12. Springer, Cham, pp 199–228

Gunawardena HP, Silva R, Sivakanesan R et al (2019) Poor glycaemic control is associated with increased lipid peroxidation and glutathione peroxidase activity in type 2 diabetes patients. Oxidative Med Cell Longev 2019, Article ID 9471697, 10 pages

Gupta P, Bansal MP, Koul A (2013) Lycopene modulates initiation of N-nitrosodiethylamine induced hepatocarcinogenesis: studies on chromosomal abnormalities, membrane fluidity and antioxidant defense system. Chem Biol Interact 206:364–374

Habib E, Linher-Melville K, Lin H-X et al (2015) Expression of xCT and activity of system xc(−) are regulated by NRF2 in human breast cancer cells in response to oxidative stress. Redox Biol 5:33–42

Hafez HA, Kamel MA, Osman MY et al (2021) Ameliorative effects of astaxanthin on brain tissues of alzheimer's disease-like model: cross talk between neuronal-specific microRNA-124 and related pathways. Mol Cell Biochem 476:2233–2249

Hajizadeh-Sharafabad F, Ghoreishi Z, Maleki V et al (2019) Mechanistic insights into the effect of lutein on atherosclerosis, vascular dysfunction, and related risk factors: a systematic review of *in vivo, ex vivo* and *in vitro* studies. Pharmacol Res 149:104477

Halabi IE, Bejjany R, Nasr R et al (2018) Ascorbic acid in colon cancer: from the basic to the clinical applications. Int J Mol Sci 19:2752

Halbwirth H (2010) The creation and physiological relevance of divergent hydroxylation patterns in the flavonoid pathway. Int J Mol Sci 11:595–621

Halliwell B, Gutteridge JMC (1989) Free radicals in biology and medicine. Clarendon Press, Oxford

Han Y, Tang X, Zhang Y et al (2021) The current status of biotechnological production and the application of a novel antioxidant ergothioneine. Crit Rev Biotechnol 41:580–593

Hano C, Tungmunnithum D (2020) Plant polyphenols, more than just simple natural antioxidants: oxidative stress, aging and age-related diseases. Medicines 7:26

Hao L, Sun Q, Zhong W et al (2018) Mitochondria-targeted ubiquinone (MitoQ) enhances acetaldehyde clearance by reversing alcohol-induced posttranslational modification of aldehyde dehydrogenase 2: a molecular mechanism of protection against alcoholic liver disease. Redox Biol 14:626–636

Harding JW, O'Fallon JV (1979) The subcellular distribution of carnosine, carnosine synthetase, and carnosinase in mouse olfactory tissues. Brain Res 173:99–109

Hargreavas IP, Sheena Y, Land JM et al (2005) Glutathione deficiency in patients with mitochondrial disease: implications for pathogenesis and treatment. J Inherit Metab Dis 28:81–88

Hargreaves I, Heaton RA, Mantle D (2020) Disorders of human coenzyme Q10 metabolism: an overview. Int J Mol Sci 21:6695

Harman D (1956) Aging: a theory based on free radical and radiation chemistry. J Gerontol 2:298–300

Harman D (2006) Free radical theory of aging: an update. Increasing the functional life span. Ann N Y Acad Sci 1067:10–21

Harrison FE, May JM (2009) Vitamin C function in the brain: vital role of the ascorbate transporter SVCT2. Free Radic Biol Med 46:719–730

Hasan WNW, Chin K-Y, Abd Ghafar N et al (2020) Annatto-derived tocotrienol promotes mineralization of MC3T3-E1 cells by enhancing BMP-2 protein expression via inhibiting RhoA activation and HMG-CoA reductase gene expression. Drug Design Dev Ther 14:969–976

Hassan AS, Bunick D, St Denis SH et al (1993) Glutathione and bile acid synthesis. II. Effect of hepatic glutathione content on the activity and mRNA levels of cholesterol 7α-hydroxylase in the rat. Biochem Pharmacol 46:555–556

Hatano T, Saiki S, Okuzumi A et al (2016) Identification of novel biomarkers for Parkinson's disease by metabolomic technologies. J Neurol Neurosurg Psychiatry 87:295–301

Hawkins RA, O'Kane RL, Simpson IA et al (2018) Structure of the blood-brain barrier and its role in the transport of amino acids. J Nutr 136:218S–226S

Hazafa A, Rehman K-U, Jahan N et al (2020) The role of polyphenol (flavonoids) compounds in the treatment of cancer cells. Nutr Cancer 72:386–397

He Y, Xia P, Jin H et al (2016) Lycopene ameliorates transplant arteriosclerosis in vascular allograft transplantation by regulating the NO/cGMP pathways and Pho-associated kinases expression. Oxidative Med Cell Longev 2016, Article ID 3128280, 9 pages

Heart Protection Study Collaborative Group (2002) MRC/BHF Heart Protection Study of antioxidant vitamin supplementation in 20,536 high-risk individuals: a randomised placebo-controlled trial. Lancet 360:23–33

Hermsdorff HHM, Barbosa KBF, Volp ACP et al (2012) Vitamin C and fibre consumption from fruits and vegetables improves oxidative stress markers in healthy young adults. Br J Nutr 107:1119–1127

Hernández-Camacho JD, Bernier M, López-Lluch G et al (2018) Coenzyme Q10 supplementation in aging and disease. Front Physiol 9:44

Hertog MG, Feskens EJ, Hollman PC et al (1993) Dietary antioxidant flavonoids and risk of coronary heart disease: the Zutphen Elderly Study. Lancet 342:1007–1011

Higashi Y, Noma K, Yoshizumi M et al (2009) Endothelial function and oxidative stress in cardiovascular diseases. Circ J 73:411–418

Hirsch IB, Atchley DH, Tsai E et al (1998) Ascorbic acid clearance in diabetic nephropathy. J Diabetes Complicat 12:259–263

Hodge AM, Calvo MS (2019) Do bioactive components in non-animal food sources contribute to the beneficial health effect of a Japanese dietary pattern? Public Health Nutr 22:2469–2471

Hoffer LJ, Robitaille L, Zakarian R et al (2015) High-dose intravenous vitamin C combined with cytotoxic chemotherapy in patients with advanced cancer: a phase I-II clinical trial. PLoS One 10:0120228

Hoffman JR, Ostfeld I, Stout JR et al (2015) Beta-alanine supplemented diets enhance behavioral resilience to stress exposure in an animal model of PTSD. Amino Acids 47:1247–1257

Holley AS, Harding SA, Sasse A et al (2016) Reduced glutathione peroxidase activity predicts increased cardiovascular risk following an acute coronary syndrome. Int Cardiovasc Forum J 6:61–65

Holmgren A (1981) Regulation of ribonucleotide reductase. Curr Top Cell Regul 19:47–76

Hosono T, Sato A, Nakaguchi N et al (2020) Diallyl trisulfide inhibits platelet aggregation through the modification of sulfhydryl groups. J Agric Food Chem 68:1571–1578

Hossain MZ, Gilbert SF, Patel K et al (2013) Biological clues to potent DNA-damaging activities in food and flavoring. Food Chem Toxicol 55:557–567

Hou YQ, He WL, Hu SD et al (2019) Composition of polyamines and amino acids in plant-source foods for human consumption. Amino Acids 51:1153–1165

Hseu Y-C, Gowrisankar YV, Chen X-Z et al (2020) The antiaging activity of ergothioneine in UVA-irradiated human dermal fibroblasts via the inhibition of the AP-1 pathway and the activation of Nrf2-mediated antioxidant genes. Oxidative Med Cell Longev 2020, Article ID 2576823, 13 pages

Hu MY, Li YI, Jiang CH et al (2008) Comparison of lycopene and fluvastatin effects on atherosclerosis induced by a high-fat diet in rabbits. Nutrition 24:1030–1038

Hu Y, Yanhong S, Jing Z et al (2009) Synthesis and biological evaluation of novel flavonoid derivatives as dual binding acetylcholinesterase inhibitors. J Enzyme Inhib Med Chem 24:372–380

Hu JY, Zhang DL, Liu XL et al (2017) Pathological concentration of zinc dramatically accelerates abnormal aggregation of full-length human Tau and thereby significantly increases Tau toxicity in neuronal cells. Biochim Biophys Acta Mol basis Dis 1863:414–427

Hua Y-F, Wang G-Q, Jiang W et al (2016) Vitamin C intake and pancreatic cancer risk: a meta-analysis of published case-control and cohort studies. PLoS One 11:e0148816

Huang C, Yin Z (2020) Highly efficient synthesis of glutathione via a genetic engineering enzymatic method coupled with yeast ATP generation. Catalysts 10:33

Huang ZZ, Chen C, Zeng Z et al (2001) Mechanism and significance of increased glutathione level in human hepatocellular carcinoma and liver regeneration. FASEB J 15:19–21

Hughes KC, Gao X, Kim IY et al (2016) Intake of antioxidant vitamins and risk of Parkinson's disease. Mov Disord 31:1909–1914

Hui C, Qi X, Qianyong Z et al (2013) Flavonoids, flavonoid subclasses and breast cancer risk: a meta-analysis of epidemiologic studies. PLoS One 8:e54318
Humphries L, Vivian B, Stuart M et al (1989) Zinc deficiency and eating disorders. J Clin Psychiatry 50:456–459
Hung CF, Huang TF, Chen BH et al (2008) Lycopene inhibits TNF-alpha-induced endothelial ICAM-1 expression and monocyte-endothelial adhesion. Eur J Pharmacol 586:275–282
Hussain T, Tan B, Murtaza G et al (2020) Flavonoids and type 2 diabetes: evidence of efficacy in clinical and animal studies and delivery strategies to enhance their therapeutic efficacy. Pharmacol Res 152:104629
Ibrahim SRM, Mohamed GA, Alshali KZ et al (2018) Lipoxygenase inhibitors flavonoids from *Cyperus rotundus* aerial parts. Rev Bras Farmacogn 28:320–324
Ibrahim SAZ, Kerkadi A, Agouni A (2019) Selenium and health: an update on the situation in the Middle East and North Africa. Nutrients 11:1457
Ide T, Tsutsui H, Ohashi N et al (2002) Greater oxidative stress in healthy young men compared with premenopausal women. Arterioscler Thromb Vasc Biol 22:438–442
Imaeda N, Tokudome Y, Ikeda M et al (1999) Foods contributing to absolute intake and variance in intake of selected vitamins, minerals and dietary fiber in middle-aged Japanese. J Nutr Sci Vitaminol (Tokyo) 45:519–522
Imran M, Saeed F, Gilani SA et al (2021) Fisetin: an anticancer perspective. Food Sci Nutr 9:3–16
Ip BC, Hu KQ, Liu C et al (2013) Lycopene metabolite, apo-10′-lycopenoic acid, inhibits diethylnitrosamine-initiated, high fat diet-promoted hepatic inflammation and tumorigenesis in mice. Cancer Prev Res (Phila) 6:1304–1316
Ismail M, Alsalahi A, Iman MU et al (2020) Safety and neuroprotective efficacy of palm oil and tocotrienol-rich fraction from palm oil: a systematic review. Nutrients 12:521
Itakura Y, Ichikawa M, Mori Y et al (2001) How to distinguish garlic from the other Allium vegetables. J Nutr 131:963s–967s
Ivey KL, Hodgson JM, Croft KD et al (2015) Flavonoid intake and all-cause mortality. Am J Clin Nutr 101:1012–1020
Jaeger BN, Parylak SL, Gage FH (2018) Mechanisms of dietary flavonoid action in neuronal function and neuroinflammation. Mol Asp Med 61:50–62
Jaganathan SK, Vellayappan MV, Narasimhan G et al (2014) Chemopreventive effect of apple and berry fruits against colon cancer. World J Gastroenterol 20:17029–17036
Jager A, Saaby L (2011) Flavonoids and the CNS. Molecules 16:1471–1485
Jahoor F, Wykes LJ, Reeds PJ et al (1995) Protein-deficient pigs cannot maintain reduced glutathione homeostasis when subjected to the stress of inflammation. J Nutr 125:1462–1472
Jahoor F, Taffet GE, Sekhar RV (2019) Glutathione deficiency and oxidative stress in aging: metabolic mechanism and targeted intervention. Innov Aging 3:S416
Jang JH, Aruoma OI, Jen LS et al (2004) Ergothioneine rescues PC12 cells from beta-amyloid-induced apoptotic death. Free Radic Biol Med 36:288–299
Janiszewska E (2014) Microencapsulated beetroot juice as a potential source of betalain. Powder Technol 264:190–196
Jardim BV, Moschetta MG, Leonel C et al (2013) Glutathione and glutathione peroxidase expression in breast cancer: an immunohistochemical and molecular study. Oncol Rep 30:1119–1128
Jayedi A, Rashidy-Pour A, Parohan M et al (2019) Dietary and circulating vitamin C, vitamin E, β-carotene and risk of total cardiovascular mortality: a systematic review and dose–response meta-analysis of prospective observational studies. Public Health Nutr 22:1872–1887
Jensen GS (2020) Improved joint mobility associated with reduced inflammation related to consumption of Nopal cactus fruit juice: results from a placebo-controlled trial using Digital Inclinometry to Objectively Document Mobility of All Major Joints. Clin Interv Aging 15:2341–2352
Jeon Y-D, Lee J-H, Lee Y-M et al (2020) Puerarin inhibits inflammation and oxidative stress in dextran sulfate sodium-induced colitis mice model. Biomed Pharmacother 124:109847

Jeremic JN, Jakovljevic VL, Zivkovic VI et al (2020) Garlic derived diallyl trisulfide in experimental metabolic syndrome: metabolic effects and cardioprotective role. Int J Mol Sci 21:9100
Ji X, Hu X, Zou C et al (2017) Vitamin C deficiency exacerbates diabetic glomerular injury through activation of transforming growth factor-β signaling. Biochim Biophys Acta 1861:2186–2195
Jiang Q (2014) Natural forms of vitamin E: metabolism, antioxidant and anti-inflammatory activities and the role in disease prevention and therapy. Free Radic Biol Med 72:76–90
Jiang Q (2017) Natural forms of vitamin E as effective agents for cancer prevention and therapy. Adv Nutr 8:850–867
Jiang W, Hu M (2012) Mutual interactions between flavonoids and enzymatic and transporter elements responsible for flavonoid disposition via phase II metabolic pathways. RSC Adv 2:7948–7963
Jiang Q, Christen S, Shigenaga MK et al (2001) γ-Tocopherol, the major form of vitamin E in the US diet, deserves more attention. Am J Clin Nutr 74:714–722
Jiang Q, Yin J, Chen J et al (2020) Mitochondria-targeted antioxidants: a step towards disease treatment. Oxidative Med Cell Longev 2020, Article ID 8837893, 18 pages
Jin CL, Yang LX, Wu XH et al (2005) Effects of carnosine on amygdaloid-kindled seizures in Sprague-Dawley rats. Neuroscience 135:930–947
Jin RC, Mahoney CE, Coleman Anderson L et al (2011) Glutathione peroxidase-3 deficiency promotes platelet-dependent thrombosis *in vivo*. Circulation 123:1963–1973
Jorat MV, Tabrizi R, Kolahdooz F et al (2019) The effects of coenzyme Q10 supplementation on biomarkers of inflammation and oxidative stress in among coronary artery disease: a systematic review and meta-analysis of randomized controlled trials. Inflammopharmacology 27:233–248
Joshipura KJ, Hu FB, Manson JE et al (2001) The effect of fruit and vegetable intake on risk for coronary heart disease. Ann Intern Med 134:1106–1114
Jung UJ, Kim SR (2018) Beneficial effects of flavonoids against Parkinson's disease. J Med Food 21:421–432
Juraschek SP, Guallar E, Appel LJ et al (2012) Effects of vitamin C supplementation on blood pressure: a meta-analysis of randomized controlled trials. Am J Clin Nutr 95:1079–1088
Kalaras MD, Richie JP, Calcagnotto A et al (2017) Mushrooms: a rich source of the antioxidants ergothioneine and glutathione. Food Chem 233:429–433
Kameda M, Teruya T, Yanagida M et al (2020) Frailty markers comprise blood metabolites involved in antioxidation, cognition, and mobility. Proc Natl Acad Sci U S A 117:9483–9489
Kanamori Y, Via LD, Macone A et al (2020) Aged garlic extract and its constituent, S-allyl-L-cysteine, induce the apoptosis of neuroblastoma cancer cells due to mitochondrial membrane depolarization. Exp Ther Med 19:1511–1521
Kanberoglu GS, Yilmaz E, Soylak M (2019) Application of deep eutectic solvent in ultrasound-assisted emulsification microextraction of quercetin from some fruits and vegetables. J Mol Liq 279:571–577
Kaneko J, Enya A, Enomoto K et al (2017) Anserine (beta-alanyl-3-methyl-l-histidine) improves neurovascular-unit dysfunction and spatial memory in aged AβPPswe/PSEN1dE9 Alzheimer's-model mice. Sci Rep 7:12571
Keli SO, Hertog MG, Feskens EJ et al (1996) Dietary flavonoids, antioxidant vitamins, and incidence of stroke: the Zutphen study. Arch Intern Med 156:637–642
Kern C, Bernards CM (1997) Ascorbic acid inhibits spinal meningeal catechol-o-methyl transferase *in vitro*, markedly increasing epinephrine bioavailability. Anesthesiology 86:405–409
Khachik F (2006) Distribution and metabolism of dietary carotenoids in humans as a criterion for development of nutritional supplements. Pure Appl Chem 78:1551–1557
Khadangi F, Azzi A (2019) Vitamin E- the next 100 years. UBMB Life 71:411–415
Khan A, Ikram M, Hahm JR et al (2020) Antioxidant and anti-inflammatory effects of citrus flavonoid hesperetin: special focus on neurological disorders. Antioxidants 9:609
Khoo NKH, White CR, Pozzo-Miller L et al (2010) Dietary flavonoid quercetin stimulates vasorelaxation in aortic vessels. Free Radic Biol Med 49:339–347

Kim Y, Je Y (2017) Flavonoid intake and mortality from cardiovascular disease and all causes: a meta-analysis of prospective cohort studies. Clin Nutr ESPEN 20:68–77

Kim JY, Paik JK, Kim OY et al (2011) Effects of lycopene supplementation on oxidative stress and markers of endothelial function in healthy men. Atherosclerosis 215:189–195

Kim JE, Clark RM, Park Y et al (2012) Lutein decreases oxidative stress and inflammation in liver and eyes of guinea pigs fed a hypercholesterolemic diet. Nutr Res Pract 6:113–119

Kim SR, Ha YM, Kim YM et al (2015) Ascorbic acid reduces HMGB1 secretion in lipopolysaccharide-activated RAW 264.7 cells and improves survival rate in septic mice by activation of Nrf2/HO-1 signals. Biochem Pharmacol 95:279–289

Kim M, Kim SH, Lim JW et al (2019) Lycopene induces apoptosis by inhibiting nuclear translocation of β-catenin in gastric cancer cells. J Physiol Pharmacol 70:605–611

Kiokias S, Proestos C, Oreopoulou V (2020) Phenolic acids of plant origin—a review on their antioxidant activity *in vitro* (O/W emulsion systems) along with their *in vivo* health biochemical properties. Foods 9:534

Kishimoto Y, Saito N, Kurita K et al (2013) Ascorbic acid enhances the expression of type 1 and type 4 collagen and SVCT2 in cultured human skin fibroblasts. Biochem Biophys Res Commun 430:579–584

Kobayashi M, Ikeda I (2017) Mechanisms of inhibition of cholesterol absorption by green tea catechins. Food Sci Technol Res 23:627–636

Kohen R (1999) Skin antioxidants: their role in aging and in oxidative stress – new approaches for their evaluation. Biomed Pharmacother 53:181–192

Kohen R, Gati I (2000) Skin low molecular weight antioxidants and their role in aging and in oxidative stress. Toxicology 148:149–157

Kohen R, Yamamoto Y, Cundy KC et al (1988) Antioxidant activity of carnosine, homocarnosine, and anserine present in muscle and brain. Proc Natl Acad Sci U S A 85:3175–3179

Komaki H, Faraji N, Komaki A et al (2019) Investigation of protective effects of coenzyme Q10 on impaired synaptic plasticity in a male rat model of Alzheimer's disease. Brain Res Bull 147:14–21

Koo SI, Noh SK (2007) Green tea as inhibitor of the intestinal absorption of lipids: potential mechanism for its lipid-lowering effect. J Nutr Biochem 18:179–183

Korde SD, Basak A, Chaudhary M et al (2011) Enhanced nitrosative and oxidative stress with decreased total antioxidant capacity in patients with oral precancer and oral squamous cell carcinoma. Oncologia 80:382–389

Korobelnik JF, Rougier MB, Delyfer MN et al (2017) Effect of dietary supplementation with lutein, zeaxanthin, and omega-3 on macular pigment: a randomized clinical trial. JAMA Ophthalmol 135:1259–1266

Krajka-Kuźniak V, Szaefer H, Ignatowicz E et al (2012) Beetroot juice protects against N-nitrosodiethylamine-induced liver injury in rats. Food Chem Toxicol 50:2027–2033

Krajka-Kuźniak V, Paluszczak J, Szaefer H et al (2013) Betanin, a beetroot component, induces nuclear factor erythroid-2-related factor 2-mediated expression of detoxifying/antioxidant enzymes in human liver cell lines. Br J Nutr 110:2138–2149

Krasinska B, Osinaka A, Osinski M et al (2017) Standardised tomato extract as an alternative to acetylsalicylic acid in patients with primary hypertension and high cardiovascular risk – a randomized, controlled trial. Arch Med Sci 14:773–780

Kuhad A, Bishnoi M, Tiwari V et al (2009) Suppression of NF-κβ signaling pathway by tocotrienol can prevent diabetes associated cognitive deficits. Pharmacol Biochem Behav 92:251–259

Kuiper C, Molenaar IG, Dachs GU et al (2010) Low ascorbate levels are associated with increased hypoxia-inducible factor-1 activity and an aggressive tumor phenotype in endometrial cancer. Cancer Res 70:5749–5758

Kumar BR (2017) Application of HPLC and ESI-MS techniques in the analysis of phenolic acids and flavonoids from green leafy vegetables (GLVs). J Pharm Anal 7:349–364

Kumar N, Goel N (2019) Phenolic acids: natural versatile molecules with promising therapeutic applications. Biotechnol Rep 24:e00370

Kumar S, Pandey AK (2013) Chemistry and biological activities of flavonoids: an overview. Sci World J 2013, Article ID 162750, 16 pages
Kvíčalová Z, Alster J, Hofmann E et al (2016) Triplet–triplet energy transfer from chlorophylls to carotenoids in two antenna complexes from dinoflagellate *Amphidinium carterae*. Biochim Biophys Acta Bioenerg 1857:341–349
Ladas EJ, Jacobson JS, Kennedy DD et al (2004) Antioxidants and cancer therapy: a systematic review. J Clin Oncol 22:517–528
Lakey-Beitia J, Doens D, Kumar DJ et al (2017) Anti-amyloid aggregation activity of novel carotenoids: implications for Alzheimer's drug discovery. Clin Interv Aging 12:815–822
Lakshmi BVS, Sudhakar M, Surya Prakash K (2015) Protective effect of selenium against aluminum chloride-induced Alzheimer's disease: behavioral and biochemical alterations in rats. Biol Trace Elem Res 165:67–74
Lane DJR, Jansson PJ, Richardson DR (2016) Bonnie and Clyde: vitamin C and iron are partners in crime in iron deficiency anaemia and its potential role in the elderly. Aging 8:1150–1152
Lang CA, Naryshkin S, Schneider DL et al (1992) Low blood glutathione levels in healthy aging adults. J Lab Clin Med 120:P720–P725
Langsjoen PH, Langsjoen AM (2014) Comparison study of plasma coenzyme Q10 levels in healthy subjects supplemented with ubiquinol versus ubiquinone. Clin Pharmacol Drug Dev 3:13–17
Laredj LN, Licitra F, Puccio HM (2014) The molecular genetics of coenzyme Q biosynthesis in health and disease. Biochimie 100:78–87
Lass A, Kwong L, Sohal RS (1999) Mitochondrial coenzyme Q content and aging. Biofactors 9:199–205
Lechner JF, Stoner GD (2019) Red beetroot and betalains as cancer chemopreventative agents. Molecules 24:1602
Lee YS, Kim AY, Choi JW et al (2008) Dysregulation of adipose glutathione peroxidase 3 in obesity contributes to local and systemic oxidative stress. Mol Endocrinol 22:2176–2189
Lee C-TC, Gayton EL, Beulens JWJ et al (2010) Micronutrients and diabetic retinopathy: a systematic review. Ophthalmology 117:71–78
Lee DH, Kim CS, Lee YJ (2011) Astaxanthin protects against MPTP/MPP+-induced mitochondrial dysfunction and ROS production *in vivo* and *in vitro*. Food Chem Toxicol 49:271–280
Lee B-J, Tseng Y-F, Yen C-H et al (2013) Effects of coenzyme Q10 supplementation (300 mg/day) on antioxidation and anti-inflammation in coronary artery disease patients during statins therapy: a randomized, placebo-controlled trial. Nutr J 12:142
Lee J, Shin A, Oh JH et al (2017) Colors of vegetables and fruits and the risks of colorectal cancer. World J Gastroenterol 23:2527–2538
Lee SJ, Jeong J-H, Lee IH et al (2019) Effect of high-dose vitamin C combined with anti-cancer treatment on breast cancer cells. Anticancer Res 39:751–758
Lei L, Liu Y (2017) Efficacy of coenzyme Q10 in patients with cardiac failure: a meta-analysis of clinical trials. BMC Cardiovasc Disord 17:196
Leopoldini M, Russo N, Toscano M (2011) The molecular basis of working mechanism of natural polyphenolic antioxidants. Food Chem 125:288–306
Letenneur L, Proust-Lima C, Le Gouge A et al (2007) Flavonoid intake and cognitive decline over a 10-year period. Am J Epidemiol 165:1364–1371
Li HT, Jiao M, Chen J et al (2010) Roles of zinc and copper in modulating the oxidative refolding of bovine copper, zinc superoxide dismutase. Acta Biochim Biophys Sin (Shanghai) 42:183–194
Li H, Wu S, Chen J et al (2013) Effect of glutathione depletion on Nrf2/ARE activation by deltamethrin in PC12 cells. Arh Hig Rada Toksikol 64:87–97
Li M, Pare PW, Zhang J et al (2018a) Antioxidant capacity connection with phenolic and flavonoid content in Chinese medicinal herbs. Rec Nat Prod 12:239–250
Li Y, Buckhaults P, Li S et al (2018b) Temporal efficacy of a sulforaphane-based broccoli sprout diet in prevention of breast cancer through modulation of epigenetic mechanisms. Cancer Prev Res 11:451–464

Li Y-Y, Cui Y, Wang H-L et al (2019a) Effect of total flavonoids of *Herba Taxilli* on osteoporotic rats induced by retinoic acid. World J Tradit Chinese Med 5:243–249
Li D, Tang H, Wei P et al (2019b) Vitamin C and vitamin E mitigate the risk of pancreatic ductal adenocarcinoma from meat-derived mutagen exposure in adults in a case-control study. J Nutr 149:1443–1450
Li Q, Wang Y, Mai Y et al (2020) Health benefits of the flavonoids from onion: constituents and their pronounced antioxidant and anti-neuroinflammatory capacities. J Agric Food Chem 68:799–807
Liang M-H, Zhu J, Jiang J-G (2018) Carotenoids biosynthesis and cleavage related genes from bacteria to plants. Crit Rev Food Sci Nutr 58:2314–2333
Liguori L, Califano R, Albanese D et al (2017) Chemical composition and antioxidant properties of five white onion (*Allium cepa* L.) landraces. J Food Qual 2017, Article ID 6873651, 9 pages
Lim JY, Wang X-D (2020) Mechanistic understanding of β-cryptoxanthin and lycopene in cancer prevention in animal models. Biochim Biophys Acta Mol Cell Biol Lipids 1865:158652
Lima GPP, Vianello F, Corrêa CR et al (2014) Polyphenols in fruits and vegetables and its effect on human health. Food Nutr Sci 5:1065–1082
Lin CC, Yin MC, Hsu CC et al (2004) Effect of five cysteine-containing compounds on three lipogenic enzymes in Balb/cA mice consuming a high saturated fat diet. Lipids 39:843–848
Lin T-Y, Cantley LC, DeNicola GM (2016a) NRF2 rewires cellular metabolism to support the antioxidant response, a master regulator of oxidative stress - the transcription factor Nrf2, Morales-Gonzalez JA, Morales-Gonzalez A, Osiris Madrigal-Santillan E. IntechOpen, London, United Kingdom. Accessed 21 December 2016
Lin J, Huang L, Yu J et al (2016b) Fucoxanthin, a marine carotenoid, reverses scopolamine-induced cognitive impairments in mice and inhibits acetylcholinesterase *in vitro*. Mar Drugs 14:67
Lin J, Yu J, Zhao J et al (2017) Fucoxanthin, a marine carotenoid, attenuates β-amyloid oligomer-induced neurotoxicity possibly via regulating the PI3k/Akt and the ERK pathways in SH-SY5Y cells. Oxidative Med Cell Longev 2017, Article ID 6792543, 10 pages
Lin S-C, Chagnaadorj A, Bayarsengee U et al (2019) The compound, diallyl disulfide, enriched in garlic, prevents the progression of doxorubicin-induced nephropathy. Food Sci Technol 39:1040–1046
Lin P, Ren Q, Wang Q et al (2020) Carotenoids inhibit fructose-induced inflammatory response in human endothelial cells and monocytes. Mediat Inflamm 2020, Article ID 5373562, 11 pages
Liñero O, Cidad M, Carrero JA et al (2017) Partitioning of nutrients and non-essential elements in Swiss chards cultivated in open-air plots. J Food Compos Anal 59:179–187
Ling P, Shan W, Zhai G et al (2020) Association between glutathione peroxidase-3 activity and carotid atherosclerosis in patients with type 2 diabetes mellitus. Brain Behav 10:e01773
Liskova A, Koklesova L, Samec M et al (2020) Flavonoids in cancer metastasis. Cancers 12:1498
Liu RH (2004) Potential synergy of phytochemicals in cancer prevention: mechanism of action. J Nutr 134:3479S–3485S
Liu Y-J, Zhan J, Liu X-L et al (2014a) Dietary flavonoids intake and risk of type 2 diabetes: a meta-analysis of prospective cohort studies. Clin Nutr 33:59–63
Liu XH, Yu RB, Liu R et al (2014b) Association between lutein and zeaxanthin status and the risk of cataract: a meta-analysis. Nutrients 6:452–465
Liu R, Wang T, Zhang B et al (2015) Lutein and zeaxanthin supplementation and association with visual function in age-related macular degeneration. Invest Ophthalmol Vis Sci 56:252–258
Liu T, Liu WH, Zhao JS et al (2017a) Lutein protects against β-amyloid peptide-induced oxidative stress in cerebrovascular endothelial cells through modulation of Nrf-2 and NF-κB. Cell Biol Toxicol 33:57–67
Liu HT, Cheng SB, Huang YC et al (2017b) Coenzyme Q10 and oxidative stress: inflammation status in hepatocellular carcinoma patients after surgery. Nutrients 9:E29
Liu F, Qu Y-K, Wang A-M et al (2019) Effects of carotenoids on the growth performance, biochemical parameters, immune responses and disease resistance of yellow catfish (*Pelteobagrus fulvidraco*) under high-temperature stress. Aquaculture 503:293–303

Liu B, Liu X, Ning Q et al (2020) Evaluation of toxicity and anti-osteoporosis effect in rats treated with the flavonoids of *Psoraleae Fructus*. J Funct Foods 75:104262

Livingstone C (2015) Zinc: physiology, deficiency, and parenteral nutrition. Nutr Clin Pract 30:371–382

Lobos P, Bruna B, Cordova A et al (2016) Astaxanthin protects primary hippocampal neurons against noxious effects of Aβ-oligomers. Neural Plast 2016, Article ID 3456783, 13 pages

López-Lluch G, Rodriguez-Aguilera JC, Santos-Ocana C et al (2010) Is coenzyme Q a key factor in aging? Mech Ageing Dev 131:225–235

Loung C-Y, Fernando W, Rupasinghe HPV et al (2019) Apple peel flavonoid fraction 4 suppresses breast cancer cell growth by cytostatic and cytotoxic mechanisms. Molecules 24:3335

Lu SC (2009) Regulation of glutathione synthesis. Mol Asp Med 30:42–59

Lu SC (2013) Glutathione synthesis. Biochim Biophys Acta 1830:3143–3153

Lu SC, Ge JL (1992) Loss of suppression of GSH synthesis at low cell density in primary cultures of rat hepatocytes. Am J Phys 263:C1181–C1189

Lubos E, Handy DE, Loscalzo J (2008) Role of oxidative stress and nitric oxide in atherothrombosis. Front Biosci 13:5323–5344

Luo JL, Hammarqvist F, Andersson K et al (1998) Surgical trauma decreases glutathione synthetic capacity in human skeletal muscle tissue. Am J Phys 275:E359–E365

Lushchak VI (2012) Glutathione homeostasis and functions: potential targets for medical interventions. J Amino Acids 2012, Article ID 736837, 26 pages

Lutchmansingh FK, Hsu JW, Bennett FI et al (2018) Glutathione metabolism in type 2 diabetes and its relationship with microvascular complications and glycemia. PLoS One 13:e0198626

Ma L, Yan S-F, Huang Y-M et al (2012a) Effect of lutein and zeaxanthin on macular pigment and visual function in patients with early age-related macular degeneration. Ophthalmology 119:2290–2297

Ma L, Dou H-L, Wu Y-Q et al (2012b) Lutein and zeaxanthin intake and the risk of age-related macular degeneration: a systematic review and meta-analysis. Br J Nutr 107:350–359

Ma L, Hao ZX, Liu RR et al (2014) A dose-response meta-analysis of dietary lutein and zeaxanthin intake in relation to risk of age-related cataract. Graefes Arch Clin Exp Ophthalmol 252:63–70

Ma J, Chen J, Bo S et al (2015) Protective effect of carnosine after chronic cerebral hypoperfusion possibly through suppressing astrocyte activation. Am J Transl 7:2706–2715

Machiela E, Southwell AL (2020) Biological aging and the cellular pathogenesis of Huntington's disease. J Huntingtons Dis 9:115–128

Maci S, Santos R (2015) The beneficial role of lutein and zeaxanthin in cataracts. Forum Nutr 14:63–69

Macielag MJ (2012) Chemical properties of antimicrobials and their uniqueness. In: Dougherty T, Pucci M (eds) Antibiotic discovery and development. Springer, Boston, pp 793–820

Mahmoud AM, Bautista RJH, Sandhu MA et al (2019) Beneficial effects of citrus flavonoids on cardiovascular and metabolic health. Oxidative Med Cell Longev 2019, Article ID 5484138, 19 pages

Maiese K, Chong ZZ, Shang YC (2007) Mechanistic insights into diabetes mellitus and oxidative stress. Med Chem 14:1729–1738

Maiti AK, Spoorthi BC, Saha NC et al (2018) Mitigating peroxynitrite mediated mitochondrial dysfunction in aged rat brain by mitochondria-targeted antioxidant MitoQ. Biogerontology 19:271–286

Maiti S, Nazmeen A, Medda N et al (2019) Flavonoids green tea against oxidant stress and inflammation with related human diseases. Clin Nutr Exp 24:1–14

Mak TW, Grusdat M, Duncan GS et al (2017) Glutathione primes T cell metabolism for inflammation. Immunity 46:675–689

Malakul W, Pengnet S, Kumchoom C et al (2018) Naringin ameliorates endothelial dysfunction in fructose-fed rats. Exp Ther Med 15:3140–3146

Malešev D, Kuntić V (2007) Investigation of metal–flavonoid chelates and the determination of flavonoids via metal–flavonoid complexing reactions. J Serb Chem Soc 72:921–939

Mangels AR, Block G, Frey CM et al (1993) The bioavailability to humans of ascorbic acid from oranges, orange juice and cooked broccoli is similar to that of synthetic ascorbic acid. J Nutr 123:1054–1061
Mannion AF, Jakeman PM, Dunnett M et al (1992) Carnosine and anserine concentrations in the quadriceps femoris muscle of healthy humans. Eur J Appl Physiol Occup Physiol 64:47–50
Manzar H, Abdulhussein D, Yap TE et al (2020) Cellular consequences of coenzyme Q10 deficiency in neurodegeneration of the retina and brain. Int J Mol Sci 21:9299
Maoka T (2009) Recent progress in structural studies of carotenoids in animals and plants. Arch Biochem Biophys 483:191–195
Maoka T (2020) Carotenoids as natural functional pigments. J Nat Med 74:1–16
Marengo B, Ciucis CD, Ricciarelli R et al (2010) DNA oxidative damage of neoplastic rat liver lesions. Oncol Rep 23:1241–1246
Mares J (2016) Lutein and zeaxanthin isomers in eye health and disease. Annu Rev Nutr 36:571–602
Markova NG, Karaman-Jurukovska N, Dong KK et al (2009) Skin cells and tissue are capable of using L-ergothioneine as an integral component of their antioxidant defense system. Free Radic Biol Med 46:1168–1176
Martinez RM, Longhi-Balbinot DT, Zarpelon AC et al (2015) Anti-inflammatory activity of betalain-rich dye of *Beta vulgaris*: effect on edema, leukocyte recruitment, superoxide anion and cytokine production. Arch Pharm Res 38:494–504
Martins TEA, de Oliveira Pinto CAS, de Oliveira AC et al (2020) Contribution of topical antioxidants to maintain healthy skin—a review. Sci Pharm 88:27
Mason SA, Rasmussen B, van Loon LJC et al (2019) Ascorbic acid supplementation improves postprandial glycaemic control and blood pressure in individuals with type 2 diabetes: findings of a randomized cross-over trial. Diabetes Obes Metab 21:674–682
Masuoka N, Lei C, Li H et al (2021) Anserine, HClO-scavenger, protected against cognitive decline in individuals with mild cognitive impairment. Aging 13:1729–1741
Mata-Granados JM, Cuenca-Acebedo R, de Castro MDL et al (2013) Lower vitamin E serum levels are associated with osteoporosis in early postmenopausal women: a cross-sectional study. J Bone Miner Metab 31:455–460
Matarrese P, Colasanti T, Ascione B et al (2011) Gender disparity in susceptibility to oxidative stress and apoptosis induced by autoantibodies specific to RLIP76 in vascular cells. Antioxid Redox Signal 15:2825–2836
Matthews RT, Yang L, Browne S et al (1998) Coenzyme Q10 administration increases brain mitochondrial concentrations and exerts neuroprotective effects. Proc Natl Acad Sci U S A 95:8892–8897
May JM (2011) The SLC23 family of ascorbate transporters: ensuring that you get and keep your daily dose of vitamin C. Br J Pharmacol 164:1793–1801
May JM (2012) Vitamin C transport and its role in the central nervous system. In: Stanger O (ed) Water soluble vitamins. Springer, Dordrecht, pp 85–103
May JM (2016) Ascorbic acid repletion: a possible therapy for diabetic macular edema? Free Radic Biol Med 94:47–54
May JM, Harrison FE (2013) Role of vitamin C in the function of the vascular endothelium. Antioxid Redox Signal 19:2068–2083
Mayland CR, Bennett MI, Allan K (2005) Vitamin C deficiency in cancer patients. Palliat Med 19:17–20
Medeiros MS, Schumacher-Schuh A, Cardoso AM et al (2016) Iron and oxidative stress in Parkinson's disease: an observational study of injury biomarkers. PLoS One 11:e0146129
Mehri A (2020) Trace elements in human nutrition (II) – an update. Int J Prev Med 11:2
Meister A (1988) On the discovery of glutathione. Trends Biochem Sci 13:185–188
Mekuria AN, Tura AK, Hagos B et al (2020) Anti-cancer effects of lycopene in animal models of hepatocellular carcinoma: a systematic review and meta-analysis. Front Pharmacol 11:1306
Melino S, Leo S, Papajani VT (2019) Natural hydrogen sulfide donors from *Allium* sp. as a nutraceutical approach in type 2 diabetes prevention and therapy. Nutrients 11:1581

Menon K, Marquina C, Hoj P et al (2021) Carnosine and histidine-containing dipeptides improve dyslipidemia: a systematic review and meta-analysis of randomized controlled trials. Nutr Rev 79:117

Messina JP, Lawrence DA (1989) Cell cycle progression of glutathione-depleted human peripheral blood mononuclear cells is inhibited at S phase. J Immunol 143:1974–1981

Mezzomo N, Ferreira SRS (2016) Carotenoids functionality, sources, and processing by supercritical technology: a review. J Chem 2016, Article ID 3164312, 16 pages

Mikolajczyk-Bator K, Czapski J (2017) Effect of pH changes on antioxidant capacity and the content of betalain pigments during the heating of a solution of red beet betalains. Pol J Food Nutr Sci 67:123–128

Mikulic-Petkovsek M, Ivancic A, Todorovic B et al (2015) Fruit phenolic composition of different elderberry species and hybrids. J Food Sci 80:C2180–C2190

Miles MV, Horn PS, Tang PH et al (2004) Age-related changes in plasma coenzyme Q10 concentrations and redox state in apparently healthy children and adults. Clin Chim Acta 347:139–144

Milton AH, Vashum KP, McEvoy M et al (2018) Prospective study of dietary zinc intake and risk of cardiovascular disease in women. Nutrients 10:38

Minich DM, Brown BI (2019) A review of dietary (phyto)nutrients for glutathione support. Nutrients 11:2073

Mirahmadi M, Azimi-Hashemi S, Saburi E et al (2020) Potential inhibitory effect of lycopene on prostate cancer. Biomed Pharmacother 129:110459

Miranda-Osorio PH, Castell-Rodriguez AE, Vargas-Mancilla J et al (2016) Protective action of Carica papaya on β-cells in streptozotocin-induced diabetic rats. Int J Environ Res Public Health 13:E446

Mirmiran P, Bahadoran Z, Tohidi M et al (2019) Higher consumption of *Allium* vegetables may modulate insulin homeostasis: a longitudinal follow-up study. J Herbal Med 17–18:100260

Mischley LK, Allen J, Bradley R (2012) Coenzyme Q10 deficiency in patients with Parkinson's disease. J Neurol Sci 318:72–75

Mishra A, Reddy IJ, Dhali A et al (2018) L-Ergothioneine improves the developmental potential of *in vitro* sheep embryos without influencing OCTN1-mediated cross-membrane transcript expression. Zygote 26:149–161

Mitchell JA, Ahmetaj-Shala B, Kirkby NS et al (2014) Role of prostacyclin in pulmonary hypertension. Glob Cardiol Sci Pract 2014:382–393

Mo M, Li S, Dong Z et al (2020) S-allylmercaptocysteine ameliorates lipopolysaccharide-induced acute lung injury in mice by inhibiting inflammation and oxidative stress via nuclear factor kappa B and Keap1/Nrf2 pathways. Int Immunopharmacol 81:106273

Moccia S, Russo M, Durante M et al (2020) A carotenoid-enriched extract from pumpkin delays cell proliferation in a human chronic lymphocytic leukemia cell line through the modulation of autophagic flux. Curr Res Biotechnol 2:74–82

Mohamad N-V, Ima-Nirwana S, Chin K-Y (2018) Effect of tocotrienol from *Bixa orellana* (annatto) on bone microstructure, calcium content, and biomechanical strength in a model of male osteoporosis induced by buserelin. Drug Des Dev Ther 12:555–564

Molyneux SL, Young JM, Florkowski CM et al (2008) Coenzyme Q10: is there a clinical role and a case for measurement? Clin Biochem Rev 29:71–82

Monacelli F, Acquarone E, Giannotti C et al (2017) Vitamin C, aging and Alzheimer's disease. Nutrients 9:670

Mondal S, Rahaman ST (2020) Flavonoids: a vital resource in healthcare and medicine. Pharm Pharmacol Int J 8:91–104

Moriya J (2019) Critical roles of inflammation in atherosclerosis. J Cardiol 73:22–27

Morsi M, Kobeissy F, Magdeldin S et al (2019) A shared comparison of diabetes mellitus and neurodegenerative disorders. J Cell Biochem 120:14318–14325

Mozos I, Luca CT (2017) Crosstalk between oxidative and nitrosative stress and arterial stiffness. Curr Vasc Pharmacol 15:446–456

Mozos I, Stoian D, Caraba A et al (2018) Lycopene and vascular health. Front Pharmacol 9:521

Mulcahy RT, Untawale S, Gipp JJ (1994) Transcriptional up-regulation of γ-glutamylcysteine synthetase gene expression in melphalan-resistant human prostate carcinoma cells. Mol Pharmacol 46:909–914

Mullan K, Williams MA, Cardwell CR et al (2017) Serum concentrations of vitamin E and carotenoids are altered in Alzheimer's disease: a case-control study. Alzheimers Dement: Transl Res Clin Interv 3:432–439

Mullan K, Cardwell CR, McGuinness B et al (2018) Plasma antioxidant status in patients with Alzheimer's disease and cognitively intact elderly: a meta-analysis of case-control studies. J Alzheimers Dis 62:305–317

Muller L, Caris-Veyrat C, Lowe G et al (2016) Lycopene and its antioxidant role in the prevention of cardiovascular diseases- a critical review. Crit Rev Food Sci Nutr 56:1868–1879

Mulvihill EE, Burke AC, Huff MW (2016) Citrus flavonoids as regulators of lipoprotein metabolism and atherosclerosis. Annu Rev Nutr 36:275–299

Murphy MP, Smith RA (2007) Targeting antioxidants to mitochondria by conjugation to lipophilic cations. Annu Rev Pharmacol Toxicol 47:629–656

Myung SK, Ju W, Cho B et al (2013) Efficacy of vitamin and antioxidant supplements in prevention of cardiovascular disease: systematic review and meta-analysis of randomised controlled trials. BMJ 346:f10

Nagase M, Yamamoto Y, Matsumoto N et al (2018) Increased oxidative stress and coenzyme Q10 deficiency in centenarians. J Clin Biochem Nutr 63:129–136

Nakamichi N, Nakayama K, Ishimoto T et al (2016) Food-derived hydrophilic antioxidant ergothioneine is distributed to the brain and exerts antidepressant effect in mice. Brain Behav 6:e00477

Nakamura A, Chieko I, Saito A et al (2017) Possible benefits of tomato juice consumption: a pilot study on irradiation human lymphocytes from healthy donors. J Nutr 16:1–11

Nam KN, Park YM, Jung HJ et al (2010) Anti-inflammatory effects of crocin and crocetin in rat brain microglial cells. Eur J Pharmacol 648:110–116

Narayanankutty A, Job JT, Narayanankutty V (2019) Glutathione, an antioxidant tripeptide: dual roles in carcinogenesis and chemoprevention. Curr Protein Pept Sci 20:907–917

Nascimento-Souza MA, de Paiva PG, Pérez-Jiménez J et al (2018) Estimated dietary intake and major food sources of polyphenols in elderly of Viçosa, Brazil: a population-based study. Eur J Nutr 57:617–627

Nasrolahi A, Mahmoudi J, Noori-Zadeh A et al (2019) Shared pathological mechanisms between diabetes mellitus and neurodegenerative diseases. Curr Pharmacol Rep 5:219–231

Navale AM, Paranjape AN (2016) Glucose transporters: physiological and pathological roles. Biophys Rev 8:5–9

Navarra M, Femia AP, Romagnoli A et al (2020) A flavonoid-rich extract from bergamot juice prevents carcinogenesis in a genetic model of colorectal cancer, the Pirc rat (F344/NTac-Apc^{am1137}). Eur J Nutr 59:885–894

Naz A, Butt MS, Sultan MT et al (2014) Watermelon lycopene and allied health claims. EXCLI J 13:650–660

Nazarov PA, Osterman IA, Tokarchuk AV et al (2017) Mitochondria-targeted antioxidants as highly effective antibiotics. Sci Rep 7:1394

Nazem A, Sankowski R, Bacher M et al (2015) Rodent models of neuroinflammation for Alzheimer's disease. J Neuroinflammation 12:74

Nelson MM, Builta ZJ, Monroe TB et al (2019) Biochemical characterization of the catecholaldehyde reactivity of L-carnosine and its therapeutic potential in human myocardium. Amino Acids 51:97–102

Ng LF, Gruber J, Cheah IK et al (2014) The mitochondria-targeted antioxidant MitoQ extends lifespan and improves healthspan of a transgenic *Caenorhabditis elegans* model of Alzheimer disease. Free Radic Biol Med 71:390–401

Nguyen D, Hsu JW, Jahoor F et al (2014) Effect of increasing glutathione with cysteine and glycine supplementation on mitochondrial fuel oxidation, insulin sensitivity, and body composition in older HIV-infected patients. J Clin Endocrinol Metab 99:169–177

Nijveldt RJ, van Nood E, van Hoorn DEC et al (2001) Flavonoids: a review of probable mechanisms of action and potential applications. Am J Clin Nutr 74:418–425

Niki E, Abe K (2019) Chapter 1: Vitamin E: structure, properties and functions. In: Vitamin E: chemistry and nutritional benefits. Royal Society of Chemistry, London, pp 1–11

Niklowitz P, Onur S, Fischer A et al (2016) Coenzyme Q10 serum concentration and redox status in European adults: influence of age, sex, and lipoprotein concentration. J Clin Biochem Nutr 58:240–245

Nimptsch K, Zhang X, Cassidy A et al (2016) Habitual intake of flavonoid subclasses and risk of colorectal cancer in 2 large prospective cohorts. Am J Clin Nutr 103:184–191

Nishikimi M, Fukuyama R, Minoshima S et al (1994) Cloning and chromosomal mapping of the human nonfunctional gene for L-gulono-gamma-lactone oxidase, the enzyme for L-ascorbic acid biosynthesis missing in man. J Biol Chem 269:13685–13688

Nita M, Grzybowski A (2016) The role of the reactive oxygen species and oxidative stress in the pathomechanism of the age-related ocular diseases and other pathologies of the anterior and posterior eye segments in adults. Oxidative Med Cell Longev 2016, Article ID 3164734, 23 pages

Noh YH, Kim K-Y, Shim MS et al (2013) Inhibition of oxidative stress by coenzyme Q10 increases mitochondrial mass and improves bioenergetic function in optic nerve head astrocytes. Cell Death Dis 4:e820

Norouzi S, Adulcikas J, Sohal SS et al (2018) Zinc stimulates glucose oxidation and glycemic control by modulating the insulin signaling pathway in human and mouse skeletal muscle cell lines. PLoS One 13:e0191727

Nørregaard R, Kwon T-H, Frøkiær J (2015) Physiology and pathophysiology of cyclooxygenase-2 and prostaglandin E2 in the kidney. Kidney Res Clin Pract 34:194–200

Nowak KL, Yaffe K, Orwoll ES et al (2018) Serum sodium and cognition in older community-dwelling men. Clin J Am Soc Nephrol 13:366–374

Nugroho A, Heryani H, Choi JS et al (2017) Identification and quantification of flavonoids in *Carica papaya* leaf and peroxynitrite-scavenging activity. Asian Pac J Trop Biomed 7:208–213

Nunes SC, Serpa J (2018) Glutathione in ovarian cancer: a double-edged sword. Int J Mol Sci 19:1882

Nweze CC, Solomon M, Haruna GH (2019) Serum lipoproteins response to twelve months changes in antioxidants: influence of nutraceuticals and functional foods intervention on age and gender. Int J Adv Biochem Res 3:7–14

Nyariki JN, Ochola LA, Jillani NE et al (2019) Oral administration of coenzyme Q10 protects mice against oxidative stress and neuroinflammation during experimental cerebral malaria. Parasitol Int 71:106–120

Okada E, Takahashi K, Nakamura K et al (2019) Dietary patterns and abnormal glucose tolerance among Japanese: findings from the national health and nutrition survey, 2012. Public Health Nutr 22:2460–2468

Olechnowicz J, Tinkov A, Skalny A et al (2018) Zinc status is associated with inflammation, oxidative stress, lipid, and glucose metabolism. J Physiol Sci 68:19–31

Omata Y, Salvador GA, Supasai S et al (2013) Decreased zinc availability affects glutathione metabolism in neuronal cells and in the developing brain. Toxicol Sci 133:90–100

Onodera K, Omoi N-O, Fukui K et al (2003) Oxidative damage of rat cerebral cortex and hippocampus, and changes in antioxidative defense systems caused by hyperoxia. Free Radic Res 37:367–372

Onur S, Niklowitz P, Fischer A et al (2014) A comparative study into alterations of coenzyme Q redox status in ageing pigs, mice, and worms. Biofactors 40:346–354

Ornoy A, Zaken V, Kohen R (1999) Role of reactive oxygen species (ROS) in the diabetes-induced anomalies in rat embryos *in vitro*: reduction in antioxidant enzymes and low-molecular-weight antioxidants (LMWA) may be the causative factor for increased anomalies. Teratology 60:376–386

Ottosson F, Smith E, Melander O et al (2018) Altered asparagine and glutamate homeostasis precede coronary artery disease and type 2 diabetes. J Clin Endocrinol Metab 103:3060–3069

Padayatty SJ, Katz A, Wang Y et al (2003) Vitamin C as an antioxidant: evaluation of its role in disease prevention. J Am Coll Nutr 22:18–35

Padayatty SJ, Riordan HD, Hewitt SM et al (2006) Intravenously administered vitamin C as cancer therapy: three cases. Can Med Assoc J 174:937–942

Pallister T, Jennings A, Mohney RP et al (2016) Characterizing blood metabolomics profiles associated with self-reported food intakes in female twins. PLoS One 11:1–16

Palloza P, Catalano A, Simone RE et al (2012) Effect of lycopene and tomato products on cholesterol metabolism. Ann Nutr Metab 61:126–134

Panahi G, Pasalar P, Zare M et al (2018) High glucose induces inflammatory responses in HepG2 cells via the oxidative stress-mediated activation of NF-κB, and MAPK pathways in HepG2 cells. Arch Physiol Biochem 124:468–474

Panat NA, Maurya DK, Ghaskadbi SS et al (2016) Troxerutin, a plant flavonoid, protects cells against oxidative stress-induced cell death through radical scavenging mechanism. Food Chem 194:32–45

Panche A, Chandra S, Diwan A et al (2015) Alzheimer's and current therapeutics: a review. Asian J Pharm Clin Res 8:14–19

Panche AN, Diwan AD, Chandra SR (2016) Flavonoids: an overview. J Nutr Sci 5:e47

Pandey KB, Rizvi SI (2009) Plant polyphenols as dietary antioxidants in human health and disease. Oxidative Med Cell Longev 2:270–278 Pandey KB, Rizvi SI (2009) Plant polyphenols as dietary antioxidants in human health and disease. Oxidative Med Cell Longev 2:270–278

Park Y, Bolpe S, Decker E (2005) Quantitation of carnosine in humans plasma after dietary consumption of beef. J Agric Food Chem 53:4736–4739

Park B, Song HS, Kwon JE et al (2017) Effects of *Salvia miltiorrhiza* extract with supplemental liquefied calcium on osteoporosis in calcium-deficient ovariectomized mice. BMC Complement Altern Med 17:545

Parrado-Fernández C, López-Lluch G, Rodríguez-Bies E et al (2011) Calorie restriction modifies ubiquinone and COQ transcript levels in mouse tissues. Free Radic Biol Med 50:1728–1736

Pastori D, Pignatelli P, Farcomeni A et al (2016) Aging-related decline of glutathione peroxidase 3 and risk of cardiovascular events in patients with atrial fibrillation. J Am Heart Assoc 5:e003682

Paul BD, Snyder SH (2010) The unusual amino acid L-ergothioneine is a physiologic cytoprotectant. Cell Death Differ 17:1134–1140

Paulsen JS (2011) Cognitive impairment in Huntington disease: diagnosis and treatment. Curr Neurol Neurosci Rep 11:474–483

Pekovic-Vaughan V, Gibbs J, Yoshitane H et al (2014) The circadian clock regulates rhythmic activation of the NRF2/glutathione-mediated antioxidant defense pathway to modulate pulmonary fibrosis. Genes Dev 28:548–560

Péneau S, Galan P, Jeandel C et al (2011) Fruit and vegetable intake and cognitive function in the SU.VI.MAX 2 prospective study. Am J Clin Nutr 94:1295–1303

Peng W, Xu S, Zhang J et al (2019) Vitamin C attenuates sodium fluoride-induced mitochondrial oxidative stress and apoptosis via Sirt1-SOD2 pathway in F9 cells. Biol Trace Elem Res 191:189–198

Peters V, Calabrese V, Forsberg E et al (2018) Protective actions of anserine under diabetic conditions. Int J Mol Sci 19:2751

Phaniendra A, Jestadi DB, Periyasamy L (2015) Free radicals: properties, sources, targets, and their implication in various diseases. Indian J Clin Biochem 30:11–26

Pierce GL, Lesniewski LA, Lawson BR et al (2009) Nuclear factor-{kappa}B activation contributes to vascular endothelial dysfunction via oxidative stress in overweight/obese middle-aged and older humans. Circulation 119:1284–1292

Piersma B, Wouters OY, de Rond S et al (2017) Ascorbic acid promotes a TGFβ1-induced myofibroblast phenotype switch. Physiol Rep 5:e13324

Pietrzkowski Z, Nemzer B, Spórna A et al (2010) Influence of betalain-rich extracts on reduction of discomfort associated with osteoarthritis. New Med 1:12–17

Pignatelli P, Pulcinelli FM, Lenti L et al (1998) Hydrogen peroxide is involved in collagen-induced platelet activation. Blood 91:484–490

Pires AS, Marques CR, Encarnação JC et al (2016) Ascorbic acid and colon cancer: an oxidative stimulus to cell death depending on cell profile. Eur J Cell Biol 95:208–218

Pires AS, Marques CR, Encarnação JC et al (2018) Ascorbic acid chemosensitizes colorectal cancer cells and synergistically inhibits tumor growth. Front Physiol 9:911

Platonova K, Kitamura K, Watanabe Y et al (2021) Dietary calcium and vitamin K are associated with osteoporotic fracture risk in middle-aged and elderly Japanese women, but not men: the Murakami Cohort Study. Br J Nutr 125:319–328

Playdon MC, Moore SC, Derkach A et al (2017) Identifying biomarkers of dietary patterns by using metabolomics. Am J Clin Nutr 105:450–465

Podda M, Grundmann-Kollmann M (2001) Low molecular weight antioxidants and their role in skin ageing. Clin Exp Dermatol 26:578–582

Polívka T, Frank HA (2010) Molecular factors controlling photosynthetic light-harvesting by carotenoids. Acc Chem Res 43:1125–1134

Ponzo V, Goitre I, Fadda M et al (2015) Dietary flavonoid intake and cardiovascular risk: a population-based cohort study. J Transl Med 13:218

Prasad AS (2020) Lessons learned from experimental human model of zinc deficiency. J Immunol Res 2020, Article ID 9207279, 12 pages

Praveen D, Puvvada RC, Vijey Aanandhi M (2020) Association of vitamin C status in diabetes mellitus: prevalence and predictors of vitamin C deficiency. Future J Pharm Sci 6:30

Pravina P, Sayaji D, Avinash M (2013) Calcium and its role in human body. Int J Res Pharmaceut Biomed Sci 4:659–668

Pravst I, Zmitek K, Zmitek J (2010) Coenzyme Q10 contents in foods and fortification strategies. Crit Rev Food Sci Nutr 50:269–280

Qi Z, Liu KJ (2019) The interaction of zinc and the blood-brain barrier under physiological and ischemic conditions. Toxicol Appl Pharmacol 364:114–119

Qi S, He J, Han H et al (2019) Anthocyanin-rich extract from black rice (*Oryza sativa* L. *Japonica*) ameliorates diabetic osteoporosis in rats. Food Funct 10:5350–5360

Raddatz-Mota D, Pérez-Flores LJ, Carrari F et al (2017) Achiote (*Bixa orellana* L.): a natural source of pigment and vitamin E. J Food Sci Technol 54:1729–1741

Rahimi P, Abedimanesh S, Mesbah-Namin SA et al (2019a) Betalains, the nature-inspired pigments, in health and disease. Crit Rev Food Sci Nutr 59:2949–2978

Rahimi P, Mesbah-Namin SA, Ostadrahimi A et al (2019b) Betalain- and betacyanin-rich supplements' impacts on the PBMC SIRT1 and LOX1 genes expression and Sirtuin-1 protein levels in coronary artery disease patients: a pilot crossover clinical trial. J Funct Foods 60:103401

Rahman SU, Huang Y, Zhu L et al (2020) Tea polyphenols attenuate liver inflammation by modulating obesity-related genes and down-regulating COX-2and iNOS expression in high fat-fed dogs. BMC Vet Res 16:234

Rahmani G, Farajdokht F, Mohaddes G et al (2020) Garlic (*Allium sativum*) improves anxiety- and depressive-related behaviors and brain oxidative stress in diabetic rats. Arch Physiol Biochem 126:95–100

Raish M, Ahmad A, Ansari MA et al (2019) Beetroot juice alleviates isoproterenol-induced myocardial damage by reducing oxidative stress, inflammatory, and apoptosis in rats. 3 Biotech 9:147

Raja SB, Rajendiran V, Kasinathan NK et al (2017) Differential cytotoxic activity of Quercetin on colonic cancer cells depends on ROS generation through COX-2 expression. Food Chem Toxicol 106:92–106

Rakic JM, Tanprasertsuk J, Scott TM et al (2021) Effects of daily almond consumption for six months on cognitive measures in healthy middle-aged to older adults: a randomized control trial. Nutr Neurosci:1–11

Rapola JM, Virtamo J, Ripatti S et al (1997) Randomised trial of alpha-tocopherol and beta-carotene supplements on incidence of major coronary events in men with previous myocardial infarction. Lancet 349:1715–1720

Ravanshad SH, Naserollahzadeh J, Sovaid M et al (2006) Effect of sour orange (*Citrus Aurantium* L.) juice consumption on blood glucose and lipid profile in diabetic patients with dyslipidemia. J Guilan Uni Med Sci 15:48–53

Ravichandran K, Saw NMMT, Mohdaly AAA et al (2013) Impact of processing of red beet on betalain content and antioxidant activity. Food Res Int 50:670–675

Raymond YCF, Glenda CSL, Meng LK (2016) Effects of high doses of vitamin C on cancer patients in Singapore: nine cases. Integr Cancer Ther 15:197–204

Reale G, Russo GI, Di Mauro M et al (2018) Association between dietary flavonoids intake and prostate cancer risk: a case-control study in Sicily. Complement Ther Med 39:14–18

Reboul E (2013) Absorption of vitamin A and carotenoids by the enterocyte: focus on transport proteins. Nutrients 5:3563–3581

Reboul E (2019) Mechanisms of carotenoid intestinal absorption: where do we stand? Nutrients 11:838

Reczek CR, Chandel NS (2015) CANCER. Revisiting vitamin C and cancer. Science 350:1317–1318

Reddy MK, Alexander-Lindo RL, Nair MG (2005) Relative inhibition of liquid peroxidation, cyclooxygenase enzymes, and human tumor cell proliferation by natural food colors. J Agric Food Chem 53:9268–9273

Remely M, Ferk F, Sterneder S et al (2017) Vitamin E modifies high-fat diet-induced increase of DNA strand breaks, and changes in expression and DNA methylation of Dnmt1 and MLH1 in C57BL/6J male mice. Nutrients 9:E607

Revathy J, Srinivasan S, Abdullah SHS et al (2018) Antihyperglycemic effect of hesperetin, a citrus flavonoid, extenuates hyperglycemia and exploring the potential role in antioxidant and antihyperlipidemic in streptozotocin-induced diabetic rats. Biomed Pharmacother 97:98–106

Ribeiro DA, Buttros JB, Oshima CTF et al (2009) Ascorbic acid prevents acute myocardial infarction induced by isoproterenol in rats: role of inducible nitric oxide synthase production. J Mol Histol 40:99–105

Rimm EB, Stampfer MJ, Ascherio A et al (1993) Vitamin E consumption and the risk of coronary heart disease in men. N Engl J Med 328:1450–1456

Rinninella E, Mele MC, Merendino N et al (2018) The role of diet, micronutrients and the gut microbiota in age-related macular degeneration: new perspectives from the gut-retina axis. Nutrients 10:1677

Rizvi S, Raza ST, Ahmed F et al (2014) The role of vitamin E in human health and some diseases. Sultan Qaboos Univ Med J 14:e157–e165

Robert C, Couëdelo L, Vaysse C et al (2020) Vegetable lecithins: a review of their compositional diversity, impact on lipid metabolism and potential in cardiometabolic disease prevention. Biochimie 169:121–132

Rodriguez-Amaya DB (2019) Update on natural food pigments—a mini-review on carotenoids, anthocyanins, and betalains. Food Res Int 124:200–205

Rodríguez-García C, Sánchez-Quesada C, Gaforio JJ (2019) Dietary flavonoids as cancer chemopreventive agents: an updated review of human studies. Antioxidants 8:137

Rogerson D (2017) Vegan diets: practical advice for athletes and exercisers. J Int Soc Sports Nutr 14:36

Rolnik A, Żuchowski J, Stochmal A et al (2020) Quercetin and kaempferol derivatives isolated from aerial parts of *Lens culinaris* Medik as modulators of blood platelet functions. Industr Crops Prod 152:112536

Rondeva T, Wolf G (2014) Reactive oxygen species in diabetic nephropathy: friend or foe? Nephrol Dial Transplant 29:1998–2003

Rosenblat M, Volkova N, Coleman R et al (2007) Anti-oxidant and anti-atherogenic properties of liposomal glutathione: studies *in vitro*, and in the atherosclerotic apolipoprotein E-deficient mice. Atherosclerosis 195:e61–e68

Ross D, Siegel D (2017) Functions of NQO1 in cellular protection and CoQ10 metabolism and its potential role as a redox sensitive molecular switch. Front Physiol 8:595

Rossman MJ, Santos-Parker JR, Steward CAC et al (2018) Chronic supplementation with a mitochondrial antioxidant (MitoQ) improves vascular function in healthy older adults. Hypertension 71:1056–1063

Ruhee RT, Ma S, Suzuki K (2019) Sulforaphane protects cells against lipopolysaccharide-stimulated inflammation in murine macrophages. Antioxidants 8:577

Ruhee RT, Roberts LA, Ma S et al (2020) Organosulfur compounds: a review of their anti-inflammatory effects in human health. Front Nutr 7:64

Russo GT, Giorda CB, Cercone S et al (2014) Factors associated with beta-cell dysfunction in type 2 diabetes: the BETADECLINE study. PLoS One 9:e109702

Ruz M, Carrasco F, Rojas P et al (2019) Nutritional effects of zinc on metabolic syndrome and type 2 diabetes: mechanisms and main findings in human studies. Biol Trace Elem Res 188:177–188

Sadikali F, Darwish R, Watson WC (1975) Carnosinase activity of human gastrointestinal mucosa. Gut 16:585–589

Saini R (2011) Coenzyme Q10: the essential nutrient. J Pharm Bioallied Sci 3:466–467

Saini RK, Keum Y-S, Daglia M et al (2020) Dietary carotenoids in cancer chemoprevention and chemotherapy: a review of emerging evidence. Pharmacol Res 157:104830

Sajjadi M, Bathaie Z (2017) Comparative study on the preventive effect of saffron carotenoids, crocin and crocetin, in NMU-induced breast cancer in rats. Cell J 19:94–101

Sakakibara T, Sawada Y, Wang J et al (2019) Molecular mechanism by which tea catechins decrease the micellar solubility of cholesterol. J Agric Food Chem 67:7128–7135

Salmonowicz B, Krzystek-Korpacka M, Noczyńska A (2014) Trace elements, magnesium, and the efficacy of antioxidant systems in children with type 1 diabetes mellitus and in their siblings. Adv Clin Exp Med 23:259–268

Samanta A, Das G, Das SK (2011) Roles of flavonoids in plants. Int J Pharm Sci Technol 6:12–35

Sanchez-Gonzalez N, Jaime-Fonseca MR, San Martin-Martinez E et al (2013) Extraction, stability, and separation of betalains from *Opuntia joconostle cv.* using response surface methodology. J Agric Food Chem 61:11995–12004

Sandoo A, Veldhuijzen van Zanten JJCS, Metsios GS et al (2010) The endothelium and its role in regulating vascular tone. Open Cardiovasc Med J 4:302–312

Sansai K, Takuathung MN, Khatsri R et al (2020) Effects of isoflavone interventions on bone mineral density in postmenopausal women: a systematic review and meta-analysis of randomized controlled trials. Osteoporos Int 31:1853–1864

Santi MD, Zunini MP, Vera B et al (2018) Xanthine oxidase inhibitory activity of natural and hemisynthetic flavonoids from *Gardenia oudiepe* (Rubiaceae) *in vitro* and molecular docking studies. Eur J Med Chem 143:577–582

Schön M, Mousa A, Berk M et al (2019) The potential of carnosine in brain-related disorders: a comprehensive review of current evidence. Nutrients 11:1196

Seddon JM, Ajani UA, Sperduto RD et al (1994) Dietary carotenoids, vitamins A, C, and E, and advanced age-related macular degeneration. Eye disease case-control study group. JAMA 272:1413–1420

Sekhar RV, Patel SG, Guthikonda AP et al (2011) Deficient synthesis of glutathione underlies oxidative stress in aging and can be corrected by dietary cysteine and glycine supplementation. Am J Clin Nutr 94:847–853

Semaming Y, Pannengpetch P, Chattipakorn SC et al (2015) Pharmacological properties of protocatechuic acid and its potential roles as complementary medicine. Evid Based Complement Alternat Med 2015, Article ID 593902, 11 pages

Seow WJ, Koh W-P, Jin A et al (2020) Associations between tea and coffee beverage consumption and the risk of lung cancer in the Singaporean Chinese population. Eur J Nutr 59:3083–3091

Sesso HD, Liu S, Gaziano JM et al (2003) Dietary lycopene, tomato-based food products and cardiovascular disease in women. J Nutr 133:2336–2341

Sesso HD, Buring JE, Christen WG et al (2008) Vitamins E and C in the prevention of cardiovascular disease in men. JAMA 300:2123–2133

Shafabakhsh R, Milajerdi A, Reiner Ž et al (2020) The effects of catechin on endothelial function: a systematic review and meta-analysis of randomized controlled trials. Crit Rev Food Sci Nutr 60:2369–2378

Shahidi F, de Camargo AC (2016) Tocopherols and tocotrienols in common and emerging dietary sources: occurrence, applications, and health benefits. Int J Mol Sci 17:1745

Shao A, Hathcock JN (2006) Risk assessment for the carotenoids lutein and lycopene. Regul Toxicol Pharmacol 45:289–298

Sharma A, Fonarow GC, Butler J et al (2016) Coenzyme Q10 and heart failure: a state-of-the-art review. Circ Heart Fail 9:e002639

Sharma P, Tripathi MK, Shrivastava SK (2020) Cholinesterase as a target for drug development in Alzheimer's disease. In: Labrou N (ed) Targeting enzymes for pharmaceutical development, Methods in molecular biology. Humana, New York, pp 257–286

Sheng Y-N, Luo Y-H, Liu S-B et al (2020) Zeaxanthin induces apoptosis via ROS-regulated MAPK and AKT signaling pathway in human gastric cancer cells. Onco Targets Ther 13:10995–11006

Shidal C, Yoon H-S, Zheng W et al (2021) Prospective study of plasma levels of coenzyme Q10 and lung cancer risk in a low-income population in the Southeastern United States. Cancer Med 10:1439–1447

Shih P-H, Yeh C-T, Yen G-C (2007) Anthocyanins induce the activation of phase II enzymes through the antioxidant response element pathway against oxidative stress-induced apoptosis. J Agric Food Chem 55:9427–9435

Shishtar E, Rogers GT, Blumberg JB et al (2020) Long-term dietary flavonoid intake and risk of Alzheimer disease and related dementias in the Framingham Offspring Cohort. Am J Clin Nutr 112:343–353

Shivavedi N, Charan Tej GNV, Neogi K et al (2019) Ascorbic acid therapy: a potential strategy against comorbid depression-like behavior in streptozotocin-nicotinamide-induced diabetic rats. Biomed Pharmacother 109:351–359

Sifuentes-Franco S, Pacheco-Moisés FP, Rodríguez-Carrizalez AD et al (2017) The role of oxidative stress, mitochondrial function, and autophagy in diabetic polyneuropathy. J Diabetes Res 2017, Article ID 1673081, 15 pages

Singh DK, Winocour P, Farrington K (2011) Oxidative stress in early diabetic nephropathy: fueling the fire. Nat Rev Endocrinol 7:176–178

Singh SK, Srivastav S, Castellani RJ et al (2019) Neuroprotective and antioxidant effect of *Ginkgo biloba* extract against AD and other neurological disorders. Neurotherapeutics 16:666–674

Siow RCM, Sato H, Leake DS et al (1998) Vitamin C protects human arterial smooth muscle cells against atherogenic lipoproteins: effects of antioxidant vitamins C and E on oxidized LDL-induced adaptive increases in cystine transport and glutathione. Arterioscler Thromb Vasc Biol 18:1662–1670

Skrajnowska D, Bobrowska-Korczak B (2019) Role of zinc in immune system and anti-cancer defense mechanisms. Nutrients 11:2273

Smirnoff N (2018) Ascorbic acid metabolism and functions: a comparison of plants and mammals. Free Radic Biol Med 122:116–129

Smith E, Ottosson F, Hellstrand S et al (2020) Ergothioneine is associated with reduced mortality and decreased risk of cardiovascular disease. Heart 106:691–697

Smyth EM (2010) Thromboxane and the thromboxane receptor in cardiovascular disease. Clin Lipidol 5:209–219

Smyth JF, Bowman A, Perren T et al (1997) Glutathione reduces the toxicity and improves quality of life of women diagnosed with ovarian cancer treated with cisplatin: results of a double-blind, randomised trial. Ann Oncol 8:569–573

Socci V, Tempesta D, Desideri G et al (2017) Enhancing human cognition with cocoa flavonoids. Front Nutr 4:19

Socha K, Klimiuk K, Naliwajko SK et al (2021) Dietary habits, selenium, copper, zinc and total antioxidant status in serum in relation to cognitive functions of patients with Alzheimer's disease. Nutrients 13:287

Soinio M, Marniemi J, Laakso M et al (2007) Serum zinc level and coronary heart disease events in patients with type 2 diabetes. Diabetes Care 30:523–528

Solis MY, Cooper S, Hobson RM et al (2015) Effects of beta-alanine supplementation on brain homocarnosine/carnosine signal and cognitive function: an exploratory study. PLoS One 10:e0123857

Somade OT, Ugbaja RN, Alli AA et al (2018) Diallyl disulfide, an organosulfur compound in garlic and onion attenuates trichloromethane-induced hepatic oxidative stress, activation of NFkB and apoptosis in rats. J Nutr Intermed Metab 13:10–19

Sousa C, Moutinho C, Vinha AF et al (2019) Trace minerals in human health: Iron, zinc, copper, manganese and fluorine. Int J Sci Res Methodol 13:57–80

Sp N, Kang DY, Joung YH et al (2017) Nobiletin inhibits angiogenesis by regulating Src/FAK/STAT3-mediated signaling through PXN in ER^+ breast cancer cells. Int J Mol Sci 18:935

Spagnuolo C, Moccia S, Russo GL (2018) Anti-inflammatory effects of flavonoids in neurodegenerative disorders. Eur J Med Chem 153:105–115

Spencer ES, Pitcher T, Veron G et al (2020) Positive association of ascorbate and inverse association of urate with cognitive function in people with Parkinson's disease. Antioxidants 9:906

Squitti R, Pal A, Picozza M et al (2020) Zinc therapy in early Alzheimer's disease: safety and potential therapeutic efficacy. Biomol Ther 10:1164

Sripanidkulchai B (2019) Benefits of aged garlic extract on Alzheimer's disease: possible mechanisms of action (review). Exp Ther Med 19:1560–1564

Stankovic MS, Niciforovic N, Topuzovic M et al (2011) Total phenolic content, flavonoid concentrations and antioxidant activity, of the whole plant and plant parts extracts from *Teucrium Montanum* L. Var. *Montanum*, F. *Supinum* (L.) Reichenb. Biotechnol Biotechnol Equip 25:2222–2227

Steele ML, Fuller S, Patel M et al (2013) Effect of Nrf2 activators on release of glutathione, cysteinylglycine and homocysteine by human U373 astroglial cells. Redox Biol 1:441–445

Stephenson J, Nutma E, van der Valk P et al (2018) Inflammation in CNS neurodegenerative diseases. Immunology 154:204–219

Stice CP, Xia H, Wang X-D et al (2018) Tomato lycopene prevention of alcoholic fatty liver disease and hepatocellular carcinoma development. Chronic Dis Transl Med 4:211–224

Stintzing FC, Carle R (2004) Functional properties of anthocyanins and betalains in plants, food and disadvantages human nutrition. Trends Food Sci Technol 15:19–38

Storelli FV, Molina AJ, Zamora-Ros R et al (2019) Flavonoids and the risk of gastric cancer: an exploratory case-control study in the MCC-Spain Study. Nutrients 11:967

Strozyk E, Kulms D (2013) The role of AKT/mTOR pathway in stress response to UV-irradiation: implication in skin carcinogenesis by regulation of apoptosis, autophagy and senescence. Int J Mol Sci 14:15260–15285

Subedi L, Cho K, Park YU et al (2019) Sulforaphane-enriched broccoli sprouts pretreated by pulsed electric fields reduces neuroinflammation and ameliorates scopolamine-induced amnesia in mouse brain through its antioxidant ability via Nrf2-HO-1 activation. Oxidative Med Cell Longev 2019, Article ID 3549274, 19 pages

Sudha K, Rao AV, Rao S et al (2003) Free radical toxicity and antioxidants in Parkinson's disease. Neurol India 51:60–62

Sugahara R, Sato A, Uchida A et al (2015) Annatto tocotrienol induces a cytotoxic effect on human prostate cancer PC3 cells via the simultaneous inhibition of Src and Stat3. J Nutr Sci Vitaminol 61:497–501

Sun L, Subar AF, Bosire C et al (2017) Dietary flavonoid intake reduces the risk of head and neck but not esophageal or gastric cancer in US men and women. J Nutr 147:1729–1738

Sun L, Zhao W, Li J et al (2021) Dietary flavonoids intake and risk of esophageal squamous cell carcinoma: a population-based case-control study. Nutrition:111235

Suresh S, Waly MI, Rahman MS et al (2017) Broccoli (*Brassica oleracea*) reduces oxidative damage to pancreatic tissue and combats hyperglycaemia in diabetic rats. Prev Nutr Food Sci 22:277–284

Suzuki T, Katsumata S-I, Matsuzaki H et al (2016) Dietary zinc supplementation increased TNFα and IL1β-induced RANKL expression, resulting in a decrease in bone mineral density in rats. J Clin Biochem Nutr 58:48–55

Tan BL, Norhaizan ME (2019) Carotenoids: how effective are they to prevent age-related diseases? Molecules 24:1801

Tan BL, Norhaizan ME (2021) Oxidative stress, diet and prostate cancer. World J Mens Health 39:195–207

Tan BL, Norhaizan ME, Liew W-P-P et al (2018) Antioxidant and oxidative stress: a mutual interplay in age-related diseases. Front Pharmacol 9:1162

Tanaka S, Yoshimura Y, Kawasaki R et al (2013) Fruit intake and incident diabetic retinopathy with type 2 diabetes. Epidemiology 24:204–211

Tang RMY, Cheah IK-M, Yew TSK et al (2018) Distribution and accumulation of dietary ergothioneine and its metabolites in mouse tissues. Sci Rep 8:1601

Tanokura M, Tasumi M, Miyazawa T (1976) ^{1}H nuclear magnetic resonance studies of histidine containing di and tripeptides. Estimation of the effects of charged groups on the p*K*a value of the imidazole ring. Biopolymers 15:393–401

Tasić A, Kureljušić J, Nešić K et al (2017) Determination of calcium content in mechanically separated meat. IOP Conf Ser: Earth Environ Sci 85:012056

Tesoriere L, Butera D, Allegra M et al (2005) Distribution of betalain pigments in red blood cells after consumption of cactus pear fruits and increased resistance of the cells to *ex vivo* induced oxidative hemolysis in humans. J Agric Food Chem 53:1266–1270

Tesoriere L, Gentile C, Angileri F et al (2013) Trans-epithelial transport of the betalain pigments indicaxanthin and betanin across Caco-2 cell monolayers and influence of food matrix. Eur J Nutr 52:1077–1087

The Alpha-Tocopherol Beta-Carotene Cancer Prevention Study Group (1994) The effect of vitamin E and beta carotene on the incidence of lung cancer and other cancers in male smokers. N Engl J Med 330:1029–1035

Thies F, Mills IM, Moir S et al (2017) Cardiovascular benefits of lycopene: fantasy or reality? Proc Nutr Soc 76:122–129

Thompson PA, Ashbeck EL, Roe DJ et al (2016) Selenium supplementation for prevention of colorectal adenomas and risk of associated type 2 diabetes. J Natl Cancer Inst 108:djw152

Thomson M, Al-Qattan KK, Bordia T (2006) Including garlic in the diet may help lower blood glucose, cholesterol, and triglycerides. J Nutr 136:800–802

Timoshnikov VA, Kobzeva TV, Polyakov NE et al (2020) Redox interactions of vitamin C and iron: inhibition of the pro-oxidant activity by deferiprone. Int J Mol Sci 21:3967

Tiwari V, Kuhad A, Bishnoi M et al (2009) Chronic treatment with tocotrienol, an isoform of vitamin E, prevents intracerebroventricular streptozotocin-induced cognitive impairment and oxidative–nitrosative stress in rats. Pharmacol Biochem Behav 93:183–189

Todoric J, Antonucci L, Karin M (2016) Targeting inflammation in cancer prevention and therapy. Cancer Prev Res (Philadelphia Pa) 9:895–905

Tolkatschevskaya N (1929) Zur Kentniss der Extraktivstoffe dur Muskeln. Uber die Extraktivstoffe des Huhnerfleisches. Hoppe Seylers Z Physiol Chem 185:28–32

Tolosa de Talamoni N, Morero R, Cañas F (1989) Vitamin D3 administration increases the membrane fluidity of intestinal mitochondria. Biochem Int 19:701–707

Tong C, Peng C, Wang I et al (2016) Intravenous administration of lycopene a tomato extract protects against myocardial ischemia reperfusion injury. Nutrients 8:138–151

Toth S, Jonecova Z, Maretta M et al (2019) The effect of betanin parenteral pretreatment on jejunal and pulmonary tissue histological architecture and inflammatory response after jejunal ischemia-reperfusion injury. Exp Mol Pathol 110:104292
Townsend DM, Tew KD, Tapiero H et al (2003) The importance of glutathione in human disease. Biomed Pharmacother 57:145–155
Traber MG, Atkinson J (2007) Vitamin E, antioxidant and nothing more. Free Radic Biol Med 43:4–15
Traber MG, Stevens JF (2011) Vitamins C and E: beneficial effects from a mechanistic perspective. Free Radic Biol Med 51:1000–1013
Traka MH, Melchini A, Coode-Bate J et al (2019) Transcriptional changes in prostate of men on active surveillance after a 12-mo glucoraphanin-rich broccoli intervention—results from the Effect of Sulforaphane on prostate CAncer PrEvention (ESCAPE) randomized controlled trial. Am J Clin Nutr 109:1133–1144
Traverso N, Ricciarelli R, Nitti M et al (2013) Role of glutathione in cancer progression and chemoresistance. Oxidative Med Cell Longev 2013, Article ID 972913, 10 pages
Travica N, Ried K, Sali A et al (2017) Vitamin c status and cognitive function: a systematic review. Nutrients 9:960
Trejo-Solís C, Pedraza-Chaverrí J, Torres-Ramos M et al (2013) Multiple molecular and cellular mechanisms of action of lycopene in cancer inhibition. Evid Based Complement Alternat Med 2013, Article ID 705121, 17 pages
Treml J, Šmejkal K (2016) Flavonoids as potent scavengers of hydroxyl radicals. Compr Rev Food Sci Food Saf 15:720–738
Tribuzi R, Crispoltoni L, Chiurchiù V et al (2017) Trans-crocetin improves amyloid-β degradation in monocytes from Alzheimer's disease patients. J Neurol Sci 372:408–412
Tronina T, Strugała P, Popłoński J et al (2017) The influence of glycosylation of natural and synthetic prenylated flavonoids on binding to human serum albumin and inhibition of cyclooxygenases COX-1 and COX-2. Molecules 22:1230
Trueba GP, Sanchez GM, Giuliani A (2004) Oxygen free radical and antioxidant defense mechanism in cancer. Front Biosci 9:2029–2044
Tu S-H, Chen L-C, Ho Y-S (2017) An apple a day to prevent cancer formation: reducing cancer risk with flavonoids. J Food Drug Anal 25:119–124
Tuzcu M, Orhan C, Muz OE et al (2017) Lutein and zeaxanthin isomers modulates lipid metabolism and the inflammatory state of retina in obesity-induced high-fat diet rodent model. BMC Ophthalmol 17:129
Uddin MS, Kabir MT, Niaz K et al (2020) Molecular insight into the therapeutic promise of flavonoids against Alzheimer's disease. Molecules 25:1267
Uesugi S, Ishihara J, Iso H et al (2017) Dietary intake of antioxidant vitamins and risk of stroke: the Japan Public Health Center–based Prospective Study. Eur J Clin Nutr 71:1179–1185
Ülger TG, Çakiroglu FP (2020) The effects of onion (*Allium cepa* L.) dried by different heat treatments on plasma lipid profile and fasting blood glucose level in diabetic rats. Avicenna J Phytomed 10:325–333
Ulla A, Mohamed MK, Sikder B et al (2017) Coenzyme Q10 prevents oxidative stress and fibrosis in isoprenaline induced cardiac remodeling in aged rats. BMC Pharmacol Toxicol 18:29
Ullah A, Khan A, Khan I (2016) Diabetes mellitus and oxidative stress—a concise review. Saudi Pharm J 24:547–553
Umigai N, Kozai Y, Saito T et al (2020) Effects of paprika carotenoid supplementation on bone turnover in postmenopausal women: a randomized, double-blind, placebo-controlled, parallel-group comparison study. Food Nutr Res 64:4565
van Gorkom GNY, Lookermans EL, Van Elssen CHMJ et al (2019) The effect of vitamin C (ascorbic acid) in the treatment of patients with cancer: a systematic review. Nutrients 11:977
Vanderslice JT, Higgs DJ (1991) Vitamin C content of foods: sample variability. Am J Clin Nutr 54:1323S–1327S
Varma V, Varma M, Sarkar PD et al (2014) Correlation of vitamin C with HbA1c and oxidative stress in diabetes mellitus with OR without nephropathy. Nat J Med Res 4:151–155

Vashum KP, McEvoy M, Shi Z et al (2013) Is dietary zinc protective for type 2 diabetes? Results from the Australian longitudinal study on women's health. BMC Endocr Disord 13:40

Vauzour D, Vafeiadou K, Rodriguez-Mateos A et al (2008) The neuroprotective potential of flavonoids: a multiplicity of effects. Genes Nutr 3:115–126

Veber DF, Johnson SR, Cheng HY et al (2002) Molecular properties that influence the oral bioavailability of drug candidates. J Med Chem 45:2615–2623

Vendrell I, Macedo D, Alho I et al (2015) Treatment of cancer pain by targeting cytokines. Mediat Inflamm 2015, Article ID 984570, 11 pages

Vidal PJ, López-Nicolás JM, Gandía-Herrero F et al (2014) Inactivation of lipoxygenase and cyclooxygenase by natural betalains and semi-synthetic analogues. Food Chem 154:246–254

Vineetha RC, Hariharan S, Jaleel A et al (2020) L-ascorbic acid and α-tocopherol synergistically triggers apoptosis inducing antileukemic effects of arsenic trioxide via oxidative stress in human acute promyelocytic leukemia cells. Front Oncol 10:65

Vomhof-Dekrey EE, Picklo MJ (2012) The Nrf 2-antioxidant response element pathway: a target for regulating energy metabolism. J Nutr Biochem 23:1201–1206

Vomund S, Schäfer A, Parnham MJ et al (2017) Nrf2, the master regulator of anti-oxidative responses. Int J Mol Sci 18:2772

von Lintig J (2012) Provitamin A metabolism and functions in mammalian biology. Am J Clin Nutr 96:1234S–1244S

Wan D, Wu Q, Ni H et al (2019) Treatments for iron deficiency (ID): prospective organic iron fortification. Curr Pharm Des 25:325–332

Wang L, Lee I-M, Zhang SM et al (2009) Dietary intake of selected flavonols, flavones, and flavonoid-rich foods and risk of cancer in middle-aged and older women. Am J Clin Nutr 89:905–912

Wang Y, Ausman LM, Greenberg AS et al (2010) Dietary lycopene and tomato extract supplementations inhibit nonalcoholic steatohepatitis-promoted hepatocarcinogenesis in rats. Int J Cancer 126:1788–1796

Wang TJ, Larson MG, Vasan RS et al (2011) Metabolite profiles and the risk of developing diabetes. Nat Med 17:448–453

Wang L, Zhang X, Liu J et al (2014) Tea consumption and lung cancer risk: a meta-analysis of case–control and cohort studies. Nutrition 30:1122–1127

Wang A, Wang S, Zhu C et al (2016a) Coffee and cancer risk: a meta-analysis of prospective observational studies. Sci Rep 6:33711

Wang X, Yang Y, Liu R et al (2016b) Identification of antioxidants in aged garlic extract by gas chromatography-mass spectrometry and liquid chromatography-mass spectrometry. Int J Food Prop 19:474–483

Wang AH, Ma Q, Wang X et al (2018) Protective effects of beef decoction rich in carnosine on cerebral ischemia injury by permanent middle cerebral artery occlusion in rats. Exp Ther Med 15:1321–1329

Warburton A, Breen G, Rujescu D et al (2015) Characterization of a REST-regulated internal promoter in the schizophrenia genome-wide associated gene MIR137. Schizophr Bull 41:698–707

Weichert W, Roske A, Gekeler V et al (2008) Histone deacetylases 1, 2 and 3 are highly expressed in prostate cancer and HDAC2 expression is associated with shorter PSA relapse time after radical prostatectomy. Br J Cancer 98:604–610

Weigand-Heller AJ, Kris-Etherton PM, Beelman RB (2012) The bioavailability of ergothioneine from mushrooms (Agaricus bisporus) and the acute effects on antioxidant capacity and biomarkers of inflammation. Prev Med 54:S75–S78

Wijesekara N, Dai FF, Hardy AB et al (2010) Beta cell-specific Znt8 deletion in mice causes marked defects in insulin processing, crystallisation and secretion. Diabetologia 53:1656–1668

Winkler C, Wirleitner B, Schroecksnadel K et al (2005) *In vitro* effects of beetroot juice on stimulated and unstimulated peripheral blood mononuclear cells. Appl Biochem Biotechnol 1:180–185

Wohlrab C, Phillips E, Dachs GU (2017) Vitamin C transporters in cancer: current understanding and gaps in knowledge. Front Oncol 7:74

Wong WL, Su X, Li X et al (2014) Global prevalence of age-related macular degeneration and disease burden projection for 2020 and 2040: a systematic review and meta-analysis. Lancet Global Health 2:e106–e116

Wong MYZ, Man REK, Fenwick EK et al (2018a) Dietary intake and diabetic retinopathy: a systematic review. PLoS One 13:e0186582

Wong SK, Chin K-Y, Suhaimi FH et al (2018b) The effects of vitamin E from *Elaeis guineensis* (oil palm) in a rat model of bone loss due to metabolic syndrome. Int J Environ Res Public Health 15:1828

Wong KW, Yap CK, Nulit R et al (2019) Zn in vegetables: a review and some insights. Integr Food Nutr Metab 6:1–7

Woo HD, Lee J, Choi IJ et al (2014) Dietary flavonoids and gastric cancer risk in a Korean population. Nutrients 6:4961–4973

World Health Organization (2020) Diabetes. Available from: https://www.who.int/health-topics/diabetes#tab=tab_1. Accessed 2 Sept 2020

Wroblewska M, Juskiewicz J, Wiczkowski W (2011) Physiological properties of beetroot crisps applied in standard and dyslipidaemic diets of rats. Lipids Health Dis 10:178

Wu G (2013) Amino acids: biochemistry and nutrition. CRC Press, Boca Raton

Wu G (2020) Important roles of dietary taurine, creatine, carnosine, anserine and 4-hydroxyproline in human nutrition and health. Amino Acids 52:329–360

Wu HC, Shiau C-Y, Chen H-M et al (2003) Antioxidant activities of carnosine, anserine, some free amino acids and their combination. J Food Drug Anal 11:148–153

Wu W, Li Y, Wu Y et al (2015) Lutein suppresses inflammatory responses through Nrf2 activation and NF-κB inactivation in lipopolysaccharide-stimulated BV-2 microglia. Mol Nutr Food Res 59:1663–1673

Wu G, Cross HR, Gehring KB et al (2016) Composition of free and peptide-bound amino acids in beef chuck, loin, and round cuts. J Anim Sci 94:2603–2613

Wu R, Feng J, Yang Y et al (2017) Significance of serum total oxidant/antioxidant status in patients with colorectal cancer. PLoS One 12:e0170003

Wu S, Liu Y, Michalek JE et al (2020) Carotenoid intake and circulating carotenoids are inversely associated with the risk of bladder cancer: a dose-response meta-analysis. Adv Nutr 11:630–643

Xiang S, Liu F, Lin J et al (2017) Fucoxanthin inhibits β-amyloid assembly and attenuates β-amyloid oligomer-induced cognitive impairments. J Agric Food Chem 65:4092–4102

Xiang W, Gu K, Wang W et al (2019) Tea consumption and risk of fractures: an updated meta-analysis. Osteoporos Int 30:1941–1951

Xie Y, Qin J, Nan G et al (2016) Coffee consumption and the risk of lung cancer: an updated meta-analysis of epidemiological studies. Eur J Clin Nutr 70:199–206

Xu M, Chen YM, Huang J et al (2016) Flavonoid intake from vegetables and fruits is inversely associated with colorectal cancer risk: a case-control study in China. Br J Nutr 116:1275–1287

Xu H, Luo J, Huang J et al (2018) Flavonoids intake and risk of type 2 diabetes mellitus: a meta-analysis of prospective cohort studies. Medicine 97:e0686

Xue Q, Yan Y, Zhang R et al (2018) Regulation of iNOS on immune cells and its role in diseases. Int J Mol Sci 19:3805

Yamagata K, Yamori Y (2020) Inhibition of endothelial dysfunction by dietary flavonoids and preventive effects against cardiovascular disease. J Cardiovasc Pharmacol 75:1–9

Yamagishi K, Ikeda A, Moriyama Y et al (2014a) Serum coenzyme Q10 and risk of disabling dementia: the Circulatory Risk in Communities Study (CIRCS). Atherosclerosis 237:400–403

Yamagishi K, Ikeda A, Moriyama Y et al (2014b) Serum coenzyme Q10 levels as a predictor of dementia in a Japanese general population. Atherosclerosis 237:433–434

Yan J-Y, Tian F-M, Hu W-N et al (2013) Apoptosis of human gastric cancer cells line SGC 7901 induced by garlic-derived compound S-allylmercaptocysteine (SAMC). Eur Rev Med Pharmacol Sci 17:745–751

Yanai R, Chen S, Uchi SH et al (2018) Attenuation of choroidal neovascularization by dietary intake of omega-3 long-chain polyunsaturated fatty acids and lutein in mice. PLoS One 13:e0196037

Yang Y, Chan SW, Hu M et al (2011) Effects of some common food constituents on cardiovascular disease. ISRN Cardiol 2011:397136

Yang NC, Lin HC, Wu JH et al (2012) Ergothioneine protects against neuronal injury induced by beta-amyloid in mice. Food Chem Toxicol 50:3902–3911

Yang S, Zhang H, Yang X et al (2015) Evaluation of antioxidative and antitumor activities of extracted flavonoids from Pink Lady apples in human colon and breast cancer cell lines. Food Funct 6:3789–3798

Yang X, Zhang Y, Xu H et al (2016) Neuroprotection of coenzyme Q10 in neurodegenerative diseases. Curr Topics Med Chem 16:858–866

Yang X, Zheng H, Liu Y et al (2020a) Puerarin for OVX-induced postmenopausal osteoporosis in murine model: systematic review and meta-analysis. Curr Stem Cell Res Ther 15:37–42

Yang C, Shi X, Xia H et al (2020b) The evidence and controversy between dietary calcium intake and calcium supplementation and the risk of cardiovascular disease: a systematic review and meta-analysis of cohort studies and randomized controlled trials. J Am Coll Nutr 39:352–370

Yang M, Lian N, Yu Y et al (2020c) Coenzyme Q10 alleviates sevoflurane-induced neuroinflammation by regulating the levels of apolipoprotein E and phosphorylated tau protein in mouse hippocampal neurons. Mol Med Rep 22:445–453

Yatoo MI, Saxena A, Deepa PM et al (2013) Role of trace elements in animals: a review. Vet World 6:963–967

Ye Q, Huang B, Zhang X et al (2012) Astaxanthin protects against MPP(+)-induced oxidative stress in PC12 cells via the HO-1/NOX2 axis. BMC Neurosci 13:156

Ye Q, Zhang X, Huang B et al (2013) Astaxanthin suppresses MPP+-induced oxidative damage in PC12 cells through a Sp1/NR1 signaling pathway. Mar Drugs 11:1019–1034

Yeum K-J, Orioli M, Regazzoni L et al (2010) Profiling histidine dipeptides in plasma and urine after ingesting beef, chicken or chicken broth in humans. Amino Acids 38:847–858

Yoon DS, Lee Y-K, Ha Y-C et al (2016) Inadequate dietary calcium and vitamin D intake in patients with osteoporotic fracture. J Bone Metab 23:55–61

Young VR, Steffee WP, Pencharz PB et al (1975) Protein turnover in the whole body. Nature 253:192–194

Yu L, Wang W, Pang W et al (2017) Dietary lycopene supplementation improves cognitive performances in Tau transgenic mice expressing P301L mutation via inhibiting oxidative stress and Tau hyperphosphorylation. J Alzheimers Dis 57:475–482

Yuan C, Fondell E, Ascherio A et al (2020) Long-term intake of dietary carotenoids is positively associated with late-life subjective cognitive function in a prospective study in US women. J Nutr 150:1871–1879

Yuan C, Chen H, Wang Y et al (2021) Dietary carotenoids related to risk of incident Alzheimer dementia (AD) and brain AD neuropathology: a community-based cohort of older adults. Am J Clin Nutr 113:200–208

Yuste S, Ludwig IA, Rubió L et al (2019) *In vivo* biotransformation of (poly)phenols and anthocyanins of red-fleshed apple and identification of intake biomarkers. J Funct Foods 55:146–155

Zaidun NH, Thent ZC, Abd Latiff A (2018) Combating oxidative stress disorders with citrus flavonoid: Naringenin. Life Sci 208:111–122

Zamora-Ros R, Agudo A, Luján-Barroso L et al (2012) Dietary flavonoid and lignan intake and gastric adenocarcinoma risk in the European Prospective Investigation into Cancer and Nutrition (EPIC) study. Am J Clin Nutr 96:1398–1408

Zamora-Ros R, Not C, Guinó E et al (2013) Association between habitual dietary flavonoid and lignan intake and colorectal cancer in a Spanish case-control study (the Bellvitge Colorectal Cancer Study). Cancer Causes Control 24:549–557

Zamora-Ros R, Alghamdi MA, Cayssials V et al (2019) Coffee and tea drinking in relation to the risk of differentiated thyroid carcinoma: results from the European Prospective Investigation into Cancer and Nutrition (EPIC) study. Eur J Nutr 58:3303–3312

Zeka K, Ruparelia K, Arroo RRJ et al (2017) Flavonoids and their metabolites: prevention in cardiovascular diseases and diabetes. Diseases 5:19

Zemrani B, Bines JE (2020) Recent insights into trace element deficiencies: causes, recognition and correction. Curr Opin Gastroenterol 36:110–117

Zhang J, Pavlova NN, Thompson CB (2017a) Cancer cell metabolism: the essential role of the nonessential amino acid, glutamine. EMBO J 36:1302–1315

Zhang S, Tomata Y, Sugiyama K et al (2017b) Mushroom consumption and incident dementia in elderly Japanese: the Ohsaki Cohort 2006 study. J Am Geriatr Soc 65:1462–1469

Zhang H-W, Hu J-J, Fun R-Q et al (2018a) Flavonoids inhibit cell proliferation and induce apoptosis and autophagy through downregulation of PI3Kγ mediated PI3K/AKT/mTOR/p70S6K/ULK signaling pathway in human breast cancer cells. Sci Rep 8:11255

Zhang D, Du M, Wei Y et al (2018b) A review on the structure-activity relationship of dietary flavonoids for protecting vascular endothelial function: current understanding and future issues. J Food Biochem 42:e12557

Zhang X-Y, Xu Z-P, Wang W et al (2018c) Vitamin C alleviates LPS-induced cognitive impairment in mice by suppressing neuroinflammation and oxidative stress. Int Immunopharmacol 65:438–447

Zhang Y, Liu J, Chen X-Q et al (2018d) Ubiquinol is superior to ubiquinone to enhance Coenzyme Q10 status in older men. Food Funct 9:5653–5659

Zhang Y, Jiang Y, Lu D (2019a) Diosmetin suppresses neuronal apoptosis and inflammation by modulating the phosphoinositide 3-kinase (PI3K)/AKT/nuclear factor-κB (NF-κB) signaling pathway in a rat model of pneumococcal meningitis. Med Sci Monit 25:2238–2245

Zhang S, Chen L, Zhou Z et al (2019b) Effects of puerarin on clinical parameters, vascular endothelial function, and inflammatory factors in patients with coronary artery disease. Med Sci Monit 25:402–408

Zhang Q, Ai Y, Dong H et al (2019c) Circulating oxidized low-density lipoprotein is a strong risk factor for the early stage of coronary heart disease. IUBMB Life 71:277–282

Zhang Y, Liu X, Ruan J et al (2020a) Phytochemicals of garlic: promising candidates for cancer therapy. Biomed Pharmacother 123:109730

Zhang S, Sugawara Y, Chen S et al (2020b) Mushroom consumption and incident risk of prostate cancer in Japan: a pooled analysis of the Miyagi Cohort Study and the Ohsaki Cohort Study. Int J Cancer 146:2712–2720

Zhao D, Kwon SH, Chun YS et al (2017) Anti-neuroinflammatory effects of fucoxanthin via inhibition of Akt/NF-κB and MAPKs/AP-1 pathways and activation of PKA/CREB pathway in lipopolysaccharide-activated BV-2 microglial cells. Neurochem Res 42:667–677

Zhitkovich A (2020) Nuclear and cytoplasmic functions of vitamin C. Chem Res Toxicol 33:2515–2526

Zhong L, Goldberg MS, Gao YT et al (2001) A population-based case-control study of lung cancer and green tea consumption among women living in Shanghai, China. Epidemiology 12:695–700

Zhou J, Chen C, Chen X et al (2020a) Vitamin C promotes apoptosis and cell cycle arrest in oral squamous cell carcinoma. Front Oncol 10:976

Zhou X, Qian H, Zhang D et al (2020b) Garlic intake and the risk of colorectal cancer: a meta-analysis. Medicine 99:e18575

Zhu Y-J, Bo Y-C, Liu X-X et al (2017) Association of dietary vitamin E intake with risk of lung cancer: a dose-response meta-analysis. Asia Pac J Clin Nutr 26:271–277

Zhu N-W, Yin X-L, Lin R et al (2020) Possible mechanisms of lycopene amelioration of learning and memory impairment in rats with vascular dementia. Neural Regen Res 15:332–341

Zielinska-Przyjemska M, Olejnik A, Dobrowolska-Zachwieja A et al (2009) In vitro effects of beetroot juice and chips on oxidative metabolism and apoptosis neutrophils from obese individuals. Phytother Res 23:49–55

Zou Z, Xu X, Huang Y et al (2011) High serum level of lutein may be protective against early atherosclerosis: the Beijing atherosclerosis study. Atherosclerosis 219:789–793

Chapter 8
Summary and Future Prospects

Aging is biological and functional changes characterized by a progressive decline of tissue renewal capability and diminish of cellular functions of an organism (Bonomini et al. 2015; Forni et al. 2019). ROS are generated as by-products by many exogenous and endogenous factors, which may induce oxidative damage to lipids, proteins, and DNA, and ultimately lead to cellular dysfunction (Harman 1956). Indeed, age-related pathologies are closely related to an elevation of oxidative stress (Liguori et al. 2018). Dysregulation of the immune response is one of the primary changes that occurred during aging, which may lead to a chronic systemic inflammatory state (Chung et al. 2019). Of all proinflammatory mediators, chemokines and cytokines are the primary contributors to chronic inflammation (Chung et al. 2019). Low-grade chronic systemic inflammation plays a crucial role in the aging process (Guarner and Rubio-Ruiz 2015). It has been demonstrated that aging is linked to the elevation of circulating proinflammatory cytokines and other inflammatory substances in the blood, such as increased levels of IL-6 and TNF-α (Bruunsgaard 2002). The production of enzymes and cytokines becomes dysregulated with increased age and possesses highly pleiotropic activities and subsequently modified the physiology of cells, tissues, and organs in the body. Indeed, many cytokines that are stimulated during aging are involved in the host inflammatory response, an innate immune response that plays a crucial role in the host defense toward infectious agents (Bernstein and Murasko 1998). Inappropriate activation of the chronic inflammatory response may lead to the pathologic condition, which plays a pivotal role in chronic and acute disease conditions (Arida et al. 2018). Substantial studies indicate that aging is linked to several pathological conditions, such as neurodegenerative diseases, cancer, CVD, obesity, and metabolic syndrome (Franceschi et al. 2018; Liguori et al. 2018; Tan et al. 2018a).

Chronic inflammation in the elderly is thought to be caused by cellular senescence, which is characterized by the onset of a multifaceted senescence-associated secretory phenotype (SASP) and an arrest of cell survival (Coppé et al. 2010;

B. L. Tan, M. E. Norhaizan, *The Role of Antioxidants in Longevity and Age-Related Diseases*, https://doi.org/10.1007/978-3-030-82859-2_8

Furman et al. 2019). A prominent characteristic of this phenotype is upregulated the secretion of other proinflammatory molecules from cells, chemokines, and proinflammatory cytokines (Coppé et al. 2010). Natural products have been suggested to counteract oxidative stress and promoting the healthy longevity of an organism (Chen et al. 2016). Such protection has been ascribed to antioxidants including polyphenols, carotenoids, and vitamins (Tan et al. 2018a). Natural antioxidants, for instance, organosulfur compounds, vitamins, carotenoids, and polyphenols are widely found in vegetables and fruits (Tan et al. 2018a). Natural antioxidants possess many biological activities, for instance, antioxidant activity, antitumor effects, and anti-inflammatory activity as well as inhibit platelet aggregation (Tan et al. 2018a, b).

Although many studies evaluating the generation and the origin of ROS, current antioxidant-based therapies show a lack of specificity for defective cells, tissues, and organelles, and hence may not achieve an effective dose at the target site. Further, antioxidant targets a high concentration of reactive species and is unable to mediate certain reactive oxygen intermediates during the oxidative reaction and thereby resulting in some therapeutic approaches are unfocused. Mitochondria-targeted antioxidants hold great promise and may provide a useful tool for the amelioration of a broad spectrum of human diseases (Jiang et al. 2020; Le Gal et al. 2021). In this regard, further studies are required to enhance the potency of antioxidant-based therapy, specifically via antioxidants-rich containing food. In addition, it is worth elucidating the role of iron-modulated oxidative damage via Fenton reaction to further verify the association of iron production and mitochondrial dysfunction in relation to the development and onset of age-related diseases. Moreover, the literature reported on the interaction between antioxidants and chelating is limited, and further elucidation may facilitate the development of novel biomarkers and potential therapeutic agents targeting certain disease organs and tissues, as well as to investigate the downstream mediators in oxidative signaling.

Oxidative stress is enhanced by other reactive species including singlet oxygen, $\bullet O_2^-$, and H_2O_2 as well as other non-radicals, which are produced continuously in the body and thereby altering basic structural components and cellular activity (Droge 2002; Pizzino et al. 2017). It has been demonstrated that ROS accumulation may contribute to the formation of mutagen compounds and increase oncogene transcription factors, and thereby induce inflammation and proatherogenic activity (Tan et al. 2018b). In fact, longevity is not merely embedded in the genes; intakes of food rich in antioxidants are crucial for the proper modulation of physiological function. The essential role of food rich in antioxidants in scavenging ROS, producing cellular energy, and regulating immune system is well-recognized (Tan et al. 2018a, b). The diverse mechanisms in which the antioxidants are involved imply that antioxidants may protect against the pathogenesis of age-related diseases and oxidative stress. Therefore, the consumption of food rich in antioxidants might be a useful strategy for promoting healthy longevity. Overall, more long-term randomized clinical studies are required to elucidate the implication of antioxidant-rich foods in relation to aging and age-related diseases.

References

Arida A, Protogerous AD, Kitas GD et al (2018) Systemic inflammatory response and atherosclerosis: the paradigm of chronic inflammatory rheumatic diseases. Int J Mol Sci 19:1890

Bernstein ED, Murasko DM (1998) Effect of age on cytokine production in humans. Age (Omaha) 21:137–151

Bonomini F, Rodella LF, Rezzani R (2015) Metabolic syndrome, aging and involvement of oxidative stress. Aging Dis 6:109–120

Bruunsgaard H (2002) Effects of tumor necrosis factor-alpha and interleukin-6 in elderly populations. Eur Cytokine Netw 13:389–391

Chen W, Jia Z, Pan M-H et al (2016) Natural products for the prevention of oxidative stress-related diseases: mechanisms and strategies. Oxidative Med Cell Longev 2016, Article ID 4628502, 2 pages

Chung HY, Kim DH, Lee EK et al (2019) Redefining chronic inflammation in aging and age-related diseases: proposal of the senoinflammation concept. Aging Dis 10:367–382

Coppé J-P, Desprez P-Y, Krtolica A et al (2010) The senescence-associated secretory phenotype: the dark side of tumor suppression. Annu Rev Pathol 5:99–118

Droge W (2002) Free radicals in the physiological control of cell function. Physiol Rev 82:47–95

Forni C, Facchiano F, Bartoli M et al (2019) Beneficial role of phytochemicals on oxidative stress and age-related diseases. Biomed Res Int 2019, Article ID 8748253, 16 pages

Franceschi C, Garagnani P, Parini P et al (2018) Inflammaging: a new immune-metabolic viewpoint for age-related diseases. Nat Rev Endocrinol 14:576–590

Furman D, Campisi J, Verdin E et al (2019) Chronic inflammation in the etiology of disease across the life span. Nat Med 25:1822–1832

Guarner V, Rubio-Ruiz ME (2015) Low-grade systemic inflammation connects aging, metabolic syndrome and cardiovascular disease. In: Yashin AI, Jazwinski SM (eds) Aging and health – a systems biology perspective. Interdiscipl Top Gerontol Basel, Karger, pp 99–106

Harman D (1956) Aging: a theory based on free radical and radiation chemistry. J Gerontol 11:298–300

Jiang Q, Yin J, Chen J et al (2020) Mitochondria-targeted antioxidants: a step towards disease treatment. Oxidative Med Cell Longev 2020, Article ID 8837893, 18 pages

Le Gal K, Wiel C, Ibrahim MX et al (2021) Mitochondria-targeted antioxidants MitoQ and MitoTEMPO do not influence BRAF-driven malignant melanoma and KRAS-driven lung cancer progression in mice. Antioxidants 10:163

Liguori I, Russo G, Curcio F et al (2018) Oxidative stress, aging, and diseases. Clin Interv Aging 13:757–772

Pizzino G, Irrera N, Cucinotta M et al (2017) Oxidative stress: harms and benefits for human health. Oxidative Med Cell Longev 2017, Article ID 8416763, 13 pages

Tan BL, Norhaizan ME, Liew W-P-P et al (2018a) Antioxidant and oxidative stress: a mutual interplay in age-related diseases. Front Pharmacol 9:1162

Tan BL, Norhaizan ME, Liew W-P-P (2018b) Nutrients and oxidative stress: friend or foe? Oxidative Med Cell Longev 2018, Article ID 9719584, 24 pages

Conclusion

This book has demonstrated clear evidence of the vital role played by dietary antioxidants in combating age-related diseases and providing optimal health in preclinical and human studies. Antioxidants exert strong potential to suppress oxidative stress and lipid peroxidation. Oxidative stress has been identified as a primary risk factor in the development of age-related diseases such as osteoporosis, vascular diseases, cancer, diabetes, metabolic syndromes, atherosclerosis, dementia, and arthritis. There is discrepant finding supporting the beneficial effects of antioxidants in preventing the progression and onset of age-related diseases, yet most clinical studies are limited to their duration of the study and sample size. Therefore, greater adherence to healthy dietary patterns is encouraging and may link to a lower risk of age-related diseases. Strikingly, some research has emerged to suggest that a combination of antioxidants exert greater beneficial potentials than the individual antioxidant, implied that the phytochemicals work in concert rather than individually. Therefore, an antioxidant might be a useful strategy for lifespan extension as well as healthspan extension. Despite antioxidants may not serve as a drug, they hold a great promise in delaying the development of chronic metabolic ailments. Accordingly, more randomized clinical studies are required to further elucidate the impact of food rich in antioxidants on age-related diseases.

Competing Interests
The authors declare that there is no conflict of interest regarding the publication of this paper.

B. L. Tan, M. E. Norhaizan, *The Role of Antioxidants in Longevity and Age-Related Diseases*, https://doi.org/10.1007/978-3-030-82859-2

Index

B. L. Tan, M. E. Norhaizan, *The Role of Antioxidants in Longevity and Age-Related Diseases*, https://doi.org/10.1007/978-3-030-82859-2

H

I

N